AF584823

Pearson Australia
(a division of Pearson Australia Group Pty Ltd)
707 Collins Street, Melbourne, Victoria 3008
PO Box 23360, Melbourne, Victoria 8012
www.pearson.com.au

First published 2014 by Pearson Australia
2019 2018 2017 2016
10 9 8 7 6 5 4 3

Publisher: Marita Tripp
Project Editor: Michelle Hessels
Editors: Nick Tapp and Writers Reign
Designer: Patrick Cannon
Copyright & Pictures Editor: Sian Bradfield
Desktop Operator: Rob Curulli
Illustrator/s: Guy Holt and Bruce Rankin
Printed in Australia by the SOS Print + Media Group

National Library of Australia Cataloguing-in-Publication entry
Kleeman, Grant.
Pearson geography student book. Year 8/Grant Kleeman, Helen Rhodes, David Hamper.
ISBN: 978 1 4886 5697 2 (pbk.)
Geography—Study and teaching (Secondary)
Dewey Number: 910

Pearson Australia Group Pty Ltd ABN 40 004 245 943

PEARSON Australian Curriculum
Writing and Development Team

We are grateful to the following people for their time and expertise in contributing to the Pearson Geography project across Years 7–10.

Grant Kleeman
Academic
Lead Author, New South Wales

Kelli Ashton
Teacher
Author, Victoria

Alan Atkinson
Teacher
Author, Western Australia

Patricia Beeton
Educator
Author, Victoria

Megan Bourke
Educator
Author, Victoria

John Butler
Educator
Author, South Australia

Andre Chadzynski
Teacher
Author, Victoria

April Cincotta
Teacher
Reviewer, Victoria

Greta Creed
Teacher
Reviewer, Queensland

David Hamper
Teacher
Author, New South Wales

Casey Hawkins
Teacher
Author, Victoria

Alon Kaiser
Teacher
Author, Victoria

Jenne King
Educator
Author, Victoria

Rod Lane
Academic
Author, New South Wales

Mick Law
Educator
Author, Queensland

Mark Manuel
Teacher
Author, South Australia

Shirley Melissas
Educator
Author, Victoria

Bec Nicholas
Teacher
Author, Queensland

Gary Passmore
Educator
Reviewer, South Australia

Andrew Peters
Teacher
Author, New South Wales

Helen Rhodes
Teacher
Author, New South Wales

Leah Truscott
Teacher
Author, Western Australia

Dan Waterworth
Teacher
Author, New South Wales

Zoe Winstanley
Teacher
Reviewer, Western Australia

Contents

PEARSON geography

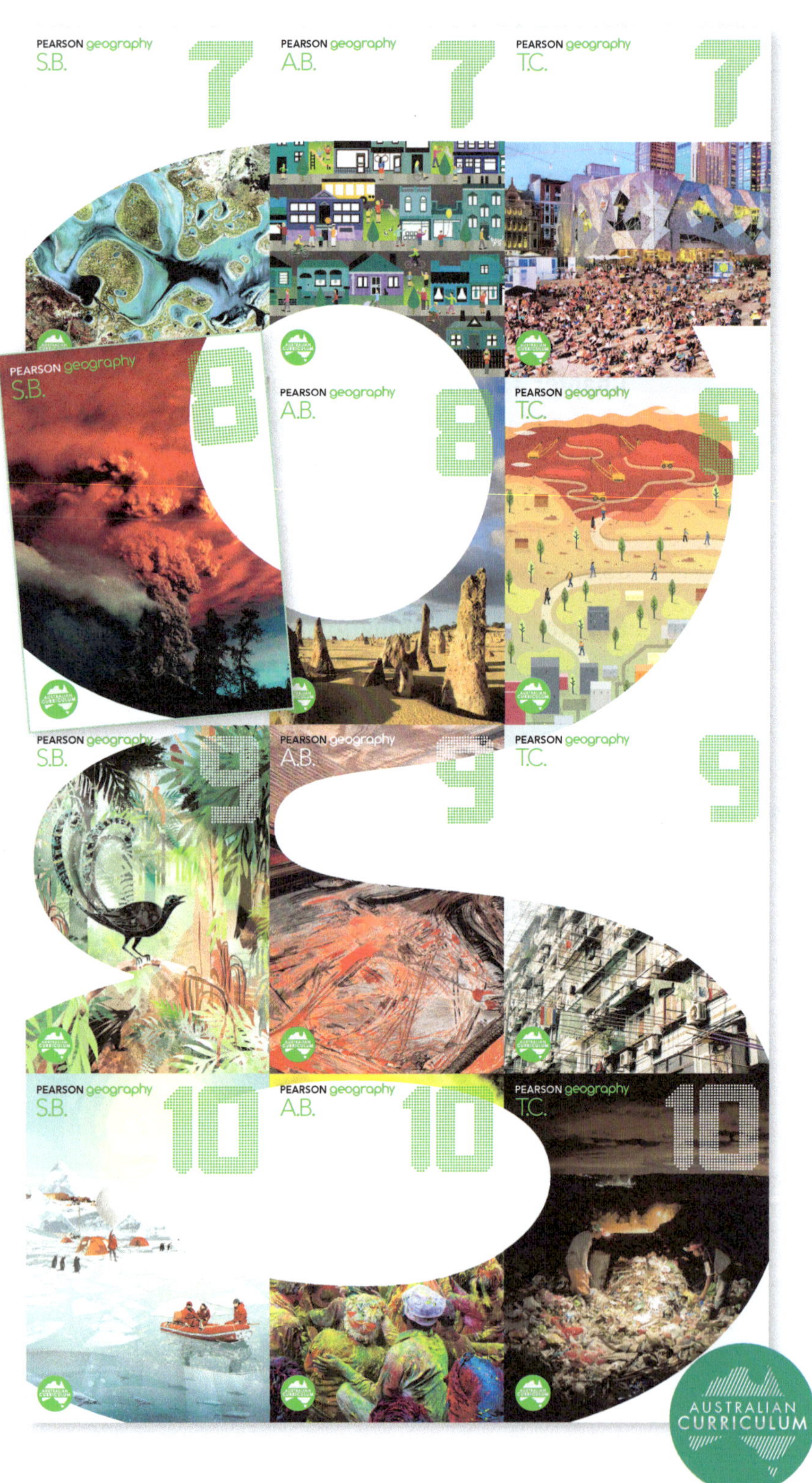

Built from the ground up

Pearson Geography has been created from the ground up for the Australian Curriculum: Geography, Years 7–10. It has been written and reviewed by a team of over 20 trained and qualified geography teachers from across Australia, ensuring its currency to teach students in the 21st century.

Professional learning, training and development

Did you know that Pearson also offer teachers a diverse range of professional learning programs? We are dedicated to supporting your implementation of **Pearson Geography**, but it doesn't stop there. We offer specific training for the Australian Curriculum and beyond, in the form of workshops, conferences and seminars. Find out more about Pearson Professional Learning at **www.pearson.com.au/pl**.

Customised content

We believe in learning for all kinds of people, delivered in a personal style. Pearson Custom is an exciting initiative that allows you to customise the content you teach with.

Schools can be involved in choosing resources specifically tailored to their school or year level. Speak to your Pearson Sales Consultant to find out more.

Student Book

Written specifically to meet the requirements of the Australian Curriculum: Geography, the Student Book acts as a guide for both students and teachers.

Full-colour presentation and contemporary design make the Student Book engaging and accessible for all students.

Activity Book

The Activity Book is a write-in resource designed to reinforce, enrich and extend the students' learning experience. The book can be used as an independent homework program or for individual classroom learning.

Teacher Companion

The Teacher Companion makes lesson planning easy by combining full-colour textbook pages, ideas for in-class activities and sample answers for the Student Book and Activity Book.

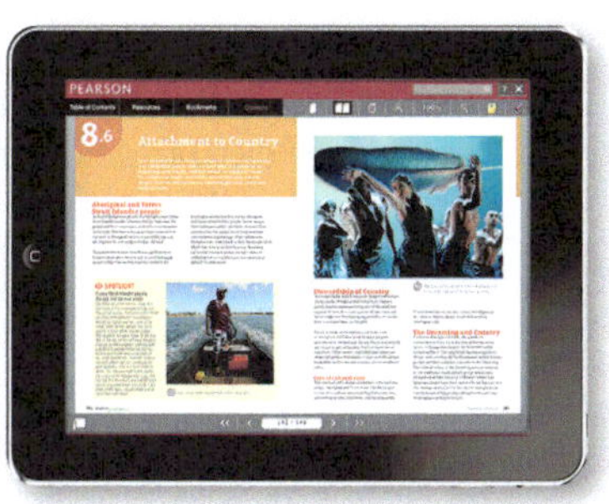

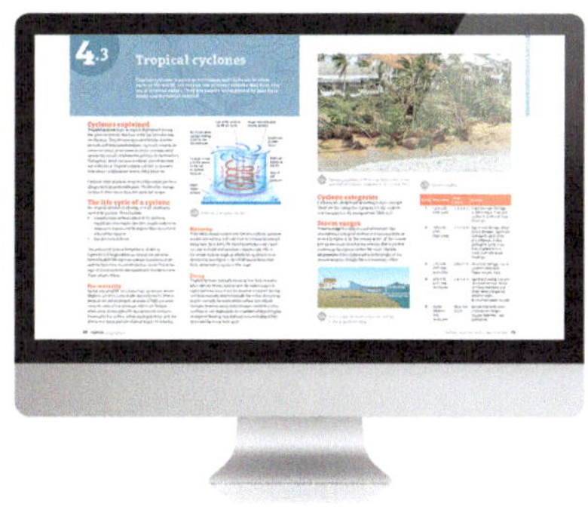

Pearson eBook 3.0

Any device, every school.

Pearson eBook 3.0 lets you use your student book online or offline on any device. Pearson eBook 3.0 retains the integrity of the printed page while offering linked interactive activities that will engage your students at school and at home.

Some activities are built with Adobe® Flash® and are not available on Android or iPad.

Pearson Reader 2.0

Collaborate. Assess. Learn.

Pearson Reader has been upgraded to provide easier and more intuitive access to content, and to all your favourite rich media resources and teacher support. Pearson Reader 2.0 also includes embedded interactive activities with tracked assessment and reporting to improve student outcomes.

Pearson Reader 2.0 is available online only.

Browse and buy at **pearson.com.au** | Access your content at **pearsonplaces.com.au**

PEARSON

How to use this book

Pearson Geography has been designed for the Australia Curriculum: Geography course. It includes content and activities that enhance the development of the Year 8 achievement standards within the two interrelated strands of Geographical Knowledge and Understanding and Geographical Inquiry and Skills.

Pearson Geography units are either two or four pages in length, designed to be completed in a lesson. Content is presented through a range of contexts to engage students and assist them. Pearson Geography has an engaging design; and it uses clear easy-to-understand language makes this a valuable resource for students of all interests and abilities.

Chapter opener

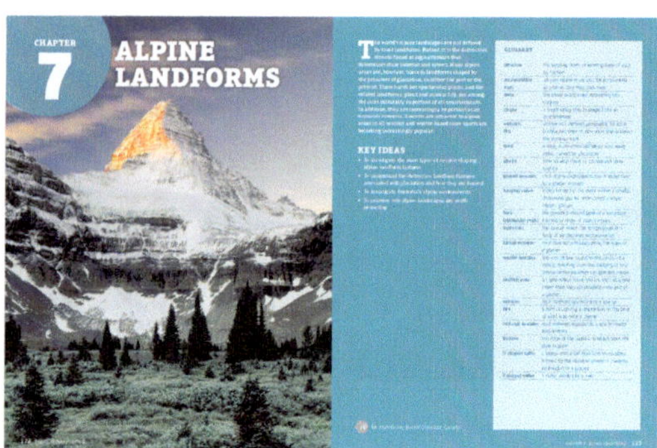

The chapter opener image is designed to engage students and provide a visual stimulus to the chapter themes. Also included is an introduction to the chapter and Key Ideas that link the chapter to the Australian Curriculum: Geography. A glossary that provides a ready reference for students to the key concepts and terms in the chapter.

Units

Each chapter of the Student Book is divided into units. Units have been written to develop students' knowledge and understanding of the concepts, skills and processes central to the study of Geography at this level. Unit are written to ensure both strands—'Knowledge and Understanding' and 'Inquiry and Skills' are interrelated as specified by the of the Australian Curriculum: Geography.

Skills builder

Skills builders are embedded in selected units and concentrate on key geographical skills.

Activity Book link

Use this icon to access an Activity Book worksheet that will consolidate and extend students' learning.

1.1

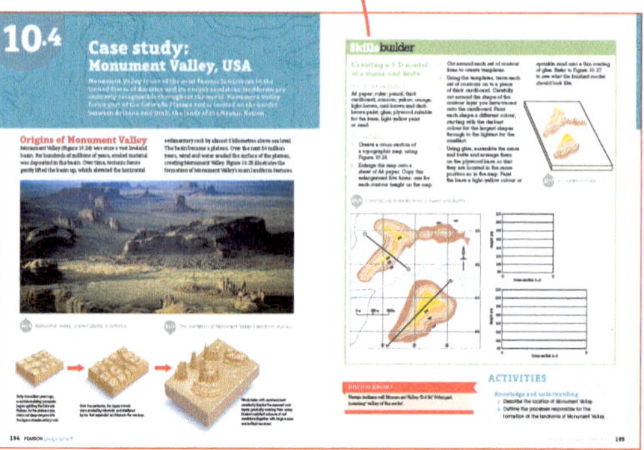

Spotlight

These features focus attention on a place, an issue or a concept relating to the unit.

Did you know?

Throughout the Student Book, these boxes give additional information and are designed to engage curious learners.

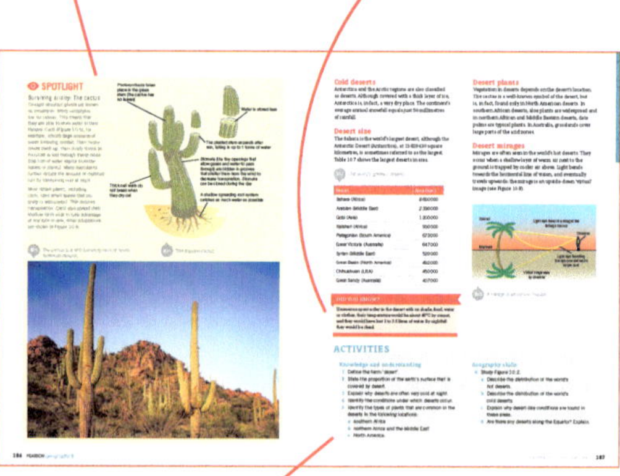

Activities

Each unit ends with a set of activities under selected headings from Bloom's Taxonomy of Cognitive Processes.

The activities include questions that guide students towards an understanding of the material covered and will extend them in a variety of learning experiences. The activities have been carefully selected to cater for the full range of students. The activities provide an opportunity for students to engage with important geographical issues from a range of perspectives.

Geoskills

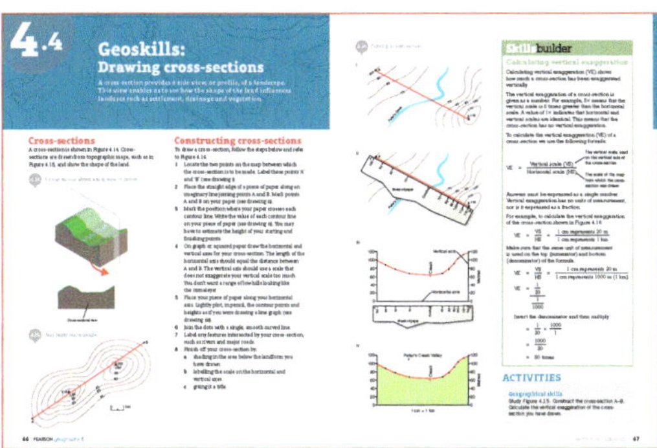

Geoskills units are designed to improve students' geographical skills: mapping, graphing, interpreting satellite images and using ABS data online. These skills relate to the Australian Curriculum: Geography Inquiry and Skills strand.

In the field

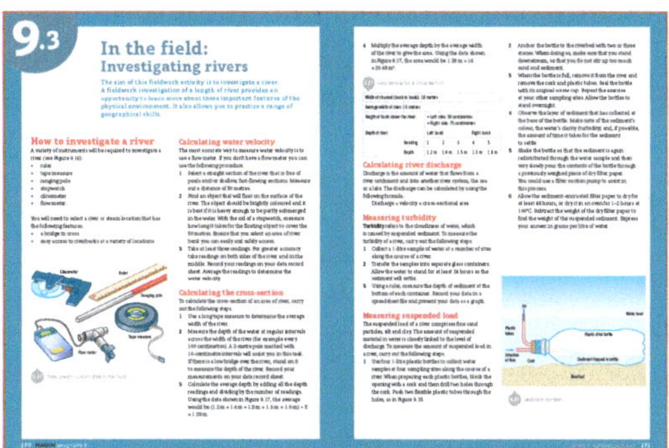

'In the field' units provide a step-by-step guide to undertaking and evaluating fieldwork. 'In the field' units have been written as a guide and are not tied to a specific location. Depending on the fieldwork task, some 'In the field' units can be conducted within school grounds.

Case studies

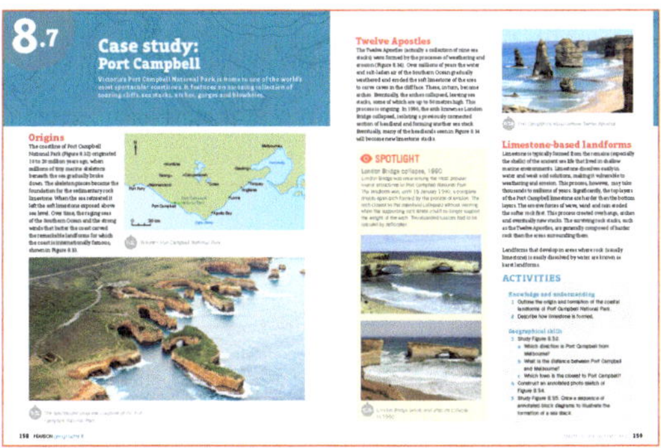

Case study units relate to a specific event or location. The units are written to extend students' knowledge and understanding. Case studies include examples from around Australia and world examples as specified in the Australian Curriculum: Geography.

Review and reflect

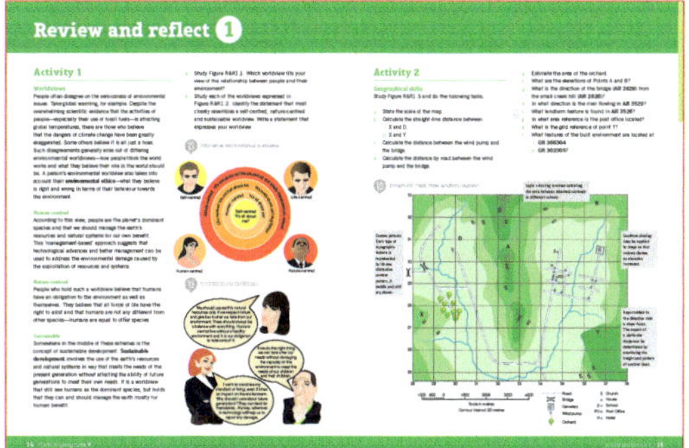

At the end of each of the three sections of the Student Book, a 'Review and Reflect' enables students to revise the key concepts and skills developed the text and complete higher order tasks. Students are encouraged to reflect on their own learning—to challenge their existing thinking and to refine their values and attitudes.

Directive terms

The following directive terms, grouped in a manner consistent with Bloom's Taxonomy, help you to identify the level or type of response required for a question or activity and provide a common language and consistent meaning in Australian Curriculum documents.

Remembering	
Define	State meaning and identify essential qualities
Label	Add annotations to a diagram or drawing
List	Write
Name	Present remembered ideas, facts and experiences
Present	Provide information for consideration
Recall	Present remembered ideas, facts and experiences
Specify	State in detail
State	Provide information without further explanation
Understanding	
Account	Account for: state reasons for, report on. Give an account of: narrate a series of events or transactions
Calculate	Ascertain/determine from given facts, figures or information
Clarify	Make clear or plain
Construct	Make; build; put together items or arguments
Describe	Provide characteristics and features
Determine	Find out the size or extent by measuring, counting or estimating
Discuss	Identify issues and provide points for and/or against
Explain	Relate cause and effect; make the relationships between things evident; provide why and/or how
Extract	Choose relevant and/or appropriate details
Gather	Collect items from different sources
Outline	Sketch in general terms; indicate the main features of
Predict	Suggest what may happen based on available information
Propose	Put forward (for example, a point of view, idea, argument, suggestion) for consideration or action
Rank	Place in order of size, age or as instructed
Recount	Retell a series of events
Summarise	Express, concisely, the relevant details

Applying	
Apply	Use, utilise, employ in a particular situation
Calculate	Ascertain/determine from given facts, figures or information
Demonstrate	Show by example
Examine	Inquire into
Identify	Recognise and name
Analysing	
Analyse	Identify components and the relationship between them; draw out and relate implications
Classify	Arrange or include in classes/categories
Compare	Show how things are similar or different
Contrast	Show how things are different or opposite
Critically analyse/evaluate	Add a degree or level of accuracy, depth, knowledge and understanding, logic, questioning, reflection and quality to analysis/evaluation
Discuss	Identify issues and provide points for and/or against
Distinguish	Recognise or note/indicate as being distinct or different from; to note differences between
Interpret	Draw meaning from
Evaluating	
Appreciate	Make a judgement about the value of
Assess	Make a judgement of value, quality, outcomes, results or size
Conclude	Come to a judgement or result based on the reasoning or arguments that you present
Critically analyse/evaluate	Add a degree or level of accuracy depth, knowledge and understanding, logic, questioning, reflection and quality to analysis/evaluation
Deduce	Draw conclusions
Evaluate	Make a judgement based on criteria; determine the value of
Extrapolate	Infer from what is known
Justify	Support an argument or conclusion
Predict	Suggest what may happen based on available information
Propose	Put forward (for example, a point of view, idea, argument, suggestion) for consideration or action
Recommend	Provide reasons in favour
Select	Select one or more items, features or objects
Creating	
Construct	Make; build; put together items or arguments
Investigate	Plan, inquire into and draw conclusions about
Synthesise	Put together various elements to make a whole

CHAPTER 1

DISCOVERING GEOGRAPHY

Geography is concerned with the changes taking place in all living and non-living elements of the earth's surface and atmosphere. The elements interact to produce the diverse landscapes that make up the world around us.

Physical geographers study the earth's climates, the formation of landforms, and the functioning and distribution of ecosystems.

Human geographers are concerned with the world's people, communities and cultures. Of particular interest are the ways in which the activities of people impact on places.

There's always something new to study in geography, for example the movements of people, including the shift to the cities (urbanisation), and the changes taking place in the distribution of economic activities. These changes are reshaping the geography of nations. Geographers are also interested in the impact of natural disasters, climate change and the ways in which new technologies are transforming the ways in which people interact.

Geography is much more than just knowing the names of countries and oceans. The study of geography allows us to better understand the changes taking place in the world in which we live.

KEY IDEAS

- To investigate change in geography
- To understand how people's worldviews affect the ways in which they interact with the natural, managed and constructed environments
- To develop the skill to identify spatial association between maps

1.0 Nabro volcano, Eritrea

GLOSSARY

biophysical environment	environments dominated by natural features such as landforms and vegetation. The biophysical environment includes the earth's soil, water, air, sunlight and all living things
change	a transformation bought about by environmental, economic, political, social and/or cultural factors
constructed environment	the earth's human-altered landscapes. It includes all those features that are normally associated with settlements, industries and agriculture
environmental ethics	what a person believes is right and wrong in terms of their behaviour towards the environment
environmental worldview	how people think the world works and what they believe their role in the world should be
geography	the study of all living and non-living elements of the earth's surface and atmosphere
managed environment	human-altered landscapes, dominated by elements of the natural environment; examples include crop and grazing lands, plantations and planted forests
perspective	a way of viewing the world
place	the human and physical characteristics of a specific location on the earth's surface
scale	the relationship between the distance on a map and the actual distance on the earth's surface
sustainable development	the managed use of the earth's resources and natural systems in a way that meets the needs of the present generation without affecting the ability of future generations to meet their own needs

1.1 Geography's focus on change

Geography is the subject in which we find answers to questions about the world around us. It is concerned with the processes that shape the earth's surface and the ways in which people interact with the environment. The study of geography helps us to better understand the world in which we live.

Change over space and time

Geographers often refer to the theme of **change** over space and time. It is the study of change that occurs in **places**. The outcomes of these changes are places that are different from other places and, in many cases, unique. The study of **geography** helps us to understand these changes and to predict what might occur.

Natural environment

Changes in the **biophysical environment** can occur very slowly or quite suddenly. The shaping of landforms by the processes of weathering and erosion can take millions of years. In contrast, earthquakes, tsunamis and volcanic eruptions can transform the landscapes within hours.

Constructed and managed environments

Change also occurs in the constructed and **managed environments**. Those of us who live in cities (an example of a **constructed environment**) are surrounded by change. Figure 1.1 provides examples of these changes.

Those of us living in rural communities are also surrounded by change. Figure 1.2 provides examples of these changes.

1.1 Changes in cities

1.2 Changes in rural communities

SPOTLIGHT

The Grand Canyon

The formation of the Grand Canyon is shown in Figure 1.3. Beginning about 200 million years ago, nearly forty layers of sedimentary material were deposited on top of volcanic rocks in the shallow seas that then existed in western North America. Uplift of the region began about 75 million years ago during a major mountain-building event. In total, the Colorado Plateau was uplifted by an estimated 3.2 kilometres. The Colorado River and its tributaries cut their channels through layer after layer of rock while the region was slowly uplifted. Because the region is relatively dry, there is little or no vegetation holding the soil together, so the Colorado River and its tributaries cut downwards at a faster rate than they cut horizontally. As a result, deep canyons rather than wide river valleys developed. Today, the canyon stretches for 446 kilometres, and is up to 29 kilometres wide and 1800 metres deep.

1.3 North America's Grand Canyon took two billion years to develop.

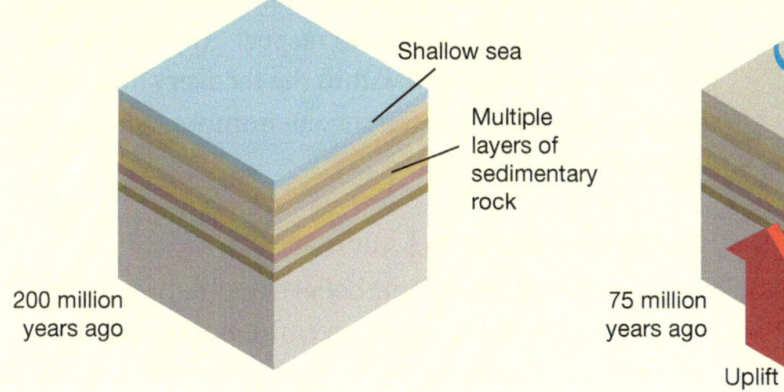

Colorado River

75 million years ago

Uplift

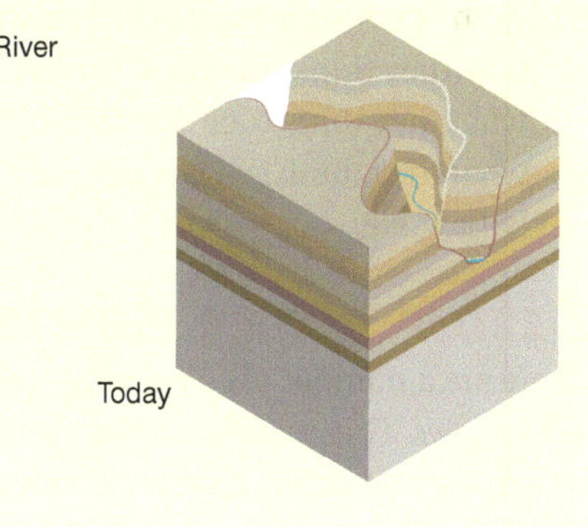

ACTIVITIES

Knowledge and understanding

1. Explain what is meant by 'change over time and space'.
2. Outline the benefits you get from studying geography.

Applying and analysing

3. Study Figures 1.1 and 1.2. Identify the changes taking place in cities and rural communities. Can you think of any other changes?
4. List the changes taking place in your own neighbourhood or community. Construct an annotated illustration to record the changes you have identified.

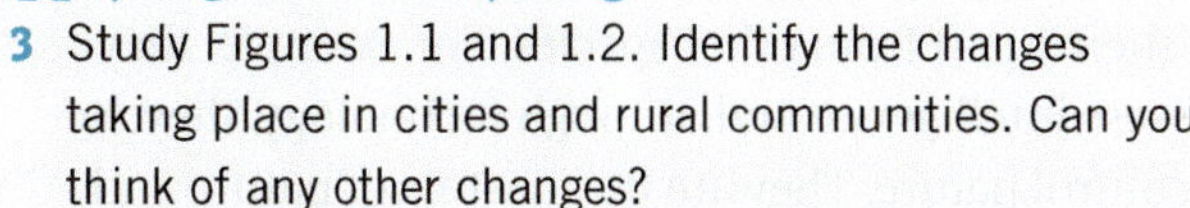

1.2 Environmental worldviews

Knowledge of people's environmental worldviews is important because it helps geographers understand the various attitudes that people have towards the natural world and its resources. Worldviews determine how people see their place in this world and are reflected in their behaviour and the decisions they make.

Perspective

A worldview is a point of view, or **perspective**. It can be held by individuals, though when accepted and shared by many it becomes a belief system that is widely acknowledged. An environmental worldview is concerned with how the earth and all its species and resources are managed.

Establishing worldviews

Conflicts arise when people hold different environmental worldviews. People will often disagree, for example about whether an environmental problem should be solved or ignored. People have different points of view or opinions about the worth of other species and the extent to which they should be protected.

A person's environmental worldview is determined by:

- how they think the world works
- what they think their role in the environment is
- what they think is the correct environmental behaviour.

Different worldviews

There are a number of competing environmental worldviews. These views may be classified as human-centred, stewardship-based and nature-centred. These views are explained below and summarised in Table 1.4.

Human-centred worldview

This worldview is based on the belief that humans are the most important species and have several traits that set them apart from other species. For example, they possess intelligence and, through technology, they try to control nature. They are the masters of nature and other species have a value only if they are considered useful to humans.

As human needs are considered the most important, the impacts of economic growth and development on the natural world are not important. Technology is seen to be the solution to problems of potential environmental damage.

Stewardship-based worldview

People with this worldview believe that humans should be caring managers, or stewards, of the natural world. They recognise that humans use resources but are aware that these resources may run out and should be managed carefully. Any development must not threaten the earth's life support systems on which humans and other species and ecosystems depend.

Indigenous people acted as stewards, as they had a close spiritual connection with the land and its creatures. They were grateful for what they were able to take and treated the natural world with respect. Any hunting or gathering was done well within the recovery limits of species and they managed the environment successfully for thousands of years.

Nature-centred worldview

People with this worldview believe that nature exists for the benefit of all species on earth, not just humans. Humans are considered equal with other species, and all species have a value separate from any material benefit humans may get from them. Every species has a right to life and does not have to be useful to humans to justify its continued existence. Humans also have a moral and ethical responsibility to guard against the extinction of any species.

The Deep Ecology movement goes even further, claiming that humans have no right to interfere with the richness and diversity of the natural world.

1.4 Comparing worldviews

Human-centred worldview	Stewardship-based worldview	Nature-centred worldview
Humans are separate from the rest of nature and can manage nature to satisfy their ever-increasing needs and wants.	Humans have an ethical responsibility to be responsible stewards of the earth and its resources.	Humans are part of nature and totally dependent on it for our wellbeing.
Technological advances will enable humanity to overcome any adverse impact on the natural environment.	The supply of natural resources is plentiful but they should be used carefully to avoid waste.	The earth's resources are limited and should not be wasted.
There is no limit on future economic growth.	Sustainable forms of economic growth should be encouraged ahead of those that are environmentally damaging.	Sustainable forms of economic activity should be encouraged. Activities that degrade the earth's resources should be discouraged.
Our future wellbeing depends on how well we manage the earth's life-support systems.	Our future wellbeing depends on how we manage the earth's life-support systems for our benefit and that of the rest of nature.	Our future wellbeing depends on our developing an understanding of how nature sustains itself.

SPOTLIGHT

Douglas Tompkins

Douglas Tompkins is a wealthy American environmentalist and former businessman who founded the clothing companies North Face and Esprit. He went on to sell these businesses when he became very concerned about the environmental impact of the fashion industry. Tompkins used his wealth to promote the importance of conservation, especially the protection of areas of exceptional environmental quality. His first project was Pumalin Park in Chile, where he purchased a 3200 square kilometre area of temperate rainforest, high peaks, lakes and rivers (see Figure 1.5). Without Tompkins's intervention, this area might have been exploited for power generation, industry or agriculture. The Chilean government has since declared it a nature sanctuary. Tompkins also established the Foundation for Deep Ecology.

1.5 Pumalin Park, Chile, is an area of exceptional beauty and unique forest ecosystems.

ACTIVITIES

Knowledge and understanding

1 Define the term 'worldview'.
2 Explain how a person's environmental worldview is determined.
3 Describe the three different types of environmental worldviews.
4 Explain why conflicts arise between people with differing environmental worldviews.

Applying and analysing

5 Determine your own worldview by considering the following: What is more important, the health of the earth's natural world or the wellbeing of people? Write a short explanation of your opinion.
6 Study Table 1.4. Which of the various descriptions most closely matches your worldview? Which of them most closely reflects the worldview of your parents?

1.3 Maps

Maps play a very important role in the study of geography. They tell us about places and help us to identify patterns and changes in the landscape. A map is a graphic representation of part of the earth's surface drawn to scale on a flat surface. The amount and type of detail shown on a map depends on the scale and purpose of the map.

Types of maps

Maps range from the very simple to the very complex. Cartographers (map makers) use symbols, shading and colour to simplify reality and show how the features of the earth's surface are arranged and distributed. Some common types of maps are:

- topographic maps
- weather maps
- thematic maps (see Figure 1.6)
- street maps.

Elements of maps

Map essentials are a title, a direction indicator, a **scale**, a legend and a border. A good way to remember the essential elements of a map is the term 'BOLTSS' (Border, Orientation, Legend, Title, Scale and Source).

Quadrants

To help us to locate features on maps, geographers sometimes express location in terms of quadrants (see Figure 1.7). The quadrants are named for the points of the compass.

1.7 The quadrants of a map

North-west quadrant	North-east quadrant
South-west quadrant	South-east quadrant

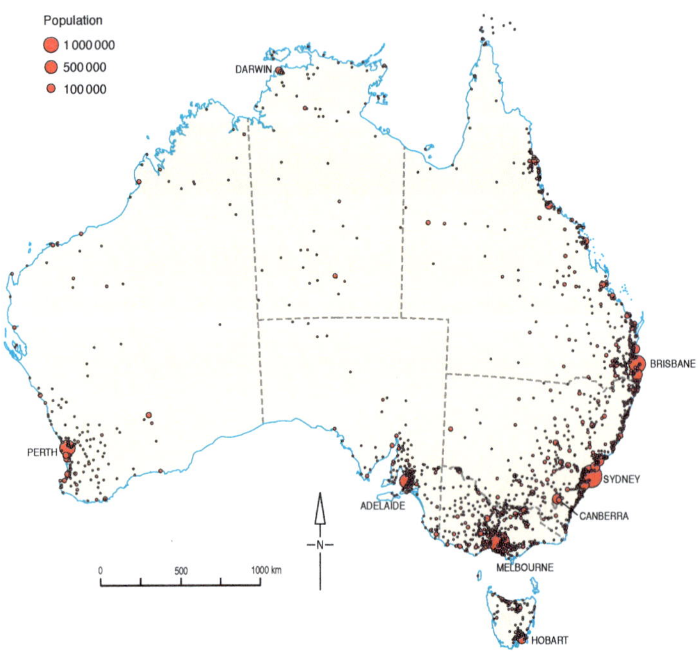

Source: *Heinemann Atlas*, 5th edition

1.6 Australia's population distribution

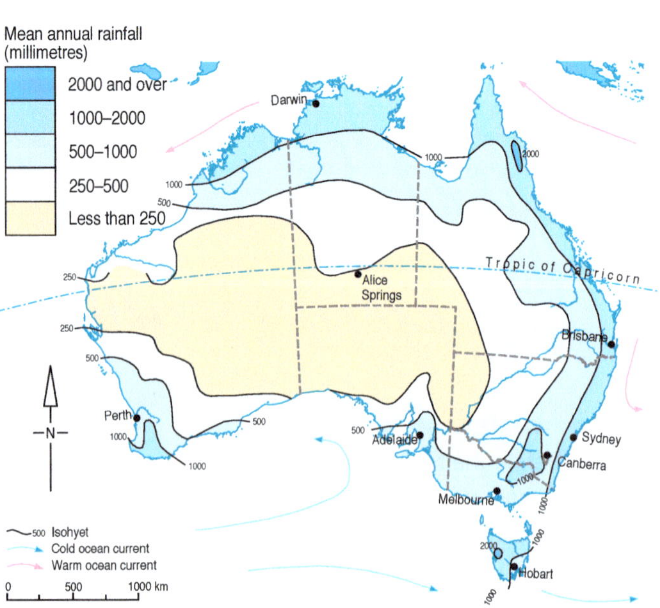

Source: *Heinemann Atlas*, 5th edition

1.8 Australia's annual rainfall

Patterns between maps

Maps help us to identify distribution patterns in landscapes. Figures 1.8 and 1.9 show a strong link between arid zones in Australia and regions that receive less than 250 millimetres of rainfall per year. Locations that receive over 2000 millimetres of rainfall per year have a tropical or maritime climate. In contrast, areas that receive less than 250 millimetres of rainfall per year have a semi-arid climate.

1.9 Climate regions of Australia

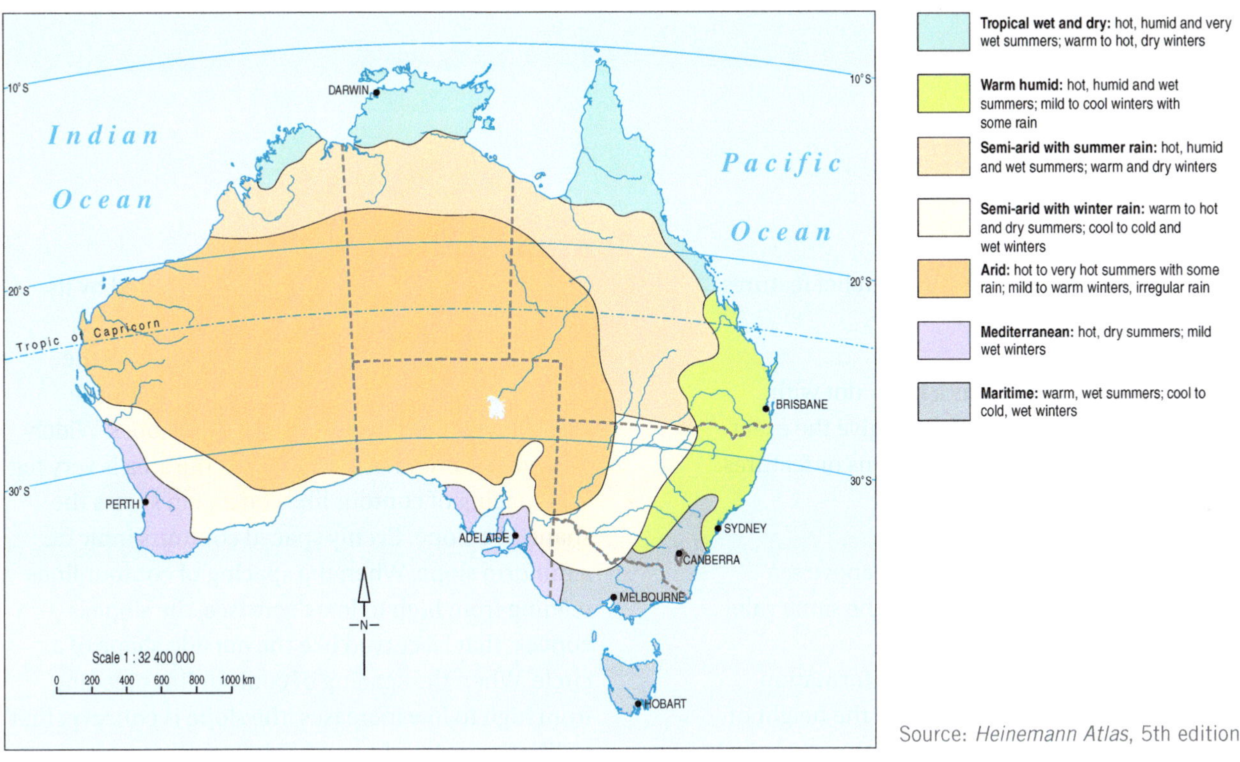

Source: *Heinemann Atlas*, 5th edition

ACTIVITIES

Knowledge and understanding

1 Name four types of maps.
2 List the elements of a map.

Geographical skills

3 Study Figure 1.6 and do the following tasks.
  a Name the quadrant/s that include cities of over 1 000 000 people.
  b Name the quadrant/s in which the least people live.
4 Study Figure 1.9 and do the following tasks.
  a Name the quadrant/s in which the maritime climate zone is located.
  b Name the quadrant/s in which the Mediterranean climate zone is located.
5 Study Figures 1.8 and 1.9 and do the following tasks.
  a Is there a link between tropical zones and areas receiving more than 1000 millimetres of rainfall per year? Explain.
  b Is there a link between semi-arid zones and areas receiving 250 to 500 millimetres of rainfall per year?
6 Study Figures 1.6 and 1.9 and answer the following questions.
  a Is there a link between areas with low population and arid zones? Explain.
  b Is there a link between areas with low population and warm-humid climate zones? Explain.
7 Study Figures 1.6 and 1.8 and discuss the following statement.
  There is a strong link between rainfall areas greater than 500 millimetres per year and the location of large cities in Australia.

1.4 Geoskills: Relief

An understanding of relief is central to the study of landscapes and landforms. 'Relief' is the term geographers use to describe the shape of the land, including height and steepness. The main techniques used by cartographers to show relief on topographic maps are spot heights, contour lines and patterns, and layer colouring and landform shading.

Topographic maps

Figure 1.10 is a topographic map showing relief features.

Spot height

A spot height is shown on a map as a black dot with the height written next to it. Spot heights give the exact height above sea level of particular locations or features.

Contour lines

A contour line joins points of equal height above sea level. Thus, every point along the line has the same value.

Contour lines provide geographers with information about the shape and slope of the land and the height of features above sea level. The contour interval, or vertical interval, is the difference in height between two adjacent contour lines. This interval is normally stated in the map's legend or near the edge of the map.

Contour patterns

Each type of topographic feature is represented by its own distinctive contour pattern.

- The spacing of the contours on a map shows the steepness of slopes. Contour lines that are close together show that the area has steep slopes. Widely spaced contour lines indicate that the area is very flat.
- The spacing of contour lines on a map shows the shape of a slope. Evenly spaced contours indicate a uniform slope. When the spacing of contour lines reading from high to low decreases, the slope is convex; that is, curved like the outside shape of a circle. When the spacing of contour lines reading from high to low increases, the slope is concave; that is, like the inside shape of a circle.

With practice, you can gain a visual impression of the shape of the land by interpreting the patterns made by the contour lines on a map.

1.10 Reading contour lines can tell us a lot about the nature of landforms.

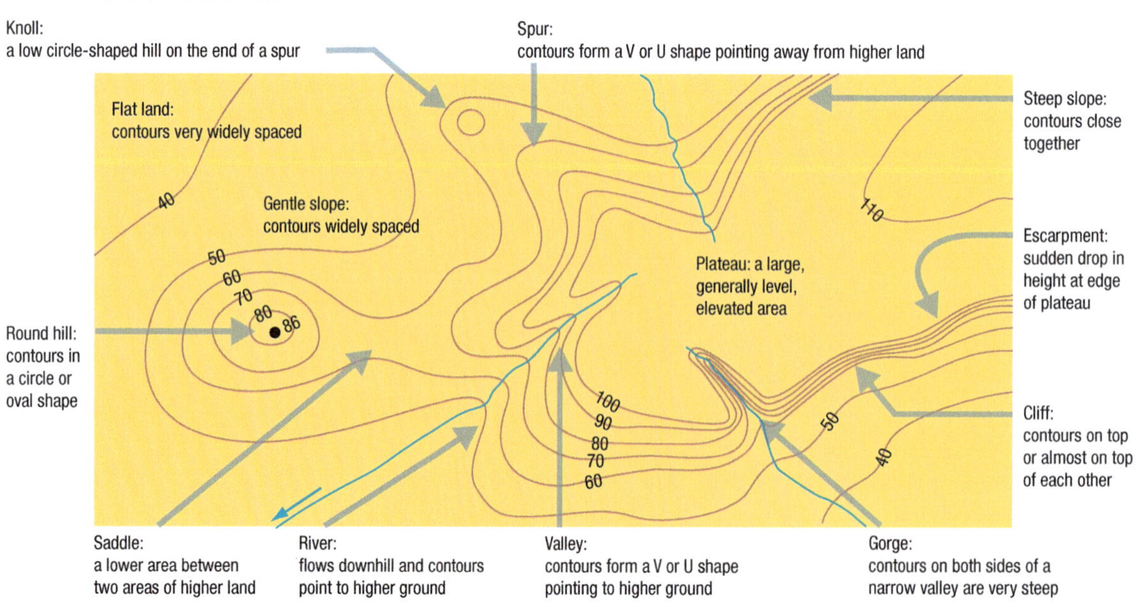

Layer colouring

Layer colouring is a simple and effective way of showing relief on maps. It involves colouring the area between selected contours in different colours. When it is used in combination with spot heights, and sometimes landform shading, layer colouring can tell you a good deal about the shape of the land.

Landform shading

Shading can be used on maps, with colours darkening as elevation increases. Shading may be used as if the light is coming from one direction, so that one side of a hill is shown in a lighter shade than the other to give it greater definition. Landform shading is sometimes used together with contour lines.

Aspect

Aspect refers to the direction that a slope faces. The aspect of a particular slope can be determined by examining the height and pattern of contour lines.

Skills builder

Estimating heights of landform features

Sometimes you will need to know the height of a map feature, such as the top of a hill or a plain. If there is no spot height on the feature, it is possible to estimate the height by studying the contour lines of the map. Use the following examples as a guide.

Example 1: Estimate the height of the hill at point A.

In Figure 1.11, point A lies more than 150 metres above sea level. However, it is obviously less than 200 metres above sea level. Your answer can be expressed in one of two ways:

- as a statement—point A is more than 150 metres, but less than 200 metres, above sea level; or
- as an estimate—point A is 175 metres (or any number between, but not including, 150 and 200 would be acceptable) above sea level.

Example 2: Estimate the height of point B.

In Figure 1.11, point B lies between the 50 and 100 metres contour lines. Your answer can be expressed in one of two ways:

- as a statement—point B is more than 50 metres, but less than 100 metres, above sea level; or
- as an estimate—point B is 75 metres (or any number between, but not including, 50 or 100 would be acceptable) above sea level.

1.11 Contour heights

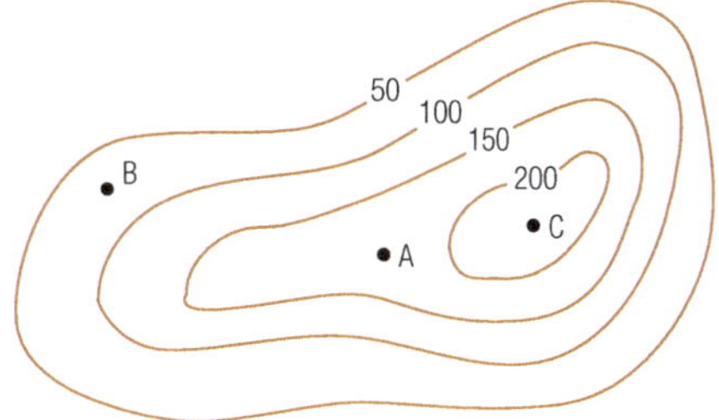

ACTIVITIES

Knowledge and understanding

1 Define the term 'relief'.
2 List the techniques to show relief on maps.
3 Explain what the spacing between contour lines tells us about relief.
4 Study Figure 1.10 and then answer the following questions.
  a Describe the difference between a cliff and an escarpment.
  b What is the spot height?

Geographical skills

5 Study Figure 1.10. What is the contour interval?
6 Study Figure 1.11. Estimate the spot height for C.

Investigating

7 Study Figure 1.10.
  a List the different landform features.
  b Find an image of each landform feature from your list.
  c Copy Figure 1.10 and annotate your diagram with the landform images you have collected.

1.5 Latitude and longitude

Most of the maps you will use include latitude and longitude. These lines allow you to quickly and accurately locate places and features on the earth's surface. Latitude and longitude also play an important role in determining times and dates.

Latitude

Lines of latitude are imaginary lines that run in an east–west direction around the earth. Because the lines of latitude are parallel to each other they are often referred to as parallels of latitude.

Important lines of latitude are shown in Figure 1.12.

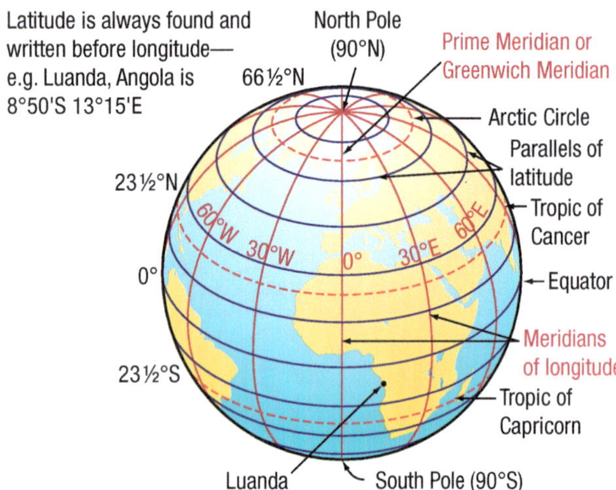

1.12 Parallels of latitude, meridians of longitude and the grid lines created by lines of latitude and longitude

Impact of latitude on climate

Latitude plays an important role in shaping the global pattern of climate. In Figure 1.13 you can see how the temperatures decrease as you move away from the Equator towards the poles. Incoming solar radiation (insolation) is the energy that reaches the earth's surface from the sun. Insolation is greatest at the Equator because the sun is directly overhead, which means that the sun's rays hit the earth at an angle of 90°. At higher latitudes the sun's rays strike the earth's surface at an angle. This means that the same amount of energy is spread over a larger area. Places near the Equator are, therefore, generally hotter than places near the poles.

1.13 Heating the earth's surface at the Equator and at higher latitudes

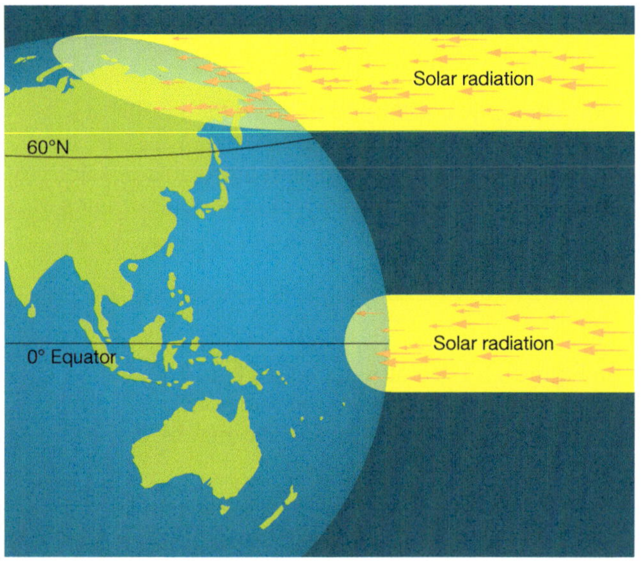

Longitude

Lines of longitude run in a north–south direction. They are not parallel to one another, but pass through both the North and South poles. Any number of these lines can be drawn but they all meet at the poles. These imaginary lines are called meridians of longitude. The most important line of longitude is the Prime Meridian (0°), which passes through Greenwich, England. All other lines of longitude are located to either the east or the west of the Prime Meridian.

Longitude and time

The Prime Meridian helps determine time. The earth's twenty-four time zones (one for each hour of the day) are organised according to Universal Coordinated Time (UTC), as shown in Figure 1.14. Places east of the Prime Meridian experience sunrise before UCT. Locations to the west of the Prime Meridian experience sunrise after UCT. Sydney, which is 115° east of Greenwich, has its sunrise before Greenwich.

Another important line of longitude is the International Date Line (IDL), which is on the opposite side of the world to the Prime Meridian, at 180°. Together, the Prime Meridian and IDL divide the earth into two halves: the Western and Eastern hemispheres.

The IDL is the point at which the change of day takes place. When you travel from east to west across the IDL, you gain a day. When you travel from east to west you lose a day. Australians travelling to the United States of America often arrive at their destination before their departure time. On the way back, however, they lose a day.

Australia has three time zones, while Russia has eleven. The zone boundaries zigzag in places so that people living in a region can operate on the same time (see Figure 1.14).

1.14 The world is divided into twenty-four time zones based on longitude.

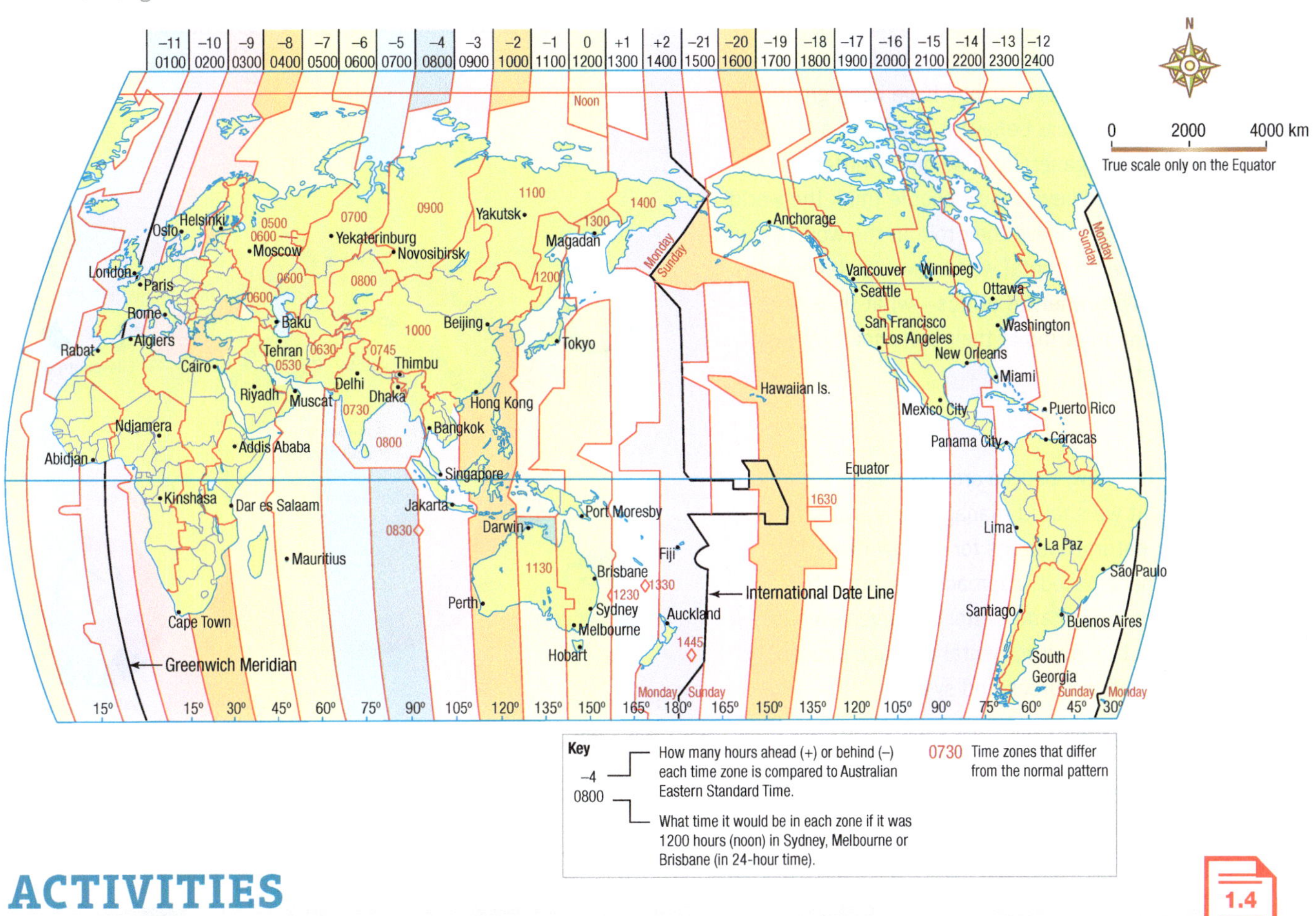

ACTIVITIES

1.4

Knowledge and understanding

1. Define the terms 'parallel of latitude' and 'meridian of longitude'.
2. Describe the location and significance of the Prime Meridian and the International Date Line.
3. How many time zones is the world is divided into? What does each zone represent?

Geographical skills

4. Study Figure 1.12. Write a paragraph describing the relationship between latitude and the global pattern of climate. Include in your response why the climate is hotter at the Equator than at higher latitudes.
5. Study Figure 1.14, showing the world's time zones.
   a. Calculate what time it will be in London, Singapore, Los Angeles and New York when it is midday in Sydney.
   b. What is the time difference between Sydney and Vancouver?
   c. If you live in Sydney and want to phone a friend in London at 8 a.m. before she leaves for school, what time of day would it be in Sydney when you call?
   d. Your Los Angeles flight leaves Sydney at 1.50 p.m. and takes 14 hours. What is the day and time when you land in Los Angeles?

Review and reflect 1

Activity 1

Worldviews

People often disagree on the seriousness of environmental issues. Take global warming, for example. Despite the overwhelming scientific evidence that the activities of people—especially their use of fossil fuels—is affecting global temperatures, there are those who believe that the dangers of climate change have been greatly exaggerated. Some others believe it is all just a hoax. Such disagreements generally arise out of differing environmental worldviews—how people think the world works and what they believe their role in the world should be. A person's environmental worldview also takes into account their **environmental ethics**—what they believe is right and wrong in terms of their behaviour towards the environment.

Human-centred

According to this view, people are the planet's dominant species and that we should manage the earth's resources and natural systems for our own benefit. This 'management-based' approach suggests that technological advances and better management can be used to address the environmental damage caused by the exploitation of resources and systems.

Nature-centred

People who hold such a worldview believe that humans have an obligation to the environment as well as themselves. They believe that all forms of life have the right to exist and that humans are not any different from other species—humans are equal to other species.

Sustainable

Somewhere in the middle of these extremes is the concept of sustainable development. **Sustainable development** involves the use of the earth's resources and natural systems in way that meets the needs of the present generation without affecting the ability of future generations to meet their own needs. It is a worldview that still see humans as the dominant species, but holds that they can and should manage the earth mostly for human benefit.

a Study Figure R&R1.1. Which worldview fits your view of the relationship between people and their environment?

b Study each of the worldviews expressed in Figure R&R1.2. Identify the statement that most closely resembles a self-centred, nature-centred and sustainable worldview. Write a statement that expresses your worldview.

Environmental worldviews

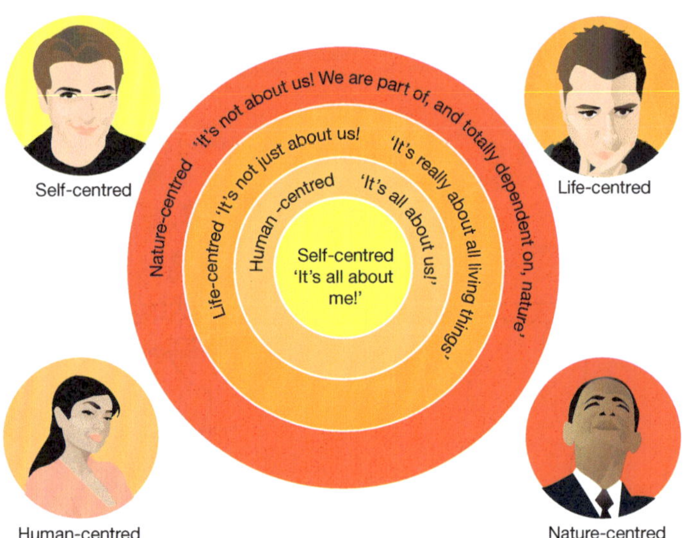

Worldviews

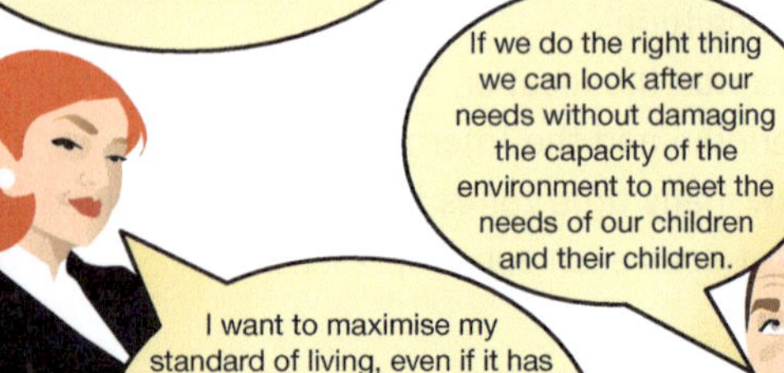

Activity 2

Geographical skills

Study Figure R&R1.3 and do the following tasks.

a State the scale of the map.

b Calculate the straight-line distance between:

  i X and D.

  ii X and Y.

c Calculate the distance between the wind pump and the bridge.

d Calculate the distance by road between the wind pump and the bridge.

e Estimate the area of the orchard.

f What are the elevations of Points A and B?

g What is the direction of the bridge (**AR 2829**) from the small creek hill (**AR 2826**)?

h In what direction is the river flowing in **AR 3529**?

i What landform feature is found in **AR 3526**?

j In what area reference is the post office located?

k What is the grid reference of point Y?

l What features of the built environment are located at:

  i **GR 366304**

  ii **GR 302305**?

1.5

R&R 1.3 Topographic maps show landform features

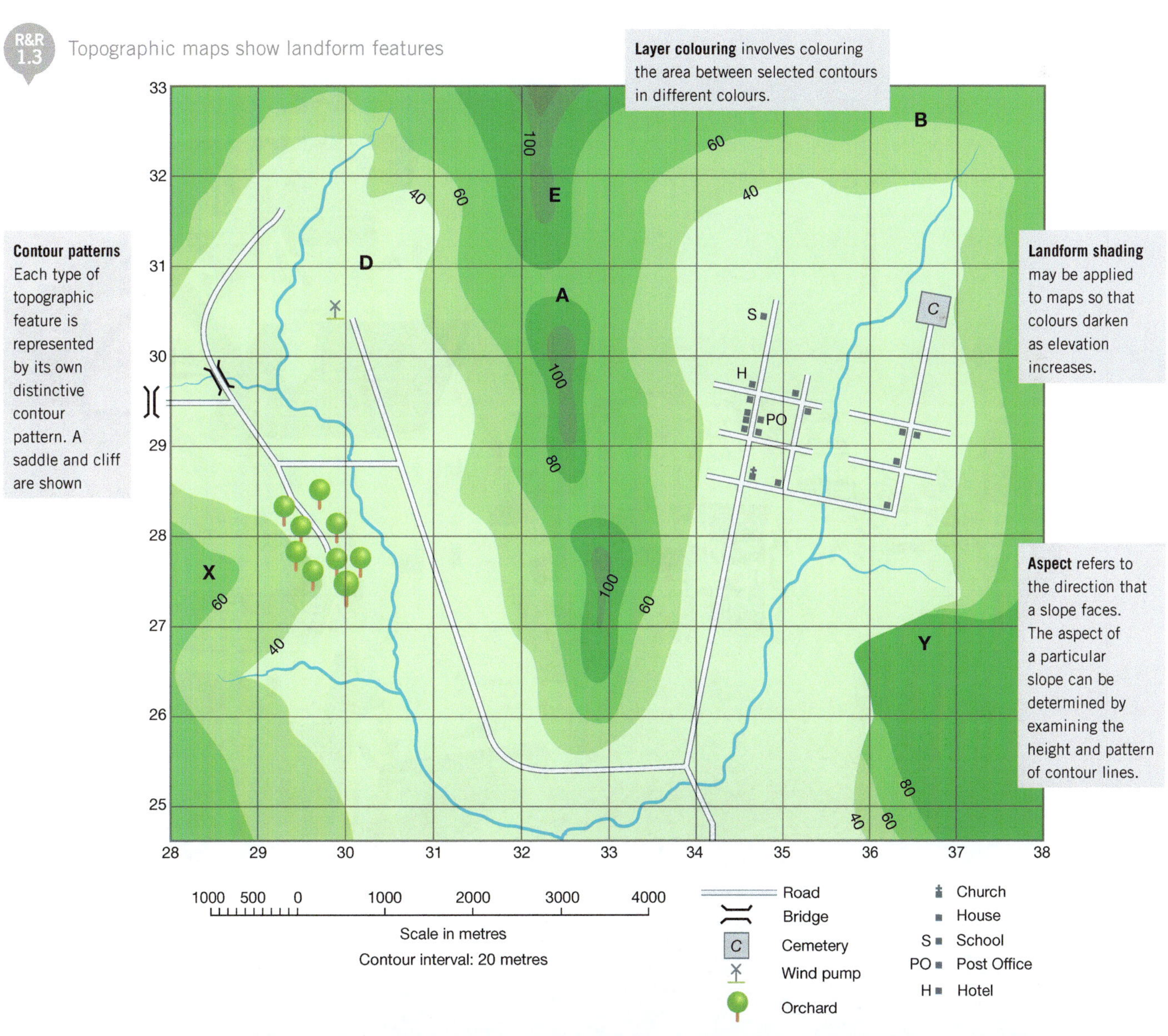

CHAPTER 2 LANDSCAPES AND LANDFORMS

The picturesque Austrian village of Hallstatt is the centrepiece of a spectacular alpine landscape. The landscape combines elements of both the biophysical and the constructed environments in a unique way. The landscape has, for 2500 years, been the setting for the mining and processing of salt. Salt has shaped all aspects of life of the region, including its economy and architecture.

In 1997, the Hallstatt-Dachstein/Salzkammergut Cultural Landscape was added to UNESCO's World Heritage List in recognition of the region's outstanding cultural importance and its magnificent natural setting.

In this chapter we look at landscapes and landforms, their link to national identity and the arts, the ways in which they are valued, and the ways in which they can be protected and conserved. We also examine indigenous explanations of landscapes and the ways in which Aboriginal and Torres Strait Islander people manage landscapes.

KEY IDEAS

- To investigate the key elements of landscapes
- To understand how national identity is shaped by landscapes
- To discuss the different ways in which landscapes are valued
- To understand the value that Aboriginal and Torres Strait Islander communities place on landscapes and landforms
- To explain how landscapes and landforms are protected

2.0 Hallstatt, Austria

GLOSSARY

aesthetic value	the value of a landscape based on its beauty or attractiveness
biophysical environment	an area dominated by natural features such as landforms and vegetation; includes the earth's soil, water, air and sunlight, and all living things
constructed environment	the earth's human-altered landscapes. It includes all those features that are normally associated with settlements, industries and agriculture
Dreaming	a key spiritual belief of Australian Aboriginal people; describes both the period of creation (the Dreaming) and the stories that come from this period (Dreaming stories)
landform	a natural feature of the earth's surface, for example a mountain, valley, lowland or volcano
landscape	the overall appearance of an area resulting from the interaction of landforms, vegetation and soils with human elements of the environment, such as transport networks, settlements, farms and factories
multicultural	relating to several or more cultural or ethnic groups in a society
national identity	a person's sense of belonging to a nation
place	the human and physical characteristics of a specific location on the earth's surface
sacred sites	places of significant cultural and spiritual meaning to Aboriginal and Torres Strait Islander people

2.1 Landscapes and landforms explained

Landscape is the term used to describe the features of an area. It includes the biophysical elements of landforms, the living elements of land cover, the human elements and changeable elements such as weather conditions.

Culture and landscapes

Landscapes combine physical features with an 'overlay' of human activity, as shown in Figure 2.1. This overlay may have accumulated over thousands of years and can often easily be identified as elements of managed and constructed environments. Sometimes, however, evidence of human activity is harder to see. Geographers seek to explain how and why places have changed over time.

Landscapes are the product of the interaction of people and **place**, and play an important role in creating what geographers refer to as 'sense of place'—the qualities that distinguish one place from another. Landscapes also play a role in shaping people's personal, local and **national identities**.

Landforms

Landforms are the natural features of the landscape. Natural physical features of the earth's surface include valleys, plateaus, mountains, cliffs, plains, hills, dunes and glaciers.

Types of landscapes

The earth has a vast range of biophysical, managed and constructed landscapes. The earth's biophysical landscapes are those largely unaffected by human activity. They include its icy polar regions; its great mountain ranges (Figure 2.2) and vast deserts; the savanna grasslands (Figure 2.3); coastal and reef landscapes (Figure 2.4); and the tropical and boreal forests. There are also a number of managed and constructed landscapes. These include the world's various agricultural landscapes (Figures 2.5 and 2.6) and its industrial and urban landscapes (Figures 2.7 and 2.8).

2.1 Elements of landscape

Natural elements

Human elements

Landscape

Living elements

Changeable elements

2.2 A mountain landscape, New Zealand

2.3 Savanna grasslands, Africa

2.4 Whitehaven Beach, Australia

2.5 Agricultural landscape, Bali

2.6 Agricultural landscape, Switzerland

2.7 Industrial landscape, China

2.8 Urban landscape, Vancouver, Canada

ACTIVITIES

Knowledge and understanding

1 State what is meant by the following terms:
   a landscape
   b overlay of human activity
   c sense of place.
2 Identify the elements of landscapes and give examples of each.
3 Explain why human activity can be difficult to observe. Give an example.
4 Define the term 'landform'.
5 Explain why geographers study landscapes.

Applying and analysing

6 Study Figures 2.2 to 2.8.
   a Describe the differences between biophysical, managed and constructed landscapes.
   b Rank the landscapes from the one you find most appealing to the one you find least appealing.
   c Write down the criteria you used when determining your ranking.
7 Identify the typical 'Australian' landscape. What are the elements of the landscape that make it uniquely Australian?

2.2 Landscapes and national identity

Landscapes can play a powerful role in the creation of national identities. People connect with an image of a landscape. This has an impact on how they see themselves as a national community. Landscapes also build a sense of belonging that helps define people's attitudes and perspectives.

Building a national identity

A national identity is developed over time. The qualities that develop represent the society and its people. These qualities foster a sense of common beliefs and ideas about identity that people feel they share with others in their society.

We are not born with a national identity. A country's history helps to develop the idea of a national identity. Our awareness and understanding of national identity are shaped by the images of our biophysical and built environment that we are exposed to in the media and the arts throughout our lives. National identity becomes the collective personality of the people of a country. It is the foundation of the values we take on as our own and it is often expressed in our behaviour.

The importance of landscape

Landscapes are important to people. Landscapes are not just scenery or interesting views; they have a lot of meaning for people. Apart from being the backdrop for day-to-day living, landscapes link people to nature and the past to the present. People create and value the concept of landscapes. For example, the Great Barrier Reef is a valued landscape and there are now laws to protect and preserve areas of the reef. Some landscapes are very special because they symbolise a nation, such as the Eiffel Tower in Paris, France.

Landscapes and national identities

Landscapes are shared by the people of a nation and often provide a living history of the past. Some national identities based on these landscapes become so popular that they last long after they are truly representative of the nation. For example, Egypt's pharaohs are long dead but the pyramids at Giza, shown in Figure 2.9, are still an important aspect of Egyptian national identity.

2.9 The pyramids at Giza, Egypt

Australia: from the bush to the beach

When the European settlers arrived in Australia, they encountered an environment very different from Europe. The vast expanses of country—from the outback to the forests and farmlands—became collectively known as 'the bush'. This landscape was uniquely Australian and it evoked many tales of the struggle for survival. Stories of drovers, outback women, bushrangers and even children lost in the bush became legends. By 1900, the landscape of the bush had become the symbol of the new nation, even though the majority of people lived in towns and cities.

The bush may have defined the national identity a century ago, but more recently the beach has become a symbolic landscape. Most Australians live near the coast and the beach culture has become a very important aspect of the Australian lifestyle (see Figure 2.10). The character attributes of the bush heroes have been transferred to the lifesavers patrolling surf beaches. They are seen as fit, courageous people who save lives.

Many people consider Australia's national identity as a work in progress, as we are still a young nation. With the majority of the population living in urban centres and the population itself becoming increasingly **multicultural**, new layers are being added to the country's national identity.

Australian beach culture as seen by cartoonist Reg Mombassa

Cartoon by Mombassa *Beachside barbecue*

ACTIVITIES

Knowledge and understanding

1 Explain the relationship between landscape and national identity.
2 Outline how a national identity is built.
3 Explain why there has been a shift in Australia's national identity.

Applying and analysing

4 Study Figure 2.10. List the points Reg Mombassa is making in his cartoon *Beachside barbecue*.
5 Select an Australian landscape and design an annotated visual display of a landscape that in your opinion characterises Australia's national identity—for example Uluru, Sydney Harbour, the Great Barrier Reef or the MCG. Your display should include a list of reasons why you chose this landscape to characterise Australia's national identity.

2.3 Landscapes and the arts

Landscapes are valued for many reasons. They may, for example, provide spectacular unique scenery that is prized by locals and visitors. Landscapes inspire people to create works of art. Such creations enrich our lives and provide a lasting record of those landscapes.

Source of inspiration

Landscapes can cause strong feelings in people and move some people to create something that expresses what they see and feel. Such creations often involve the telling of a story in a range of mediums or art forms. Artists, poets and novelists have long used landscapes as a source of inspiration. More recently, music, film, television and the various forms of multimedia have captured elements of landscapes.

Creative impressions of Australia

Images of landscape have featured strongly in the arts in Australia, reflecting the changing nature of the land and its people.

Traditional Indigenous art

Aboriginal and Torres Strait Islander people have a rich artistic tradition stretching back tens of thousands of years. Their connection to the landscape is expressed in stories and art, as well as in performance through ceremonies, music, songs and dance. Sacred ground mosaics in Central Australia are the most elaborate of these artworks, each representing an identifiable geographic location and many depicting the mythological incident that occurred there.

Most recognisable today are the dot paintings, in which the various colours such as red, yellow, brown and white represent the desert sand, sun, soil and the clouds or sky respectively. Many such paintings present aerial views of the landscape and are often titled simply *Country*.

2.11 In his painting *Wood Splitters*, Tom Roberts captures the true colours of eucalypt trees in a misty, winter light.

The painters

For most of the first century after colonisation (late 1800s to 1900s), artists struggled to paint an unfamiliar environment. Their paintings and drawings were awkward and the trees in particular looked unrealistic, mainly because the artists' training had taken place in England. It was not until the emergence of artists such as Tom Roberts and the Australian-born Arthur Streeton and Frederick McCubbin that the colour, light and mood of the Australian landscape were truly captured in painting. This can be seen in Figure 2.11. More recent artists such as Fred Williams, Russell Drysdale, Sidney Nolan, Margaret Olley and Brett Whitely produced some highly acclaimed paintings of Australian landscapes.

Many Australian films draw on biophysical environment settings.

Beach landscapes are popular settings for many film and television productions in Australia.

Poets and writers

Poets and novelists such as Banjo Paterson and Henry Lawson were inspired by the Australian bush. In his classic poem *The Man from Snowy River*, Paterson immortalised the stockman as a hero in the setting of the rugged high country of the Snowy Mountains.

Other media

Creative people find many ways to express their ideas about landscapes. Music has always been an important part of Australian life. Many children today are familiar with the songs *Home Among the Gum Trees* and *Waltzing Matilda*, which both discuss the Australian bush landscape. Australia's national anthem makes reference to the landscape: 'Our home is girt by sea' and 'Our land abounds in nature's gifts' (verse 1); and 'We've boundless plains to share' (Verse 2).

Australia's natural landscapes have featured prominently as the setting for many notable films. Baz Luhrmann's epic film *Australia* (see Figure 2.12) was an example of storytelling that drew heavily on the splendour and uniqueness of the vast open spaces of Australia's Top End.

The impression that many people overseas have of Australia comes from Australian television shows. *Home and Away*, filmed at Sydney's Palm Beach, is hugely popular in the United Kingdom, where many viewers believe it never rains in this seaside community (see Figure 2.13).

ACTIVITIES

Knowledge and understanding

1. List the art forms for which landscapes have provided inspiration.
2. Trace the history of landscape painting in Australia.
3. Describe how landscapes have featured in other artistic productions in Australia.

Applying and analysing

4. Study Figures 2.11, 2.12 and 2.13 and compare the images of Australia. Describe the similarities and differences.

2.4 Valuing landscapes

Landscapes are important to us for many reasons. They are a shared resource—they belong to everyone. They are a living record of our past, and an inspiration for our culture. They provide a wide range of social and health benefits. The value placed on landscapes changes over time.

The different values

Landscapes are said to have aesthetic, emotional, spiritual and economic value.

Aesthetic value

The term '**aesthetic value**' refers to the idea of attractiveness or beauty. The aesthetic value of all landscapes is not the same, as not all landscapes are equally appealing. People can, for example, appreciate and interpret the same landscape quite differently. What may appeal to one person may not appeal to another. For US retirees seeking a warm, sunny climate, California's Palm Springs is an attractive desert landscape. Others view desert landscapes as hostile places. International tourists travel thousands of kilometres to see the landscapes of Australia's Red Centre, while many Australians head overseas to see the managed and constructed landscapes of South-East Asia (see Figure 2.14), Western Europe and North America.

Whether we find a landscape personally appealing depends on a range of factors. These include our emotions or feelings, our attitudes and personal values, and our preferences, experiences and memories.

Emotional value

People often develop an emotional attachment to a place or landscape. This attachment usually results from a long-term connection with that place or landscape. While this is especially strong for indigenous peoples, it also applies to people more generally. We all remember, often with fondness, places we went for holidays, and we often develop an emotional attachment to the places in which we live or have lived.

This attachment is different from a simple aesthetic response, such as recognising that a certain landscape or place is special because it is beautiful. For a deeper and lasting emotional attachment to develop, a long-term relationship with a place is normally required. This relationship may be physical or emotional. It might be the place in which we live or a place our mind wanders to from time to time. A snowboarder might, for example, find their thoughts drifting to the mountains on which they snowboard. A bushwalker often develops an emotional attachment to their favourite national park, such as Wilsons Promontory, shown in Figure 2.15.

Many people find landscapes inspiring. Landscape artists and photographers, for example, seek to capture or portray the beauty of landscapes in their paintings and photographs. These provide an opportunity for people who can't observe the landscapes directly to share the experience.

2.14 Singapore

DID YOU KNOW?

In 2012, approximately 1 billion people travelled internationally, generating more than US$930 billion in economic activity. This represents 5 per cent of the world's total gross domestic product and one in twelve jobs.

2.15 Wilsons Promontory, Victoria

Spiritual value

Landscapes hold special spiritual significance for some people. Many Aboriginal and Torres Strait Islander people, for example, recognise that features of the landscape such as rivers, mountains and even individual trees have a spiritual value. Through these features, Indigenous Australians connect, by means of song and ritual, to the **Dreaming** and ancestral creator beings. **Sacred sites** are places of special spiritual importance to Indigenous Australians.

Economic value

Landscapes also have an economic value. Some, such as agricultural, industrial and urban landscapes, are in fact the product of economic activity. Others generate economic activity even though they remain in a near-natural state.

Landscapes that are spectacular and/or unique are important tourist destinations. Historic Venice in Italy, for example, attracts nearly 22 million tourists a year; California's Yosemite National Park more than 4 million; Machu Picchu in Peru, more than 700 000; and Australia's Great Barrier Reef, more than 1.6 million people. Catering for the needs of these visitors creates economic activity and employment. The economic wellbeing of many communities depends on tourism.

Landscapes provide a wide range of opportunities for people to enjoy the outdoors. They vary from the local park through to coastal national parks and remote mountain wilderness areas. All offer relaxation, challenge, inspiration and an opportunity to experience first-hand our natural and cultural heritage.

ACTIVITIES

Knowledge and understanding

1 Explain what we mean when we talk about a landscape's 'aesthetic value'.
2 Name a place or places that have aesthetic value for you.
3 Explain the economic value of landscapes.

Geographical skills

4 Construct a photo sketch of Figure 2.14 or 2.15.
5 Annotate the sketch, naming as many natural elements of the landscape as you can. Are there any human elements evident?
6 List the possible aesthetic, emotional, spiritual and/or economic value of such a landscape.

2.5 Indigenous explanations of landscapes

The relationship with landscape is central to the life of indigenous peoples worldwide. Over thousands of years, indigenous peoples have developed deep spiritual links with their traditional lands.

Creation myths

Indigenous peoples have their own stories about the origins of the world, their people and all the living things. This can be seen in their paintings and carvings. An example is shown in Figure 2.16. Stories are told and retold and have been passed down from generation to generation over thousands of years. The stories are also expressed in song and dance. The rich oral and artistic history of indigenous peoples has created a sense of social continuity and harmony, and taught people how to continue the traditions of their ancestors.

Supreme beings

Many of the stories passed down from generation to generation involve supernatural beings. This is especially the case in the stories dealing with the creation of the world. They are used to explain the origins of people, plants, animals and even landform features. Supreme beings, in one form or another, appear in ancient myths about creation in most cultures.

Animals

Indigenous peoples see themselves as part of nature and many view animals as their equals. Many ancient stories tell of a time when both humans and animals lived together peacefully, without any fear of each other. Indigenous explanations of the creation of the landscape feature animals in very important roles. In many cases they are credited with saving the human race. Such beliefs reflect the high regard indigenous peoples have for animals. While they may have hunted them for food and skins, they acknowledged how greatly they relied on them to support their needs.

2.16 Totem poles in Vancouver's Stanley Park recount the legends of the First Nations peoples of Canada.

SPOTLIGHT

The Iroquois creation myth

The Iroquois Native Americans (see Figure 2.17) believed the world was formed on the back of a giant turtle. The first woman, Sky Woman, dropped out of the sky and the water animals saved her from falling into the ocean that covered the earth. After saving her, the animals built an island for her to live on. Without their help, the Sky Woman might have perished and people might have never existed.

Source: Creation Myths—the Relationships of Animals and People, http://www.cs.williams.edu/~lindsey/myths/myths_8.html

2.17 The Iroquois

The Dreaming

According to Aboriginal belief, the landscape can be explained by the Dreaming—especially the time known by many Aboriginal people as Tjukurpa. This was when the great Ancestor Spirits roamed the earth. Before this time, the earth was flat and bleak and empty of any life. When the Ancestor Spirits came up from their dwelling places below the ground, they took the form of humans and animals. They created the features of the landscape, such as rivers, waterholes and mountains. By transforming themselves into these landforms they left evidence of their presence in the landscape. These spirits then created all life on earth—the plants, animals and people.

The tracks of the Dreaming cover Australia. The whole of the landscape is embedded with evidence of the Dreaming, which remains a powerful spiritual force for many Aboriginal and Torres Strait Islander people. For them, it still exists today and the network of tracks and sacred sites link the physical world to the Dreaming. Dreaming stories map out significant landscape features, the location of waterholes, places to camp, and places to gather food and hunt animals.

2.18 Aboriginal artist Turkey Tolson Tjupurrula at work, Central Desert, Australia

ACTIVITIES

Knowledge and understanding

1 Identify the link that an indigenous person has with their land.
2 Explain how these beliefs were maintained over very long periods of time.
3 Outline the role of the Ancestor Spirits in the Dreaming.
4 Discuss the importance of animals to indigenous peoples.

Applying and analysing

5 Compare the Iroquois creation myth and the Aboriginal Dreaming.

2.6 Case study: Aboriginal and Torres Strait Islander people

Australia's Aboriginal and Torres Strait Islander people have their own explanations of how the world and its landscapes and landforms were created. Aboriginal people believe that everything around them exists because of the Dreaming, a past that continues to exist in the present and in the future.

The origins of landscapes and landforms

At the beginning of time, spirit ancestors wandered the earth in the form of animals. While some of these ancestors went into the sky to create the sun, moon and stars, others stayed and created the natural world. They shaped the land, provided its resources and created the spirits that made plants, animals and human life. All living things, including Aboriginal people themselves, are descendants of all these original spirit beings, who did not die but live on in the different features of the landscape, flora and fauna.

The land and its features are therefore sacred to Aboriginal people, and a spiritual link exists between the land and the people who were placed on it. This means that separate language groups have a spiritual connection to particular areas and features of the landscape. This attachment is the basis for the recognition of different homelands. Aboriginal people believe that the spirit beings also gave to each group its own distinctive spiritual practices, beliefs, customs and languages.

The Dreaming Stories are a central element of the sacred Aboriginal beliefs that have been handed down from one generation to the next for over forty thousand years. Two examples of the spiritual relationship between Indigenous Australians and landforms are provided below: the Gagudju People of Kakadu, in the Northern Territory, and the South Australian Ngarrindjeri Dreaming Story known as Ngurunderi.

The Gagudju of Kakadu

Archaeological evidence suggests that the Kakadu region was one of the first parts of the Australian continent to be occupied by people. If this is correct, the descendants of the Gagudju may have occupied the area for at least 50 000 years. For much of this time they recorded Kakadu's history on the walls of the sandstone escarpment that dominates the landscape. By examining this art and the other evidence of Aboriginal occupation we can build up our knowledge of how the Aboriginal people interacted with the landscape, its landforms and the environment more generally.

2.19 Warramurrungundje came from the sea and created Kakadu's spectacular landscape.

DID YOU KNOW?

Originating more than 50 000 years, or 1600 generations, ago, Australia's Aboriginal heritage is one of the oldest continuous cultural traditions on earth.

According to a Gagudju Dreaming Story, a female ancestral being called Warramurrungundje came out of the sea and created Kakadu's landscape. During her travels she left many spirit children and taught them the different languages they were to speak. When her work was completed she turned into a rock, which remains her Dreaming place. There are many such features throughout Kakadu. They are the sacred sites of the Gagudju People. These sites include caves, hills, rocks and waterholes (see Figure 2.19).

The close spiritual bond that exists between the Gagudju People and the land means that they do not seek to dominate nature. Instead they live in harmony with it. They believe that they belong to the land and that they are part of it. Without the land, they believe they are nothing. The land and the life forms it supports are seen as a sacred trust, to be cared for and passed on to future generations.

The Dreaming Story of Ngurunderi

In the Dreaming, Ngurunderi and his two sons, in search of his two wives who had run away from him, followed a massive Murray cod down the Murray from where the Murray and the Darling rivers meet. As the huge fish swam, its tail swept the water aside, creating billabongs and swamps. Long Island, near Murray Bridge, is said to be a spear thrown by Ngurunderi. Ngurunderi eventually speared the giant cod and proceeded to cut it into small portions. As he threw the small pieces into the river they became the many different species of fish now found in the lakes and streams of the Murray–Darling.

When Ngurunderi discovered his two wives cooking a silver bream, a fish forbidden to women, he was very angry. The women sought to escape on a raft they had built. Ngurunderi pursued them down through the Coorong, creating the natural features of the landscape. When he caught up with the women, who were crossing to Kangaroo Island, he caused the sea to rise. The women drowned and became the rocky islands known as the Pages. Ngurunderi crossed to Kangaroo Island, removed his old skin of life and went to heaven. The Dreaming Story is depicted in Figure 2.20.

The fate of Ngurunderi was re-enacted in the traditional funeral ceremonies of the Ngarrindjeri People. The skin of the dead was removed before the remains were cremated on a raised platform.

2.20 A mural showing a Dreaming Story of Ngurunderi about the origins of the Murray River, Berri, South Australia

ACTIVITIES

Knowledge and understanding

1 Explain what the Dreaming is and explain its role in Aboriginal and Torres Strait Islander culture.

Applying and analysing

2 In small groups, discuss and list how Indigenous Australians' relationship with the land influences how they use it. Share the key points of your group's discussion with the class.

3 Outline the Gagudju People's explanation for the creation of Kakadu and describe their relationship with the physical environment.

4 Reread 'The Dreaming Story of Ngurunderi'.

- a What does the story seek to explain?
- b List the features of the physical environment that have their origins explained by the Dreaming Story of Ngurunderi.
- c Describe how the Dreaming Story of Ngurunderi influenced the cultural practices of the Ngarrindjeri People.

Investigating

5 Select a prominent feature of the biophysical environment and look for an Aboriginal Dreaming Story that explains its creation. Present an oral report to the class outlining the Dreaming Story and its connection to the physical environment.

2.7 Case study: Squamish Canada

The landscape is the dynamic backdrop to people's lives. In traditional indigenous cultures, the landscape not only influences people's way of life, it is often central to their spiritual wellbeing.

Their lands

The Squamish are the indigenous people of south-west British Columbia, Canada. Their territory extends northwards from north Vancouver up through the Howe Sound to the Squamish Valley and then towards Whistler—an area of 6732 square kilometres (see Figure 2.21). Today, the majority of Squamish people live in several communities located in west Vancouver, north Vancouver and the town of Squamish. In 2012, nearly 3900 people were identified as belonging to the Squamish Nation. According to traditional custom the Squamish Nation have occupied these lands since the beginning of time (see Figure 2.22).

In some places, the Squamish lands overlap with the territories of neighbouring indigenous peoples. In the Whistler area, for example, the Squamish and Lil'wat territories overlap. Rather than fight for control of these lands the two peoples agreed to share the territory. Today, a modern A$31 million cultural centre stands as tribute to this cooperative spirit.

2.21 Squamish territory

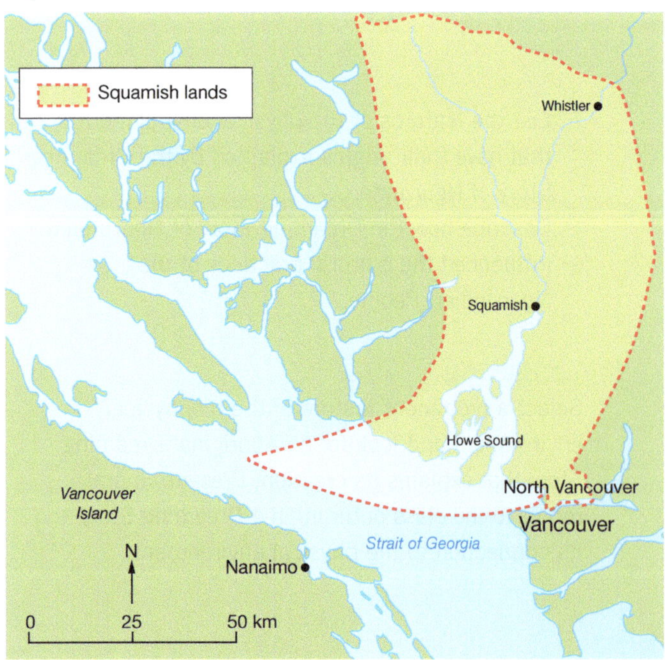

2.22 Squamish people in traditional dress

SPOTLIGHT

A landscape shaped by ice

Before the decline of the last Ice Age some 10 000 years ago, glacial ice covered the territory of the Squamish people to a depth of about 2000 metres. As the climate changed and the ice began to recede, the landscape was revealed. The glaciers had carved out deep valleys, depositing vast moraines of rock, the rivers were steeper and faster, and the Pacific Ocean extended into the upper Squamish River valley.

As the ice retreated, vegetation began to flourish in the valleys, and animals migrated northward as food became available. It is not known when the first humans entered Howe Sound in search of food and shelter, but experts estimate that is was at least 5000 years ago.

Oral tradition

Traditionally, the history, culture and customs of the Squamish people were passed onto future generations by spoken word rather than writing. This oral tradition describes and explains the Squamish's relationship to the land and provides important insights into the location of resource sites, hunting grounds, cedar-bark gathering areas, rock quarries, clam-processing camps, and the spiritual and ritual places of their ancestors.

Squamish society

Traditional Squamish society was sophisticated and organised, with complex laws and rules governing all forms of social relations, economic rights and relations with other First Nations. They did not believe in ownership as we understand it today. They believed that the land belonged by natural right to everyone and that they held it in custody for the next generation. For the Squamish today, to live on the land their ancestors occupied for thousands of years is still a powerful bond.

Biophysical environment

Like indigenous peoples worldwide, the Squamish have a very close relationship with the **biophysical environment**. Traditionally, they took from it only what they needed to survive. They used it in a way that was sustainable; that is, in a way that would enable them to maintain their traditional way of life for generations to come.

They hunted and gathered just enough salmon, herring, shellfish, seals, berries and plant roots to meet their needs. They also managed the forests sustainably. For example, the cedar bark used to make baskets was stripped from trees in a way that did not threaten the survival of the tree. A vertical strip, just two hands wide, was taken from a tree only once in its lifetime.

2.23 Black Tusk, Garibaldi National Park, Canada

2.24 The Twin Sisters, Vancouver, Canada

The landscape also has special meaning to the Squamish nation. The volcanic landform known as Black Tusk was, according to Squamish tradition, created from the Thunderbird's lightning (see Figure 2.23). The landform feature known as the Twin Sisters (also known as the Lions) is said to be the daughters of a war chief who were immortalised in stone for achieving peace among the coastal tribes. They stand as a reminder that each generation can live together in peace (see Figure 2.24).

ACTIVITIES

Knowledge and understanding

1 Outline the role played by landscape in traditional indigenous societies.
2 Describe the location of the lands occupied by the Squamish people.
3 Outline the role of the oral tradition in Squamish society.
4 Describe the relationship the Squamish people had with the land.
5 Describe a sustainable practice by the Squamish people.

Applying and analysing

6 Compare the following aspects of Squamish society with your way of life:
   a relationship with the natural environment
   b an oral tradition.

2.8 Case study: Plains Indians USA

For generations, nomadic Indian tribes such as the Sioux, the Comanche and the Crow roamed the Great Plains of the midwest of North America in search of buffalo. Hunting was a fundamental part of their life, and they had a deep spiritual connection to the land.

Relationship with the landscape

The Plains Indians believed that all things had spirits. They held deep beliefs about the creation and sacredness of the natural world and saw creation as an ongoing process. Through their guardian spirits they felt joined to the familiar shapes of their land, the sky and the wild animals they depended on for survival. They did not seek to dominate other creatures. They recognised the powers of nature and through their daily rituals they sought connection with the spirits. This connection was sustained and nourished by the landscape.

The rituals and observances of the tribes celebrated what was special about their land. On the Great Plains, elaborate ceremonies were held to honour the sun and the big sky that were so important to their daily lives.

The Great Spirit

The Plains Indians' worship centred on the Great Spirit, or Wakan Tanka, their creator, who reigned supreme over everything that had ever existed, including animals, trees, stones and clouds. The Plains Indians believed that worshipping the Great Spirit would make them stronger.

The Sun Dance

The sun was considered to possess great power because of the warmth and light it shed on the earth. The Indians performed the ritual of the Sun Dance to demonstrate love and respect for the Great Spirit. It was their way of showing gratitude for the good things that may have happened to the tribe or the Great Spirit's help in protecting them or healing a sick person.

The Sun Dance was the most important group gathering of the Plains Indians.

The Sun Dance ritual would last for four days from dawn to dusk, during which time no eating or drinking was allowed. The men would dance to drums in a circle around a sacred tree that had been cut down and set up as a pole in the middle. Offerings were made to the Great Spirit under the pole. The dancers stared at the sun and whistled through pipes. Some tribes practised self-torture by piercing the skin on their chests with ropes of hair or leather that were tied to the pole (see Figure 2.25). This personal sacrifice was made for the benefit of all the tribe.

SPOTLIGHT

Geronimo

The great Apache chief Geronimo (see Figure 2.27) emphasised the importance of the land to his people.

> For each tribe of men Usen created, He also made a home. In the land for any particular tribe He placed whatever would be best for the welfare of that tribe … Thus it was in the beginning: the apaches and their homes each created for the other by Usen Himself. When they are taken from these homes they sicken and die.

Miller, L (ed.), *From the Heart, Voices of the American Indian*, Pimlico, 1997

2.26 Geronimo emphasised the importance of the land to his people.

2.27 Geronimo was one of the fiercest Apache chiefs in recorded history.

Vision quests

Considerable importance was attached to visions. The Plains Indians believed that a person's success in life depended on the intervention of a spirit being, which could occur only through a vision.

A young man would go on a quest to seek this spirit being. He would go to a lonely spot to pray and fast for many days, and would lapse into a trance. Whatever animal or bird he saw in his dreams became his spirit being, who would be his guardian for life (see Figure 2.28).

Medicine bundles

Plains Indians carried medicine bundles, which contained objects suggested by a guardian spirit, such as feathers, animal teeth or claws, pipes and tobacco. The Indians believed that the bundle possessed powers that would bring them good luck, protection, successful hunting and even healing when needed. When a person died their medicine bundle was buried with them, unless they had chosen to pass it on to a special friend.

2.28 An Indian warrior's shield was his most prized possession. It was decorated with paintings and feathers. The shield usually showed the Indian's guardian spirit being, who would protect him from harm.

ACTIVITIES

Knowledge and understanding

1. Explain the role of the guardian spirits in a person's life.
2. Explain why the Sun Dance is performed.
3. Discuss the importance of visions to the Plains Indians.
4. Outline the importance of a person's medicine bundle.

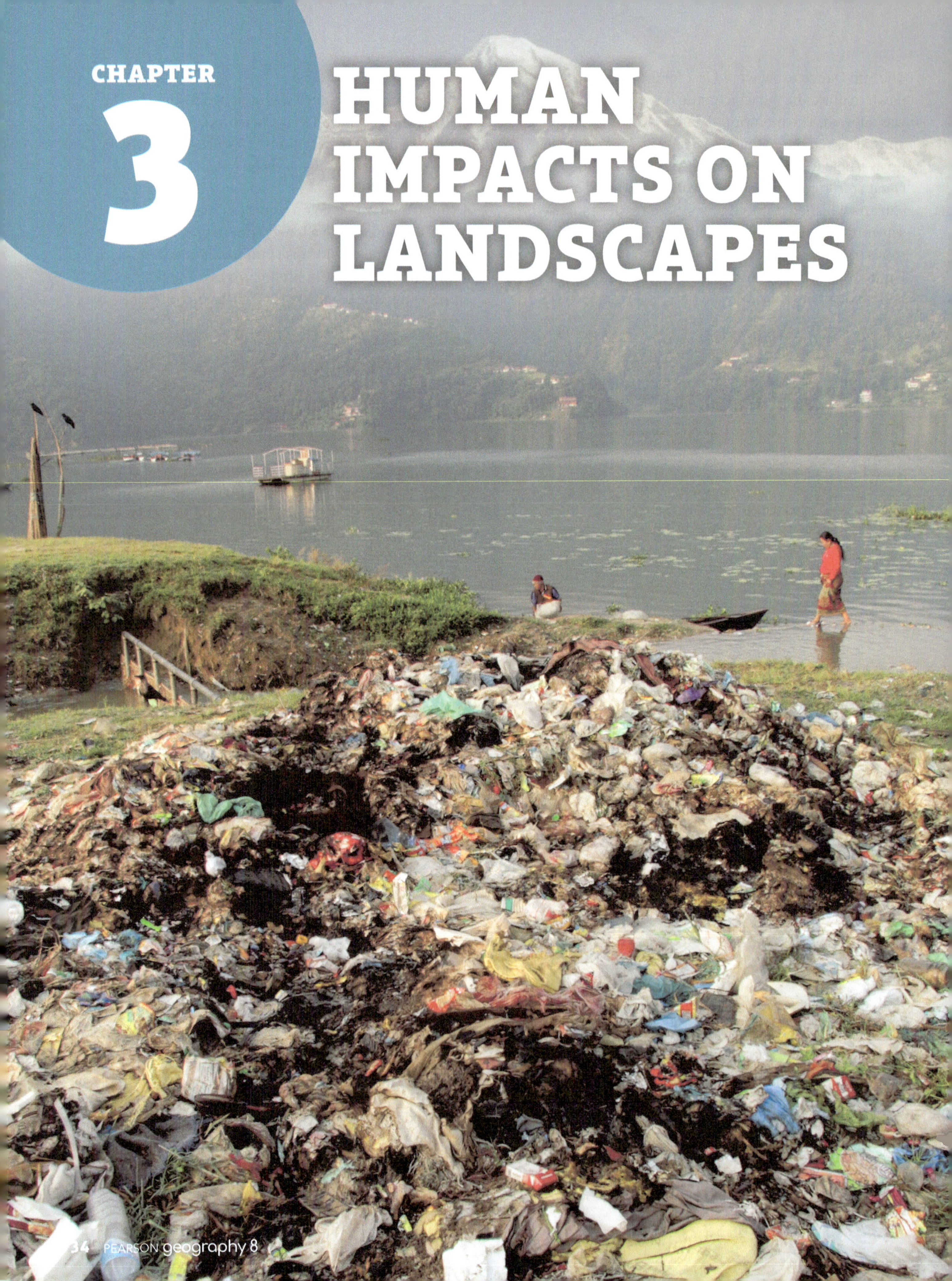

CHAPTER 3

HUMAN IMPACTS ON LANDSCAPES

Landscapes comprise the physical elements of the earth's surface and the cultural overlay of human activity, some of which stretches back for thousands of years. Landscapes reflect the interactions of place and people over time and are important in shaping local and national identity. Landscapes contribute to the development of our 'sense of place' and form the dynamic (ever-changing) backdrop to our lives.

People's impacts on landscapes vary from the indirect (for example the long-term effects of human-induced climate change) to those that transform landscapes (for example our large cities and agricultural systems). In this chapter we examine the nature of these impacts and the ways in which landscapes can be protected. We also look at the ways in which Aboriginal and Torres Strait Islander people managed the land.

KEY IDEAS

- To understand the variety of ways in which humans impact on landscapes
- To recognise the need to protect landscapes and the means by which this can be dome
- To appreciate the land management skills of Aboriginal and Torres Strait Islander people
- To understand the role of national parks and World Heritage listing in protecting places of natural and cultural significance

3.0 Rubbish at the edge of Lake Machhapuchhre, Nepal

GLOSSARY

conservation	the development and application of management practices to conserve, protect and, where possible, restore landscapes
Country	as used by Aboriginal and Torres Strait Islander people, refers to family origins and associations with particular parts of Australia
habitat	an area inhabited by a particular species of animal or plant
megafauna	giant marsupials that once roamed the Australian continent
preservation	protection of landscapes (and environments and habitats) from loss or damage
wilderness	extensive, largely undisturbed, areas of the original habitat
World Heritage sites	natural and cultural sites that have been identified by UNESCO as having international significance and being worthy of protection

3.1 Landscapes—human impacts

Many of the world's landscapes show the effects of human activity. It can now be argued that there are no longer any truly 'natural' landscapes left. The images of landscapes in this unit show the extent to which the 'cultural overlay' of human activity can transform landscapes.

Major human impacts

Agriculture

Since farming began about 10 000 years ago, whole landscapes have been transformed to create fields for crops and the raising of animals. Figure 3.1 shows irrigated crop circles on land that was once a grass prairie. Swamps and coastal marshes have been drained. Forests have been clearfelled and grasslands ploughed. The removal of trees destroys the roots that help to bind the soil together. Once exposed, the soil can be blown or washed into rivers. In some places, soil erosion has turned once-fertile farmland into barren wasteland.

3.1 Irrigated crop circles cover what was once a grass prairie in south-western Kansas, USA. The most common crops grown in this region are corn, wheat and sorghum.

Transport

Transport infrastructure (such as railways, roadways and airports) occupies large areas of the earth's land surface. Think about, for example, the amount of land devoted to roads and car parks in Australia's big cities. In China, the scale of such infrastructure is enormous. By 2025, China will have added five billion square metres of road paving and 170 mass-transit systems. In Europe, the 2.4 kilometre Viaduc de Millau (see Figure 3.2) dominates the landscape of the Tarn Valley.

3.2 The giant Viaduc de Millau, France

3.3 The giant Argyle diamond mine in Western Australia illustrates the impact that mining can have on landscapes.

Mining

Mining involves the extraction of valuable minerals and ores from the earth. Open-cut mining is especially noticeable in terms of its impact on landscapes. Mountains can be mined or deep pits dug to reach an ore body. Figure 3.3 shows the giant Argyle diamond mine in Western Australia.

Forestry

Forestry can be conducted in two ways. Selective logging involves removal of specific of trees from a forest. The trees are selected according to specific criteria (species, diameter or height). Selective logging leaves some trees standing, which allows for natural regrowth, has less impact on the wildlife, and leaves fewer ugly patches in the landscape.

Clearfelling is the practice of cutting down all the trees in an area. It is a practice that can destroy whole habitats. Figure 3.4 shows the impact of clearfelling in Tasmania.

3.4 The impact of clearfelling in Tasmania

Industry

The Industrial Revolution transformed methods of manufacturing and made them more efficient. Since then, factories have been built all over the world. Figure 3.5 shows the giant Auchi Prefecture Toyota Motor Plant at Toyota City. Factories consume huge amounts of natural resources and energy, and many give off chemical wastes, resulting in air and water pollution.

3.5 The Auchi Prefecture Toyota Motor Plant at Toyota City

Urban areas

The world's urban areas cover just 3 per cent of the earth's land surface but house 50 per cent of the world's population. It is also estimated that there are 75 000 distinct urban settlements worldwide, although many are better regarded as 'clumps' of settlements. Tokyo, the world largest city, has, for example, an urban area spreading over 30 000 square kilometres and takes in more than 500 connected settlements.

Energy

The infrastructure needed to produce and transport the energy we use is a major feature of many landscapes. This infrastructure includes wind farms (see Figure 3.7) and solar collectors, coal- and gas-burning power stations, nuclear power stations and electricity transmission lines that transport the energy to consumers.

3.6 Tokyo

3.7 Albany wind farm

China to flatten 700 mountains

In late 2012, the UK-based *Guardian* newspaper reported that the China Pacific Construction Group, one of China's largest construction firms, had announced plans to spend $3.5 billion to flatten 700 mountains, making room for a brand new city. Called the Lanzhou New Area, the planned city will be some 70 kilometres from the north-west province's capital city, Lanzhou. Featuring high-rises apartments and office towers, parks and even beaches, the futuristic city will cost billions to construct.

DID YOU KNOW?

Every year humans move about 0.8 Gt (Gt = 1 billion tonnes) of earth in house construction, 3.2 Gt in mineral production and 3 Gt in road construction. If all this earth were dumped into the Grand Canyon, it would fill the canyon in about 400 years, or about 0.01 per cent of the time it has taken the Colorado River to carve it!

ACTIVITIES

Knowledge and understanding

1 Define the term 'cultural overlay' in your own words.

2 Explain why 700 mountains are to be flattened.

Applying and analysing

3 Rank each image in this unit according the extent to which humans have transformed the landscape. Compare your ranking with a partner and then with another pair of students. Reach agreement on ranking for your group and share this ranking with the rest of the class.

4 Select three of the landscapes shown in this unit. Describe the main element of the 'cultural overlay' evident in each image. Suggest what the nature of the biophysical environment was prior to the human transformation.

5 Use the internet to gather images of landscapes dominated by the imprint of humans. Create a wall display using the following organisers: agriculture, transport, mining, forestry, industry, urban areas, energy.

3.2 Protecting landscapes

The terms 'preservation' and 'conservation' have different meanings. Preservation involves protecting landscapes (and environments and habitats) from loss or damage. Conservation involves developing and applying management practices that seek to conserve, protect and, where possible, restore landscapes.

Biophysical landscapes

Australia's most highly valued biophysical landscapes are protected by nature conservation reserves, national parks and **wilderness** areas. These are areas of land set aside for conservation purposes, such as the protection of wildlife and habitats, and the preservation of areas with natural features of scientific or recreational value. National parks are large areas of scenic or other natural significance open to the general public. Wilderness areas are defined as the most intact, undisturbed natural areas. Wilderness areas and national parks are considered important for the survival of certain species, ecological processes, conservation and recreation. In total, 7.9 per cent of the Australian landmass is set aside for conservation purposes. Marine parks and reserves cover a further 380 200 square kilometres.

Constructed landscapes

Constructed landscapes of cultural importance can include whole streetscapes, individual buildings such as the historic Coolamine Homestead, shown in Figure 3.8, the work of leading garden designers and architects, notable parks, gardens and avenues of trees, and in some cases, whole towns and cities. In Australia, important landscapes are protected by heritage legislation. In Victoria, for example, the Heritage Act (1995) protects places of cultural and heritage significance by listing them on the Victorian Heritage Register. Once on the Register, they are legally protected from inappropriate alteration and eligible for financial assistance for restoration work through state funding programs. Local governments also protect constructed landscapes through the role they play in town planning.

3.8 Coolamine Homestead, New South Wales, built in 1883, is under the care of the National Parks and Wildlife Service and the Kosciuszko Huts Association.

SPOTLIGHT

Iconic landscapes

The Victorian Government released legislation in 2012 to prohibit the building of wind farms to protect iconic Victorian landscapes such as the Great Ocean Road, the Mornington Peninsula (see Figure 3.9), the Macedon Ranges, the Yarra Ranges and Wilsons Promontory.

There has been both support for and criticism of the legislation. Supporters argue that these are important landscapes that need to be protected. Critics argue that Victoria will find it difficult to meet its clean energy targets and billions of dollars of investment will go elsewhere.

3.9 Mornington Peninsula, Victoria

ACTIVITIES

Knowledge and understanding

1 Distinguish between the terms 'preservation' and 'conservation'.
2 Outline the role of national parks.
3 State what a wilderness is.
4 State how important elements of the constructed environment are protected in Australia.
5 Outline the arguments against the Victorian Government legislation about the building of wind farms.

Geographical skills

6 Construct a photo sketch of Figure 3.8 or 3.9. Annotate your sketch with the reasons why such a landscape is worth protecting.

Investigating

7 Investigate a constructed environment protected by heritage legislation in your local area and create an annotated visual display.
   a Go to the National Trust website for your state to start your research and select an environment or building to investigate.
   b Include the following information in your display:
      - a map showing the location of the environment or building
      - images or illustrations (including historical and new)
      - arguments for and against the preservation and significance of the environment or building.

3.3 Indigenous land management

Australia's Aboriginal and Torres Strait Islander people have managed their land for tens of thousands of years. The land and its natural resources provided for their needs, shaped their history and were fundamental to their culture and spiritual beliefs.

The importance of 'country'

Land (or '**Country**', as it is often called) is central to the wellbeing of Aboriginal and Torres Strait Islander people. The land is not just the soil and rocks—it is the whole environment. It is at the centre of all spirituality and, together with the 'spirit of country', is central to the issues important to Indigenous Australians.

Indigenous ways of life

Traditionally, Australia's Aboriginal and Torres Strait Islander language groups had their own territory that sustained their way of life. They understood and cared for their different environments and adapted to them (see the quote in Figure 3.10). Indigenous Australians were skilled in managing the resources on which they depended. Geographic features such as rivers, lakes and mountains defined the boundaries of their territory or Country. In areas where there was plenty of water and food, Aboriginal people were semi-permanent, and planted crops such as the yam daisy. In more arid areas, people were semi-nomadic, moving according to the changing seasons and availability of food.

> We cultivated our land, but in a way different from the white man. We endeavoured to live with the land; they seemed to live off it. I was taught to preserve, never to destroy.

Tom Dystra, Aboriginal elder

Environmental impacts

The idea that Aboriginal and Torres Strait Islander people lived in harmony with nature for tens of thousands of years, without significantly changing the Australian environment, while an attractive one, especially for Indigenous Australians, has been challenged by scientists in recent years.

Loss of megafauna

Megafauna were large land mammals and birds living in Australia that evolved after the dinosaurs. There were over 60 species of megafauna, including 450-kilogram kangaroos, 7-metre long goanna-like creatures, 180-kilogram flightless birds, 130-kilogram marsupial lions, small-car-sized tortoises. Megafauna became extinct 16000 to 50000 years ago.

It is now thought that Indigenous Australians may have played a role in the extinction of the continent's megafauna, through hunting and the use of fire. Megafauna were slow moving and found it difficult to hide from hunters armed with spears. The larger animals reproduced slowly. The use of fire changed vegetation and mean the megafauna's food supply was changed or destroyed.

Other explanations for megafauna extinction include climate change. The loss of the megafauna transformed the landscape. The extinction of the large, plant-eating megafauna resulted in an increase in the amount of vegetation. In turn, this provided fuel for fire, which led to an increase in the frequency and intensity of fires.

Aboriginal people practised regular burning. Over time, these events changed the vegetation of the continent. Plant species that found it difficult to survive this use of fire were replaced by fire-tolerant species, such as eucalyptus.

After bringing about these changes, Aboriginal and Torres Strait Islander peoples settled into a way of life that established a balance with the environment they had created. This continued for tens of thousands of years until Europeans arrived in 1788.

 SPOTLIGHT

Indigenous use of fire

Aboriginal and Torres Strait Islander people made frequent use of fire (see Figure 3.11) to keep the country more open and easy to travel through; to promote the growth of fresh green grass, which would attract animals; to signal and hunt; and for the more obvious purposes of warmth and cooking.

There is considerable evidence to suggest that the purposeful use of fire extended from the earliest days of Aboriginal settlement. While fires had always occurred in Australia (due to lightning strikes), it is argued by some scientists that after the arrival of Indigenous people the fires became more frequent and intense. The use of fire transformed the Australian landscape because it advantaged those plants best adapted to fire.

Today, we use fire to reduce fuel loads, just like Aboriginal and Torres Strait Islander people did for thousands of years.

3.11 *Working the Land*, Joseph Lycett (1817). This painting shows Aboriginal people using fire to hunt kangaroos.

ACTIVITIES

Knowledge and understanding

1 Describe the Aboriginal and Torres Strait Islander way of life.
2 Outline how the thinking about the environmental impact of Aboriginal and Torres Strait Islander people has changed in recent times.
3 State the possible causes for the extinction of Australia's megafauna.
4 Explain how the extinction of the megafauna affected the Australian landscape.
5 Outline the reasons why Aboriginal and Torres Strait Islander people used fire.
6 Describe the impact that regular burning has on the landscape.

Applying and analysing

7 Study the quote by Tom Dystra in Figure 3.10. Discuss the nature of the relationship between Aboriginal and Torres Strait Islander people and 'Country'.

3.4 Australia's national parks

Australia has thousands of national parks and other conservation reserves. They protect a huge variety of biophysical environments and landscapes. These protected areas are managed according to principles of sustainability.

National parks

National parks are areas that are protected for the benefit of present and future generations. These diverse areas are largely untouched by human activity and occupation (see Figure 3.12). Many are home to unique plant and animal species, and habitats that are of scientific, educational and recreational value. Many include unique landform features or landscapes of great beauty.

The impact of visitors in national parks is carefully managed. Visitors are able to enter under special conditions (see Table 3.13).

3.13 Rules to manage the impact of people visiting national parks

Rules	Reason
Vehicles and mountain bikes are restricted to main roads	Vehicles and mountain bikes destroy plants and cause erosion
Pets are prohibited	Pets can scare or kill native animals
Rubbish must be disposed of properly or taken home	Rubbish can be unsightly and pollute the environment
Tread lightly—keep on designated tracks	Avoids damage to plant and animal habitats
Do not pick wild flowers	Flowers are food for birds and insects and produce the seeds from which new plants grow
Use the toilet facilities provided, not the bush	Protects water quality
Take care not to touch historic artefacts (e.g. Aboriginal rock art)	Such sites are fragile and easily damaged
Use the barbecues provided and don't light fires in the open	Fire can destroy vegetation and kill animals
Leave plants, animals, rocks and soil and shells as you find them	Disturbing these elements of the natural environment threatens plant and animal habitats
Avoid releasing introduced species of plants and animals either accidentally or deliberately (e.g. illegal dumping)	Introduced species overwhelm native plants and animals, changing habitats

3.12 Australia's diverse natural landscapes (clockwise from top left): Daintree National Park, North Queensland; Kakadu National Park, Northern Territory; Freycinet National Park, Tasmania; Kata Tju̲ta National Park, Northern Territory

Wilderness areas

Wilderness areas are large, undisturbed biophysical areas. They are the last truly wild places, where the impact of people has been minimal. Wilderness is considered important for the survival of certain species and habitats. It is also valued for cultural, spiritual, moral and aesthetic reasons. It is also a valued recreational source.

Wilderness areas are given the highest level of protection. Their wilderness values and pristine condition are protected by the limiting of activities likely to damage flora, fauna and cultural heritage.

SPOTLIGHT

Wollemi National Park

The Wollemi National Park is located just 150 kilometres north-west of Sydney. It is an area of untamed wilderness, wild rivers and spectacular landscapes. This maze of canyons, cliffs and undisturbed forest is the largest wilderness area in New South Wales and forms part of the Greater Blue Mountains World Heritage Area. It is also home to the Wollemi pine (see Figure 3.14). Discovered in 1994, this ancient tree species grows in a deep gorge in a small pocket of temperate rainforest.

The Wollemi pine is classified as critically endangered. A recovery plan has been developed to ensure that this species remains viable in the long term.

3.14 Wollemi pine, Wollemi National Park, New South Wales

Australia's landscapes

Australia has some of the world's most distinctive and spectacular natural landscapes. While many are already protected through recognition as national parks and World Heritage Areas, the National Landscape initiative gives many of these landscapes more protection, closely linked to their economic value as a tourist destination. The purpose of the initiative is to promote tourism and improve visitor experiences; increase the role of protected areas in local economies; and build broader community support for protecting our natural and cultural assets.

There are currently fifteen national landscapes: the Wet Tropics, the Great Barrier Reef, Australia's Red Centre, Flinders Ranges, the Australian Alps, the Great Ocean Road, Australia's Coastal Wilderness, Kakadu, Australia's Green Cauldron, Greater Blue Mountains, the Kimberley, Kangaroo Island, Great South West Edge and Ningaloo-Shark Bay. Sydney Harbour is under consideration.

Australia's biosphere reserves

The world's system of biosphere reserves is an initiative of the United Nations Educational, Scientific and Cultural Organization (UNESCO). United Nations member countries nominate areas of outstanding environmental importance, that meet a set of guidelines established by UNESCO. The area is then recognised under UNESCO's Man and the Biosphere Programme, which aims to promote sustainable development. The reserves include protected areas, and are managed with a view to both conservation and sustainable use of natural resources. Australia currently has fourteen biosphere reserves, including NSW's Kosciuszko National Park, Victoria's Croajingolong and Wilsons Promontory National Parks; Queensland's Great Sandy Biosphere Reserve; South Australia's Mamungari Conservation Park; and Western Australia's Fitzgerald River National Park, shown in Figure 3.15.

Protected marine areas

Australia has the world's third-largest area of protected marine territory—an area larger than the Australian landmass (see Figure 3.16). Australia's oceans provide food and recreation, and much of Australia's oil and natural gas production is drawn from the rock strata below these waters. There is also an extraordinarily high level of biodiversity. With this great natural wealth comes responsibility. We need to ensure that our oceans, and the life in them, remain healthy and productive so that future generations can enjoy them as we do.

3.15 Fitzgerald River National Park, Western Australia

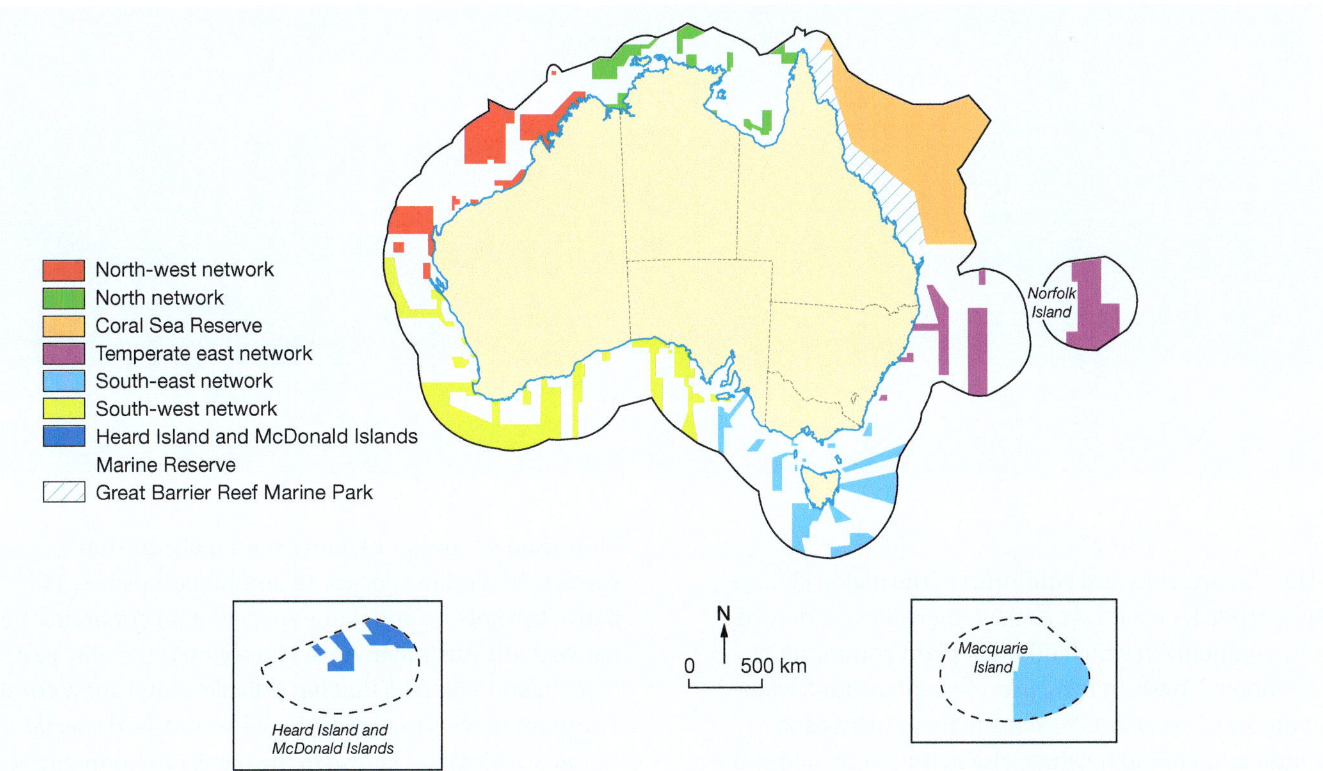

Source: Commonwealth of Australia

3.16 Australia's protected marine areas

ACTIVITIES

3.1

Knowledge and understanding

1 State the qualities of landscapes nominated as national parks that make them worth protecting.

2 Explain what wilderness is.

3 Outline how the criteria for recognition as an Australian National Landscape differ from the criteria for other protected areas.

Applying and analysing

4 Complete the following table with arguments for and against the following statement: *To fully protect our most fragile landscapes we should close the protected area to recreational land uses.*

For	Against

Geographical skills

5 Study Figure 3.16 and do the following tasks.

a Name three marine-protected areas.

b Compare the size of the Coral Sea Reserve to the other protected areas.

c Name the protected area that extends the furthest from the Australian landmass (including Tasmania).

d There are large areas not included in the proposed protection zones. List reasons why this might be the case.

Investigating

6 Select one of the biosphere reserves identified in the text or another Australian national park. Construct a multimedia presentation outlining the qualities that make these landscapes worth protecting. Include in your presentation:

a a map of the reserve or park

b qualities of the reserve or park that warrant this level of protection, giving:

i flora and fauna information

ii topography of the area

iii Aboriginal and Torres Strait Islander links

c an overview of the history of the park or reserve

d activities permitted and not permitted in the park

e information about the management of the reserve or park, noting:

i particular environmental issues (feral animals, invasive weeds, etc.)

ii development demands (building hotels)

iii number of tourists who visit.

3.5 Australia's alpine national parks

Australia's unique alpine areas are protected by system of eleven national parks and reserves covering 1.6 million hectares and stretching south from Canberra through the Brindabella Range to the Snowy Mountains of New South Wales, and along the Great Divide through eastern Victoria.

The alpine environment

The climate, soils and landforms of the region change as altitude increases. As a result, there are a variety of environments in which different plant communities dominate. These, in turn, provide habitats for a wide range of native animals. Many of these plants and animals are found nowhere else in the world, and some are threatened or endangered. Many species are found both within and outside the alpine national parks and reserves (shown in Figure 3.17).

3.17 Australian Alps national parks

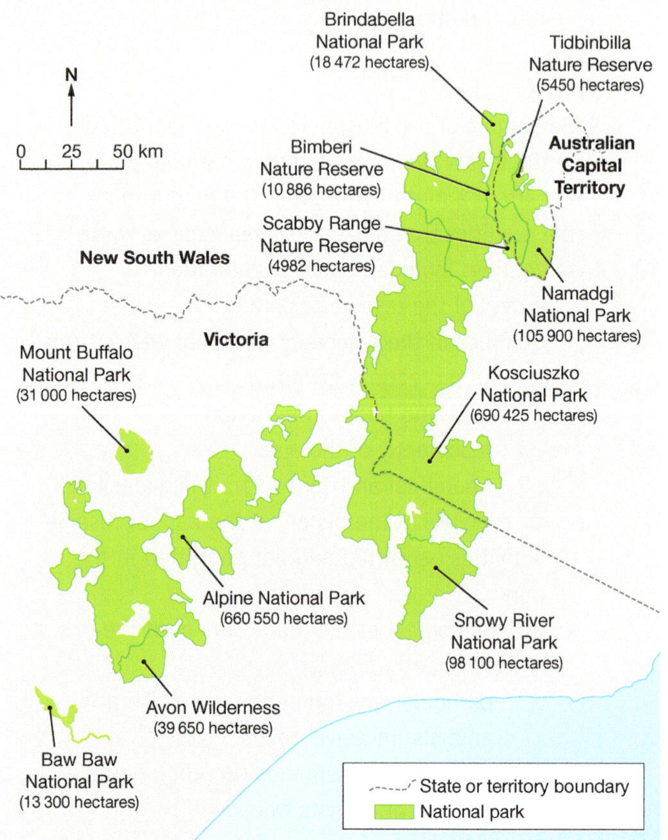

More than 40 species of native mammals, 200 bird species, 30 reptiles species, 15 amphibian species, 14 native fish species and many species of invertebrates are native to the Australian Alps. The region is the only part of mainland Australia that has reliable winter snow cover. The plant species that grow there have evolved special characteristics in response to the harsh environment. This includes sub-zero temperatures, burial by snow for months at a time and a short growing season.

Figure 3.18 shows the highest part of the Alps above the treeline. Here alpine heaths, herbfields, bogs and fens are found. Plants include the yellow billy buttons, pink trigger plants, and white snow and silver daisies. Below the treeline, the alpine woodlands are dominated by snow gums, montane and wet sclerophyll forest, alpine ash and mountain gum.

3.18 Lake Cootapatamba, New South Wales, formed by ice spilling from Mount Kosciuszko's southern face

Landforms

There is a range of distinctive landform features in the Australian Alps, some of which date from the last glacial age, which peaked about 20 000 years ago. The climate of the highest peaks of the main range near Mount Kosciuszko was cold enough for glaciers to form. Evidence of this glaciation can still be found. Figure 3.19 shows how the Australian Alps were formed. The process started about 130 million years ago when magma pushed up to create an uplifted plain. About 100 million years ago, a rift valley was created and the landmass began to spread apart. The sea floor spreading continued, a narrow sea became the Tasman Sea and the Australian Alps were formed.

Human impacts

Indigenous Australians visited the Alps for thousands of years, leaving little obvious evidence of their presence. The impact of Europeans has, however, been much more obvious and severe. The first European explorers visited in the early 1800s and graziers followed in the 1820s. Gold miners arrived in the 1850s, hoping to strike it rich. Skiing was introduced to Australia at Kiandra goldfields in New South Wales in the 1860s. The past 60 years has seen the development of large hydroelectric and water supply schemes, extensive road infrastructure, ski resorts and other tourist facilities. Tourism is central to the economy of the region and Kosciuszko National Park receives more than 3 million visitors each year.

While the alpine landscape may appear rugged and, at times, a harsh environment, it is actually very fragile and sensitive to disturbance. Careful management is needed if this unique environment is to be protected for the benefit of future generations.

3.19 The formation of the Australian Alps

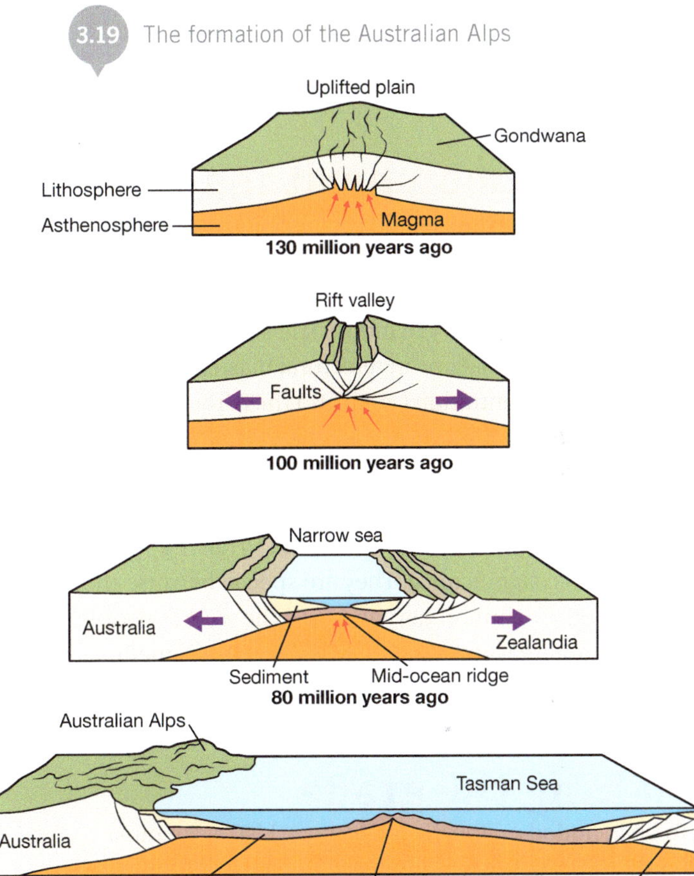

ACTIVITIES

Knowledge and understanding

1 Describe the biodiversity of the alpine region and outline the factors that that influence this diversity.

2 State what is special about the landform features surrounding Mt Kosciuszko.

3 Outline the nature of the human impacts on the alpine environment over time.

4 Explain why Australia's alpine environments need protecting.

Applying and analysing

5 Hypothetical: It has recently been proposed that local graziers be allowed to graze their cattle in the alpine areas of the Kosciuszko National Park. The NSW State Government is considering the request. Consider this statement and list the arguments for and against allowing farmers to graze their stock in Australia's alpine region.

Geographical skills

6 Study Figure 3.17.

a Describe the shape of the national park network through the Australian Alps.

b List and name the number of national parks in the Australian Alps.

c List the smallest and largest national parks.

3.6 Case study: World Heritage listing

One of the most important international treaties for protecting and managing globally significant natural and cultural sites is the United Nations Educational, Scientific and Cultural Organization's (UNESCO) World Heritage Convention.

Our common heritage

Africa's Serengeti National Park is important for its wildlife. Peru's Machu Picchu and Egypt's pyramids are archaeological treasures. Germany's Volklingen Ironworks tells the story of industrialisation. Italy's Venice and India's Taj Mahal are architectural masterpieces. Auschwitz Concentration Camp and the Hiroshima Peace Memorial remind us of the horrors of war.

These special places are just a few of the natural and cultural sites that have been identified as having international significance. They are special because they represent the common heritage of all the world's people and they are treasures to be protected for the benefit of present and future generations. These special places have been identified as **World Heritage sites**.

Making the grade

To be included on the World Heritage List, a site must have global natural or cultural significance and satisfy the strict criteria outlined by UNESCO. A cultural site could, for example, be a masterpiece of creative genius, have great architectural merit, be associated with ideas or beliefs of universal significance, or be an outstanding example of a traditional way of life representative of a particular culture. A natural site may represent a significant stage in the earth's history, represent ongoing ecological and biological processes, contain the natural habitat of endangered species, or be a place of great scenic beauty.

The UNESCO World Heritage List began in 1979. By mid-2012 there were 936 sites on the list (including 183 natural, 725 cultural and 28 mixed sites), spread across 153 countries. When a country nominates a site it must agree to conserve it for future generations.

By mid-2012 Australia had nineteen World Heritage sites.

Heritage listing

The main benefit of a World Heritage listing is that people become aware of the special nature of the site, and pressure is exerted on governments to protect the site. This means that when a site is threatened there is enough public interest to make it an issue for community debate and action.

Some sites may be endangered by war, pollution, poachers, lack of maintenance, the damming of rivers, illegal logging or similar activities. Such sites are often placed on a temporary danger list and financial assistance may be given to help protect the site. A list of endangered sites is given in Table 3.21. The Ngorongo Crater, in Tanzania, is shown in Figure 3.22. It had

SPOTLIGHT

UNESCO

The United Nations Educational, Scientific and Cultural Organization's role is to contribute to peace and security in the world by promoting cooperation between nations through education, science, culture and communication. Their aim is increase respect for justice, for the rule of law and for the human rights and freedoms to which everyone is entitled without discrimination based on race, sex, language or religion.

United Nations
Educational, Scientific and
Cultural Organization

3.20 UNESCO's World Heritage logo

been on the temporary danger list due to lack of controls on land uses. It was taken off the list when management plans were put in place to protect the crater.

Listing, however, does not always guarantee protection. Historic Dubrovnik in Croatia, for example, was partially destroyed during the war in the former Yugoslavia in the early 1990s.

Protecting sites

Governments that sign up to the World Heritage Convention agree to protect listed cultural and natural sites located within their borders.

Such protection includes:

- developing community-based planning programs that protect sites
- establishing a body whose responsibility it is to protect a site
- developing scientific and technical studies and research that help to protect sites against potential dangers
- putting in place the appropriate legal, scientific, administration and financial measures to protect sites
- a recognition that the protection of sites is the responsibility of the international community
- an undertaking not to damage, directly or indirectly, any World Heritage site.

3.22 Ngorongo Crater, Tanzania

3.21 UNESCO's World Heritage Endangered list, June 2012

Country	World Heritage Site
Afghanistan	Minaret and archaeological remains of Jam; cultural landscape archaeological remains of the Bamiyan Valley
Belize	Belize Barrier Reef Reserve System
Central African Republic	Manovo-Gounda St Floris National Park
Chile	Humberstone and Santa Laura Saltpeter Works
Colombia	Los Katios National Park
Côte d'Ivoire	Comoé National Park and Mount Nimba Strict Nature Reserve
Democratic Republic of the Congo	Virunga National Park; Garamba National Park; Kahuzi-Biega National Park; Okapi Wildlife Reserve; Salonga National Park
Egypt	Abu Mena
Ethiopia	Simien National Park
Georgia	Bagrati Cathedral and Gelati Monastery; historical monuments of Mtskheta
Guinea	Mount Nimba Strict Nature Reserve
Honduras	Rio Plátano Biosphere Reserve
Indonesia	Tropical rainforest heritage of Sumatra
Iran	Bam and its cultural landscape
Iraq	Ashur (Qal'at Sherqat); Samarra archaeological city
Jerusalem	Old City of Jerusalem and its walls
Madagascar	Rainforests of the Atsinanana
Mali	Timbuktu; tomb of Aski
Niger	Air and Ténéré Natural Reserves
Palestine	Birthplace of Jesus: Church of the Nativity and the Pilgrimage Route, Bethlehem
Panama	Fortifications on the Caribbean side of Panama: Portobelo-San Lorenzo (2012)
Peru	Chan Chan archaeological zone
Senegal	Niokolo-Koba National Park (2007)
Serbia	Medieval monuments in Kosovo
Tanzania	Ruins of Kilwa Kisiwani and ruins of Songo Mnara
Uganda	Tombs of Buganda Kings at Kasubi (2010)
United Kingdom of Great Britain and Northern Ireland	Liverpool—Maritime Mercantile City (2012)
United States of America	Everglades National Park
Venezuela	Coro and its port
Yemen	Historic town of Zabid

Source: UNESCO World Heritage Endangered List, June 2012

Wood Buffalo National Park, Canada

Listed: 1983

Criteria: Natural

Description: Located on the plains in the north-central region of Canada, the park (which covers 44 807 square kilometres) is home to North America's largest population of wild bison. Another of the park's attractions is the world's largest inland delta, located at the mouth of the Peace and Athabasca rivers.

Palace and Park of Fontainebleau, France

Listed: 1981

Criteria: Cultural

Description: Used by the kings of France as royal hunting lodge, the palace is surrounded by a huge park. The Italianate structure combines Renaissance and French artistic traditions.

Venice, Italy

Listed: 1994

Criteria: Cultural

Description: Founded in the second century BC in northern Italy, Venice prospered under Venetian rule from the early fifteenth to the end of the eighteenth century.

The Pyramids of Giza, Egypt

Listed: 1979

Criteria: Cultural

Description: Built to house the remains of Egypt's rulers, these vast structures are considered one of the Seven Wonders of the (ancient) World.

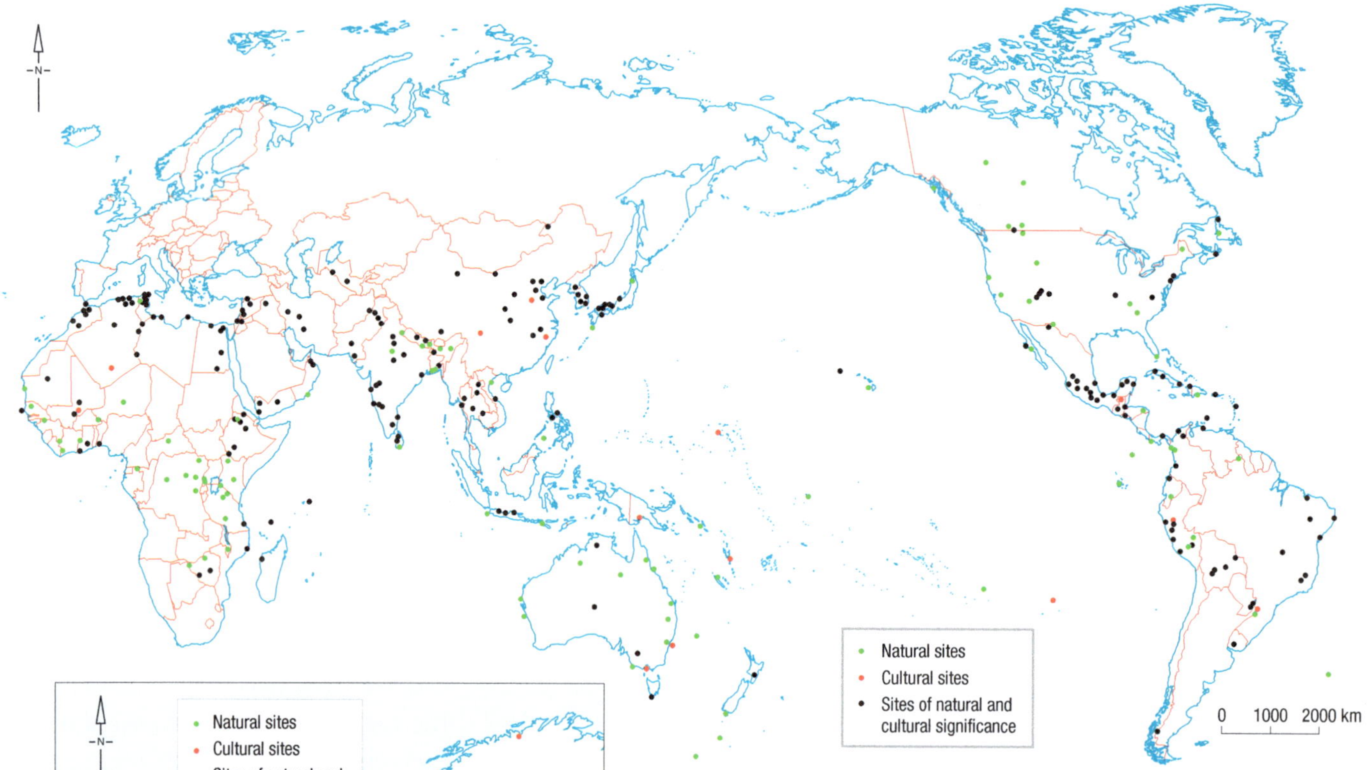

N

Natural sites
Cultural sites
Sites of natural and cultural significance

0 1000 2000 km

N

Natural sites
Cultural sites
Sites of natural and cultural significance

0 500 1000 km

3.23 Distribution of World Heritage sites

Auschwitz Concentration Camp, Poland

Listed: 1979

Criteria: Cultural

Description: The fortified walls, barbed wire, platforms, barracks, gallows, gas chambers and cremation ovens show the conditions within which the Nazi genocide took place in the former concentration and extermination camp of Auschwitz–Birkenau, the largest in the Third Reich. One and a half million people, among them a great number of Jews, were put to death in the camp. Protected as a reminder of humanity's cruelty to fellow human beings.

Grand Canyon National Park, Arizona, USA

Listed: 1979

Criteria: Natural

Description: Carved by the Colorado River, the Grand Canyon (nearly 1500 metres deep) is the most spectacular gorge in the world. Its horizontal strata retrace the geological history of the past two billion years. There are also prehistoric traces of human occupation.

Machu Picchu, Peru

Listed: 1983

Criteria: Cultural

Description: Machu Picchu is located 2430 metres above sea level, in the middle of a tropical mountain forest. It was probably the most amazing urban creation of the Inca Empire; its giant walls, terraces and ramps seem as if they have been cut naturally in the continuous rock escarpments.

Taj Mahal, India

Listed: 1983

Criteria: Cultural

Description: An immense mausoleum of white marble, built in Agra between 1631 and 1648 by order of the Mughal emperor Shah Jahan in memory of his favourite wife. The Taj Mahal is universally admired as a masterpiece of the world's cultural heritage.

The Great Wall, China

Listed: 1987

Criteria: Cultural

Description: In c. 220 BC, sections of earlier fortifications were joined to form a united defence system against invasions from the north. Construction continued up to the Ming dynasty (1368–1644), when the Great Wall became the world's largest military structure.

Iguazú National Park, Argentina

Listed: 1984

Criteria: Natural

Description: At the heart of this park is one of the world's most spectacular waterfalls. The semicircular falls (80 metres high and 2700 metres in diameter) feature numerous cascades that produce vast sprays of water.

ACTIVITIES

Knowledge and understanding

1. Explain the factors that determine whether a site qualifies for World Heritage listing.
2. Explain why World Heritage sites are special.
3. Describe the process to protect endangered World Heritage sites.

Applying and analysing

4. Do you think the responsibilities of national governments under the World Heritage Convention are sufficient to protect the sites?

Geographical skills

5. Study Figure 3.23 and answer the following questions.
   - a Name the continent/s with the most World Heritage sites.
   - b Name the continent/s with the most World Heritage natural sites.
   - c Name the continent/s with the most World Heritage cultural sites.
   - d Is there a pattern for the location and type of heritage sites? Describe the pattern.
6. Study Table 3.21 and do the following tasks.
   - a Draw up a table and with seven columns. Label each column with the seven continents: Africa, North America, South America, Europe, Asia, Australia and Antarctica.
   - b Place each of the World Heritage locations under threat in the correct column.
   - c List the continent/s that have the most number of World Heritage areas under threat.
   - d List the continent/s that have no World Heritage areas under threat.
   - e Describe the pattern to the location of World Heritage sites under threat.

Investigating

7. Select one place from Table 3.21 to investigate further. Present your research as an annotated visual display. Include the following information:
   - a a map of the site
   - b an explanation on why the site is considered important
   - c an overview of the threats to the site
   - d a description of programs in place to protect the site from the threats, and an analysis of these programs.

3.7 Case study: Machu Picchu

Built as a retreat for the rulers of the Inca Empire, Machu Picchu is one of the 'New Seven Wonders' and a popular tourist destination. World Heritage listing ensures that this amazing contribution to the world's cultural heritage will be protected for the benefit of future generations.

Location

Machu Picchu is located on a Peruvian mountain ridge, 2430 metres above sea level, and 80 kilometres to the north-west of Cusco, the historic capital of the Inca Empire. Access to Machu Picchu is via narrow mountain path. Today, tens of thousands of tourists walk the Inca Trail to visit Machu Picchu (see Figure 3.24).

History

The construction of Machu Picchu began around AD 1450, when the Inca Empire was at its strongest. It is now believed that the city was built as a secret religious or ceremonial site by the Inca emperor Pachacuti. Machu Picchu was abandoned as an official site for Inca rulers in 1572 after the Spanish conquest of South America.

Historically, the location of Machu Picchu was a closely guarded military secret, and its position was able to be readily defended. A rope bridge across the Urubamba River provided a secret entrance for the Inca army. Another removable bridge to the west of the city spanned a 6-metre gap in a stone path cut into the cliff face 570 metres above the valley floor.

Because the site remained unknown to the Spanish invaders it is the most intact cultural site of the Incas.

Construction

Machu Picchu was built in the classical Inca style, with polished dry-stone (granite) walls. Many of the stone blocks weigh 50 tonnes or more yet are so precisely sculptured and fitted together that a thin knife blade cannot be inserted into the mortarless joints.

3.24 Machu Picchu

Layout

Machu Picchu was divided into three distinct land-use zones—agricultural, urban and religious—as shown in Figure 3.25. The urban and religious sectors are divided into two (the east and west sectors) by wide plazas. The agricultural sector, which is located outside the city's wall, was subdivided into upper and lower sectors. The city's structures—terraces, walls, palaces, baths, temples, storage rooms and some 150 houses—were arranged so that the function of the buildings matched the form of their surroundings. The agricultural terracing and aqueducts, for example, take advantage of the natural slopes, and the lower areas contain buildings occupied by farmers. The most important religious areas are located at the top of the ridge, overlooking the Urubamba Valley far below. The agricultural terraces used to grow crops to feed the population were watered by natural springs, making the city largely self-sufficient.

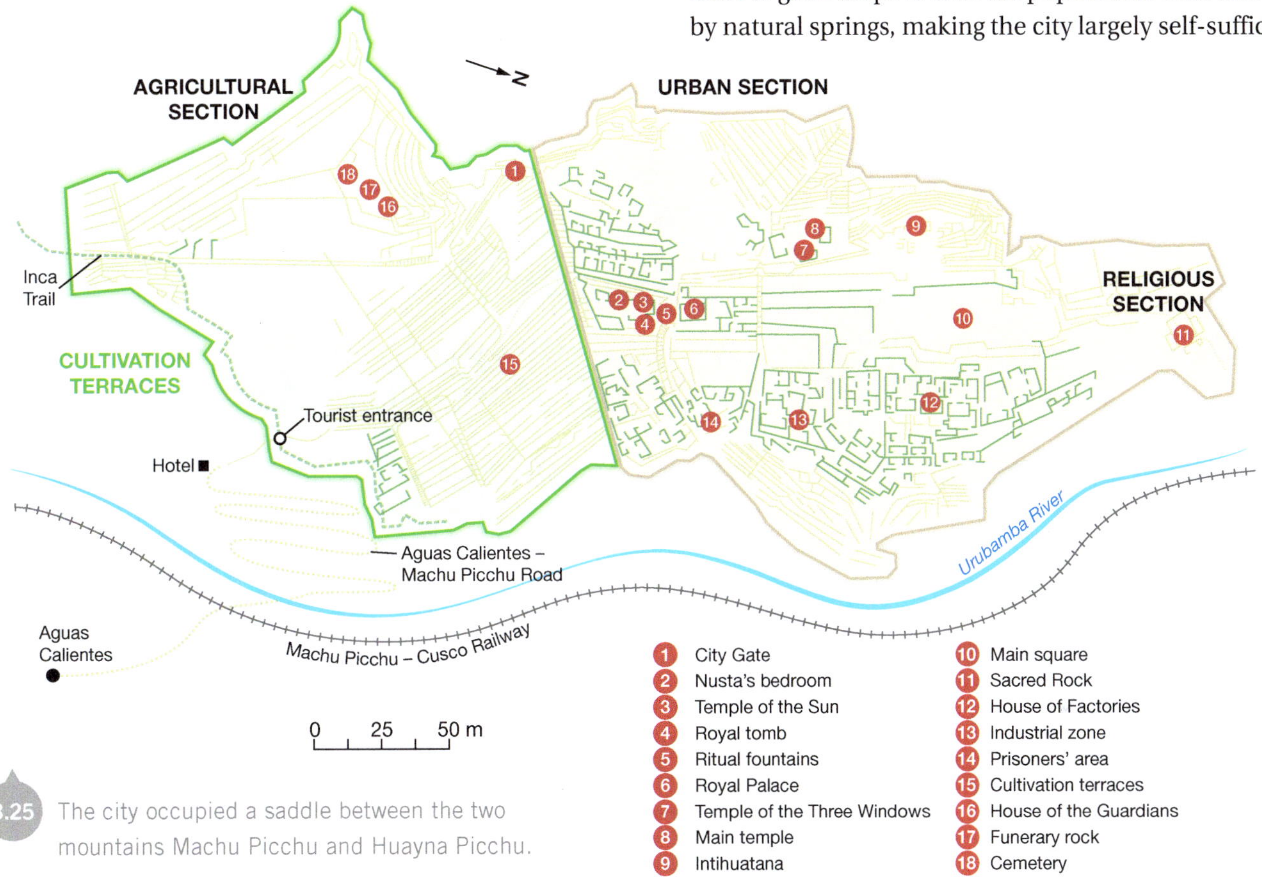

3.25 The city occupied a saddle between the two mountains Machu Picchu and Huayna Picchu.

ACTIVITIES

Knowledge and understanding

1 Explain why Machu Picchu is referred to as the 'lost city' of the Incas.
2 State the role of Machu Picchu in the time of the Inca Empire.
3 Identify the features of the site of Machu Picchu that made it the ideal location for a 'secret' city.

Geographical skills

4 Study Figure 3.24 and describe the landscape surrounding Machu Picchu.
5 Study Figure 3.24 and discuss the advantages and disadvantages of living in such a location.
6 Study Figure 3.25 and describe the layout of Machu Picchu.

Investigating

7 Undertake research into the Inca civilisation. Present your findings as an illustrated extended report that includes:
   a history and geographical boundaries of the Inca Empire
   b population distribution
   c economy
   d social structure
   e religion
   f architecture
   g urban centres (including Cusco)
   h the Inca civil war and the Spanish Conquest.

CHAPTER 4

RESTLESS EARTH

The earth is a restless, dynamic planet. Powerful forces deep below the earth's crust create volcanoes and earthquakes, build mountains and pull continents apart. Rocks, and the minerals they contain, provide clues about how these forces have shaped and changed the planet over billions of years.

The surface of the earth is made up of massive plates that rest on the earth's molten interior. The movement of these plates is caused by convection currents beneath their surface. This process is called plate tectonics. In this chapter we examine this process and how it produces the earth's great mountain ranges and deep ocean trenches.

KEY IDEAS

- To investigate particular landforms found in different landscapes
- To understand the role that plate tectonics plays in the development of the earth's landforms
- To demonstrate how tectonic forces, such as folding, faulting and volcanism, produce landforms
- To understand how the processes of weathering, erosion and deposition shape landforms
- To develop the skill to draw cross-sections

4.0 Mt Batok, Mt Bromo (both foreground) and Mt Semeru, Java, Indonesia

GLOSSARY

crust	the thin outer layer of the earth (the lithosphere)
deposition	the accumulation of sediment by the action of erosional agents, such as water and wind
differential erosion	the wearing away of softer rock at a different (faster) rate than harder, more resistant rock
earthquake	a sudden movement of the earth's crust caused by the release of pressure
erosion	the wearing down, transportation and deposition of material by water, wind and ice
exfoliation	the process of rock breakdown that occurs when rock expands and contracts due to temperature changes, then cracks, peeling off in layers like an onion
faulting	the fracturing of rock along lines of physical weakness
fold mountain	a mountain resulting from folding of the earth's crust
folding	the buckling of rock due to pressure
mantle	the layer between the earth's core and its crust
mid-ocean ridge	an underwater ridge formed when continental plates move apart, allowing molten material to fill the gap created
ocean trench	an underwater depression created when oceanic plates are drawn down into the earth's mantle
plate margins	the areas where the earth's plates meet
plate tectonics	the study of the movement of the earth's plates by currents deep within the earth's liquid mantle
rift valley	a large elongated depression with steep walls, formed by the downward movement of a block of the earth's surface between nearly parallel faults
rock cycle	the recycling of material in the earth's crust
seafloor spreading	the separation of oceanic plates
tectonic plate	a large segment of the earth's crust that is slowly moving due to convection currents in the mantle
volcano	an opening in the earth's surface through which molten rock, lava and ash erupt
weathering	the physical or chemical breakdown of rocks into smaller pieces

4.1 The changing face of the earth

Forces deep within the earth cause the movement of tectonic plates, which in turn create new landforms. Without these new landforms, the earth's surface would have long ago been reduced to a flat, featureless plain—worn down by the processes of weathering and erosion.

Plate tectonics

The earth's thin **crust** is broken into eight vast segments or **plates** (and several smaller plates) that travel slowly across the face of the planet at a rate of about 15 centimetres per year. This movement is caused by currents deep within the earth's liquid **mantle** (shown in Figure 4.1). This process is known as **plate tectonics**, or continental drift.

Continents on the move

Scientists believe that all the earth's continents were once part of one large supercontinent, known as Pangaea (a Greek word meaning 'all lands'). Pangaea consisted of two main areas: Gondwanaland (Australia, Antarctica, Africa, India and South America) and Laurasia (Asia, Europe, Greenland and North America). These two main areas began to move apart and break up about 200 million years ago. Over time they 'drifted' to their present locations.

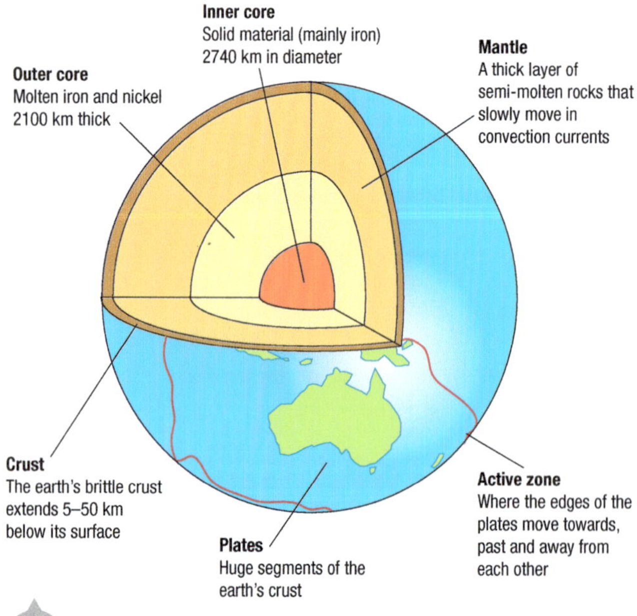

4.1 The internal structure of the earth

Types of plate movements

Each of the earth's plates moves in a different way, as:

- a convergent plate boundary(towards each other)
- a divergent plate boundary (away from each other) or
- a transform plate boundary (past each other).

The places where the plates meet are known as **plate margins**. These are often areas of great stress and activity. Many of the earth's earthquakes, volcanoes and **fold mountains** are located at the plate margins. Figure 4.2 shows the location of the earth's plates and the directions in which they are moving.

Convergent plate boundary

Collision plate margins

When two plates made of continental crust move towards one another they create a collision zone, as shown in Figure 4.3. Because neither plate can sink beneath the other, their crusts crumple upwards to form fold mountains. The Himalaya (see Figure 4.4), formed as a result of the collision between the Indian and Eurasian plates, is an example of a fold mountain system. Sometimes pressure builds up over time. Eventually the crust breaks, sending out shockwaves in the form of an earthquake.

DID YOU KNOW?

At 8880 metres above sea level, Mt Everest is the highest point on the earth's surface. It is not, however, the world's highest mountain. From base to peak, Mauna Kea on Hawaii measures 10 023 metres, 5818 metres of which is below the Pacific Ocean.

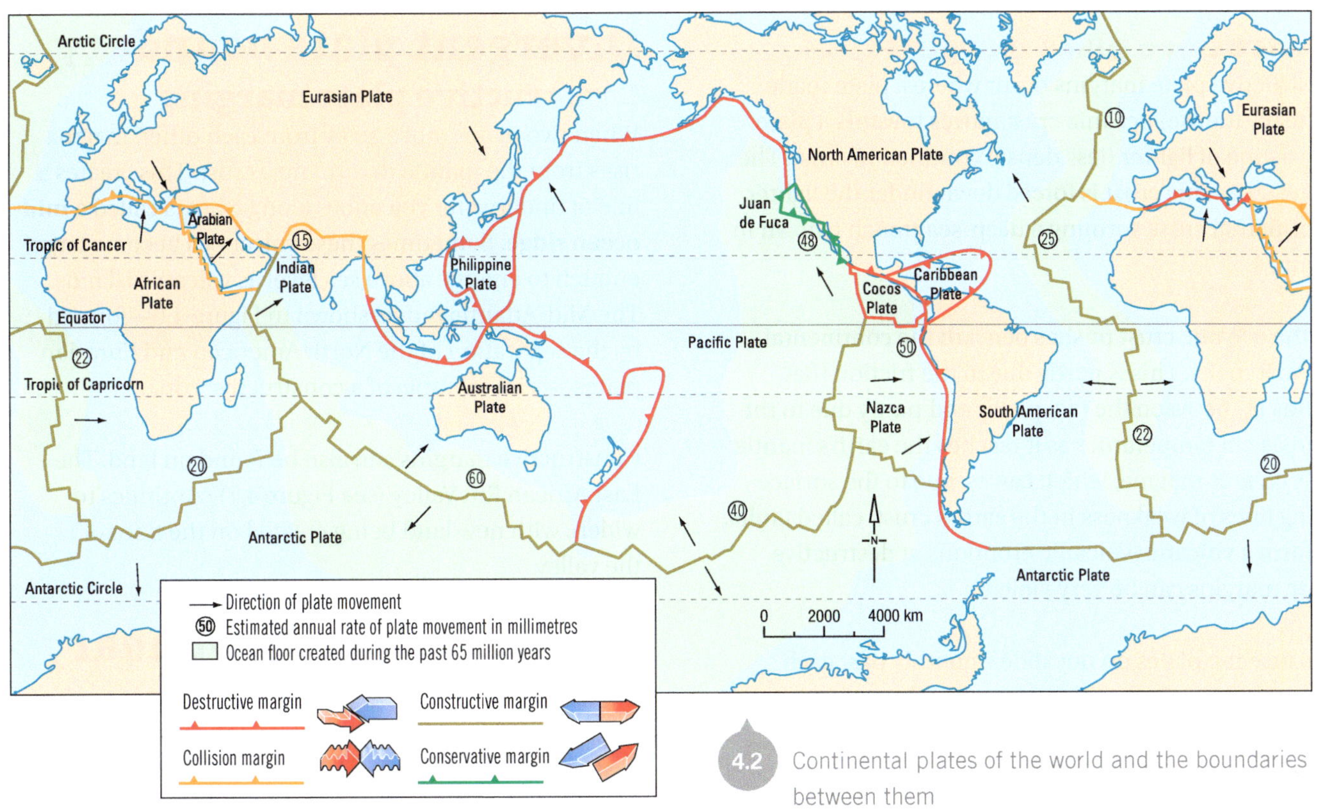

4.2 Continental plates of the world and the boundaries between them

4.3 A collision plate margin

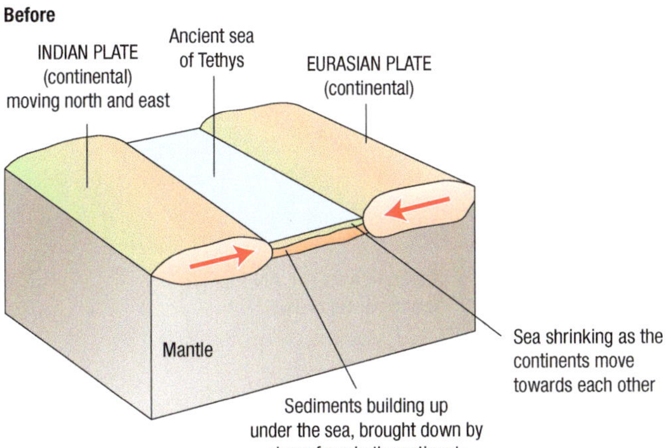

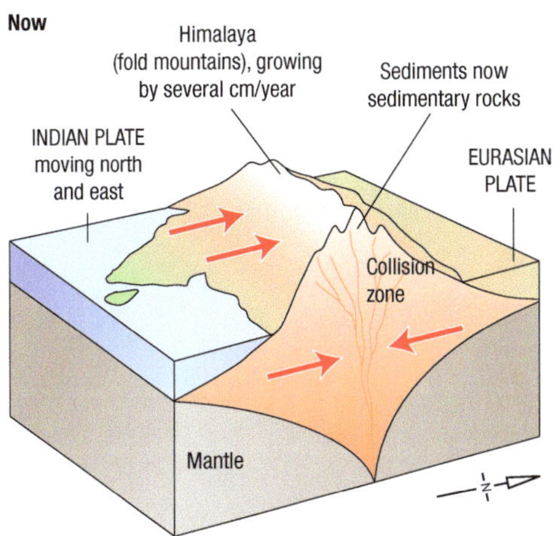

4.4 The Himalayan Mountains are still increasing in height as the Indian Plate moves into the Eurasian Plate at a rate of about 5 centimetres a year.

Destructive plate margins

Destructive plate margins occur where a plate made of heavy (dense) oceanic crust moves towards a plate consisting of lighter (less dense) continental crust. The heavier oceanic crust is forced down under the lighter continental crust, forming a deep-sea trench (shown in Figure 4.5).

As the oceanic crust pushes beneath the continental crust, it melts. This is partly due to the friction that builds up between the two plates and partly due to the increase in temperature as it reaches the earth's mantle. This creates magma, which can escape to the surface along lines of weakness in the earth's crust, called faults, to form a **volcano**. Volcanic eruptions at destructive plate margins can be very violent.

Because the plates do not slide smoothly past each other, there is often an enormous build-up of pressure. If the crust breaks, shock waves are sent out in all directions, causing an **earthquake**.

Divergent plate boundary

Constructive plate margins

When two plates move away from each other, magma rises from the mantle to form new crust. This creates a line of underwater volcanoes along what is called a **mid-ocean ridge**. Sometimes these volcanoes become large enough to emerge above sea level as volcanic islands. The Mid-Atlantic Ridge, shown in Figure 4.6—formed by the separation of the North American and Eurasian plates—is an example of a constructive zone.

Constructive margins can also be found on land. The East African Rift Valley (see Figure 4.7) continues to widen, with new land being created on the floor of the valley.

Transform plate boundary

Conservative plate margins

Conservative plate margins occur where two plates move past one another. The San Andreas Fault in California, for example, marks the point at which the North American

4.5 A destructive plate margin

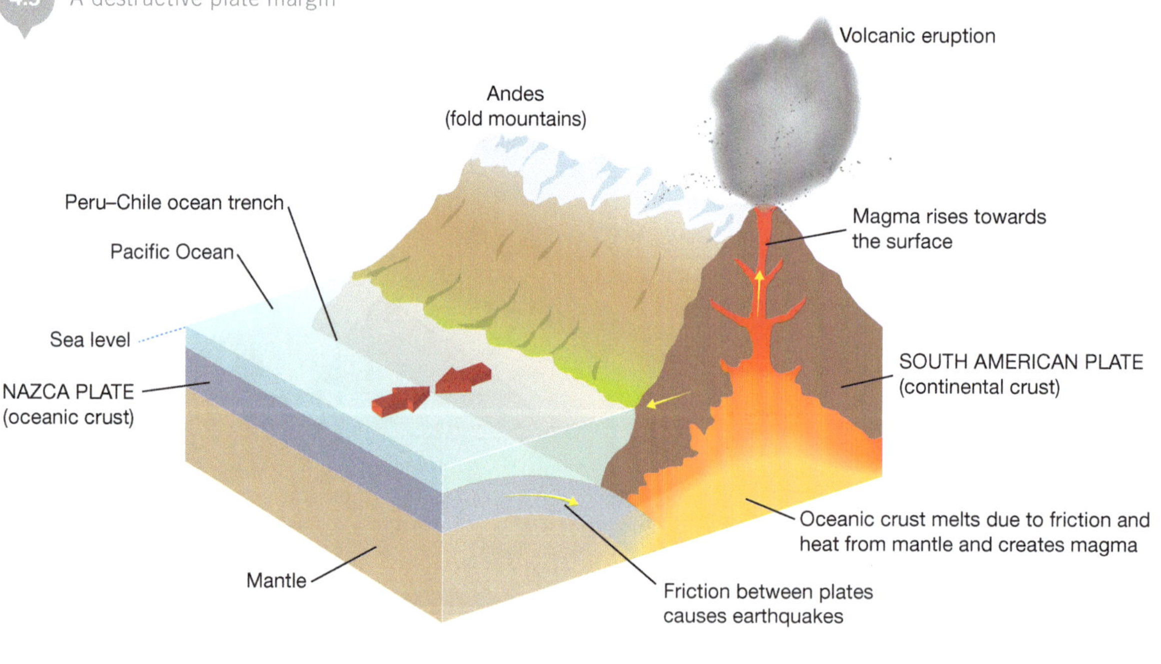

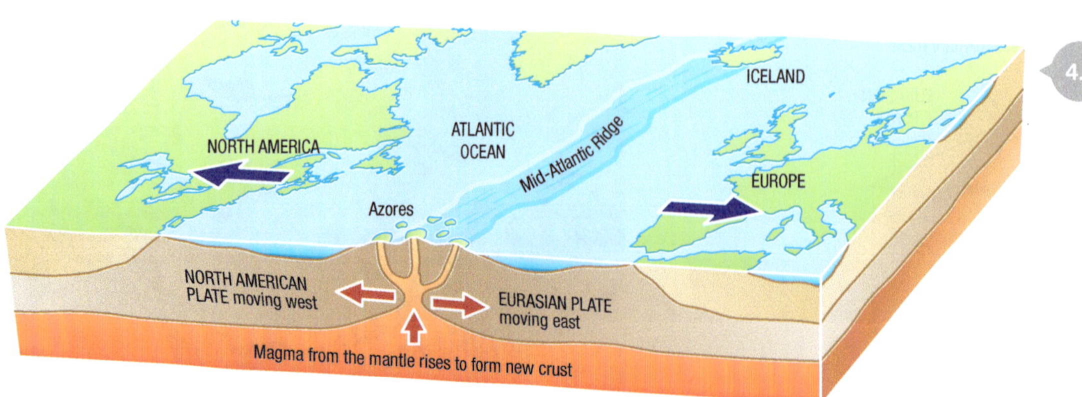

4.6 The Mid-Atlantic Ridge was formed by the separation of the North American and Eurasian plates. Each year the Atlantic Ocean widens by about 3 centimetres.

and Pacific plates meet. Although the two plates are moving in the same direction, the Pacific Plate is moving faster, causing the plate margins to 'grind' past each other. The tensions between the two plates build up over time. When this tension is suddenly released an earthquake occurs. Minor earthquakes occur almost daily along the fault, but major earthquakes, causing loss of life and destruction of property, are less frequent. The last major earthquake was the 1994 Northbridge earthquake. It measured 6.6 on the old Richter scale and killed 60 people (see Figure 4.8).

4.7 The Great Rift Valley in eastern Africa was formed through the rifting (tearing apart) and separation of the African, Arabian and Indian tectonic plates.

4.8 California's San Andreas Fault: a conservative plate margin

Pacific Ocean
PACIFIC PLATE moving north-west by 6 cm per year
0 300 km
KEY
Large earthquakes near the San Andreas Fault
(1) San Francisco 1906
(2) San Fernando 1971
(3) Coalinga 1983
(4) San Francisco/ Loma Prieta 1989
(5) Northridge 1994
San Andreas Fault
San Francisco Bay area
CALIFORNIA
Los Angeles
San Diego
MEXICO
USA
NORTH AMERICAN PLATE moving north-west by 1 cm per year

ACTIVITIES

4.1

Knowledge and understanding

1 State the name given to the processes involved in the movement of the earth's crust. Explain the causes of this movement.
2 Name the types of plate margins.
3 Describe what happens to the earth's crust in a collision zone.

Geographical skills

4 Study Figure 4.2.
  a Name two plates that are colliding, moving towards each other, moving away from each other and moving past each other.
  b Discuss the plate that is moving the greatest number of millimetres per year.
5 Study Figures 4.3 and 4.5. Write a paragraph comparing the formation of the Andes Mountains and the Himalayan mountain range.
6 Study Figure 4.8. Explain why parts of California experience earthquakes.

4.2 Rocks and the rock cycle

To understand the formation and evolution of the earth's landform features you need to understand the rock cycle and the characteristics of the different rock types.

Rock types

Rocks are classified into three different types: igneous, sedimentary and metamorphic.

Igneous rocks

Igneous rocks form when molten material (magma) cools and becomes solid, or solidifies. There are two main types of igneous rocks. They have the same chemical composition but differ in structure depending on where the magma cooled. Intrusive igneous rocks are rocks that form slowly in the earth's crust (underground). Extrusive igneous rocks form when the magma erupts from a volcano and then cools quickly above ground. Igneous rocks are distinguished by their crystal-based structure. The crystals present in intrusive igneous rocks are large due to the slow rate of cooling, while those in extrusive igneous rocks are small because of their faster rate of cooling. Granite (which features relatively large crystals) is an example of an intrusive igneous rock. Basalt (characterised by small crystals) is an example of an extrusive igneous rock.

Sedimentary rocks

Sedimentary rocks form when eroded material carried in water settles to form sediment. Over millions of years, successive layers of sediment build up on top of one another. These layers are slowly compressed and bond to form sedimentary rocks. These layers of rocks can be folded and faulted by forces within the earth's crust.

Sedimentary rocks have a layered appearance and often contain fossils of plants and animals that were trapped in the sediment as it was deposited. Sandstone is an example of a sedimentary rock.

Metamorphic rocks

Metamorphic rocks are formed when existing igneous and sedimentary rocks are compressed and heated without melting. They have the same chemical composition as the rocks from which they were made. Metamorphic rocks are often shiny and hard, and sometimes peel off in layers. Slate and marble are examples of metamorphic rock.

Rate of erosion

Rocks erode at different rates, depending on their hardness. **Differential erosion** occurs when softer rocks erode faster than more resistant, harder rock. Figure 4.9 shows a landform feature characteristic of differential erosion. The softer sedimentary rock has been eroded at a faster rate than the hard caps of these stone pillars.

4.9 Rock pillars (known as hoodoos) in Alberta, Canada. The pillars consist of soft sedimentary rock capped by a piece of harder rock. Hoodoos form in semi-arid regions of sedimentary rock that experience occasional heavy downpours. The rain erodes away the softer material, leaving behind columns of rock under hard caps.

The rock cycle

Figure 4.10 shows the **rock cycle**. The rock cycle is the process in which material from the earth's crust is recycled to form new rocks.

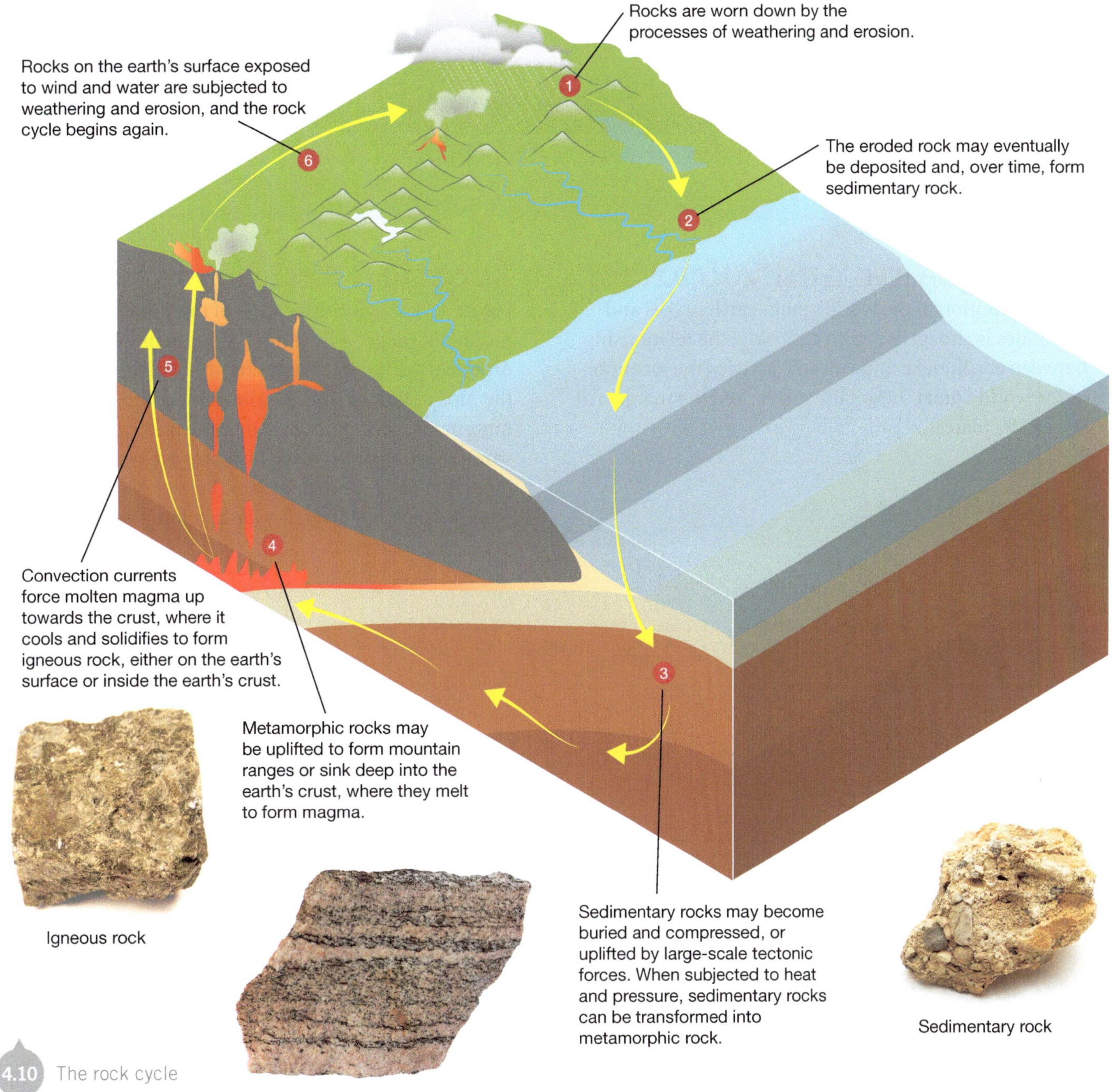

4.10 The rock cycle

ACTIVITIES

Knowledge and understanding

1 Distinguish between intrusive and extrusive igneous rocks.
2 Explain how sedimentary and metamorphic rocks are formed.
3 State what differential erosion is.
4 Name the different stages of the rock cycle.

Geographical skills

5 Study Figure 4.10. Draw your own annotated illustration of the rock cycle.

4.3 Mountain building

Mountains are defined as landform features that, when compared with surrounding landscapes, rise abruptly, and are impressive and noticeable. Factors taken into account when defining such landforms include elevation, relief, steepness and mass.

Earthquakes and volcanoes

The distribution of the world's main earthquakes and volcanoes is shown in Figure 4.11. Note the relationship between earthquake and volcanic activity, the location of the world's great mountain ranges, and the margins of the earth's plates.

4.11 There is a relationship between the location of the earth's great mountain ranges and areas of earthquake and volcanic activity.

Location and formation

Figure 4.11 shows the location of the world's great mountain ranges—the Andes Mountains of South America, the Rocky Mountains of North America, the Alps of Europe and the Himalaya of Asia. These mountains are found along the active margins of the world's tectonic plates. As the plates press against each other, the pressure increases and layers of rock are compressed and forced upwards, folding and faulting as pressure is released.

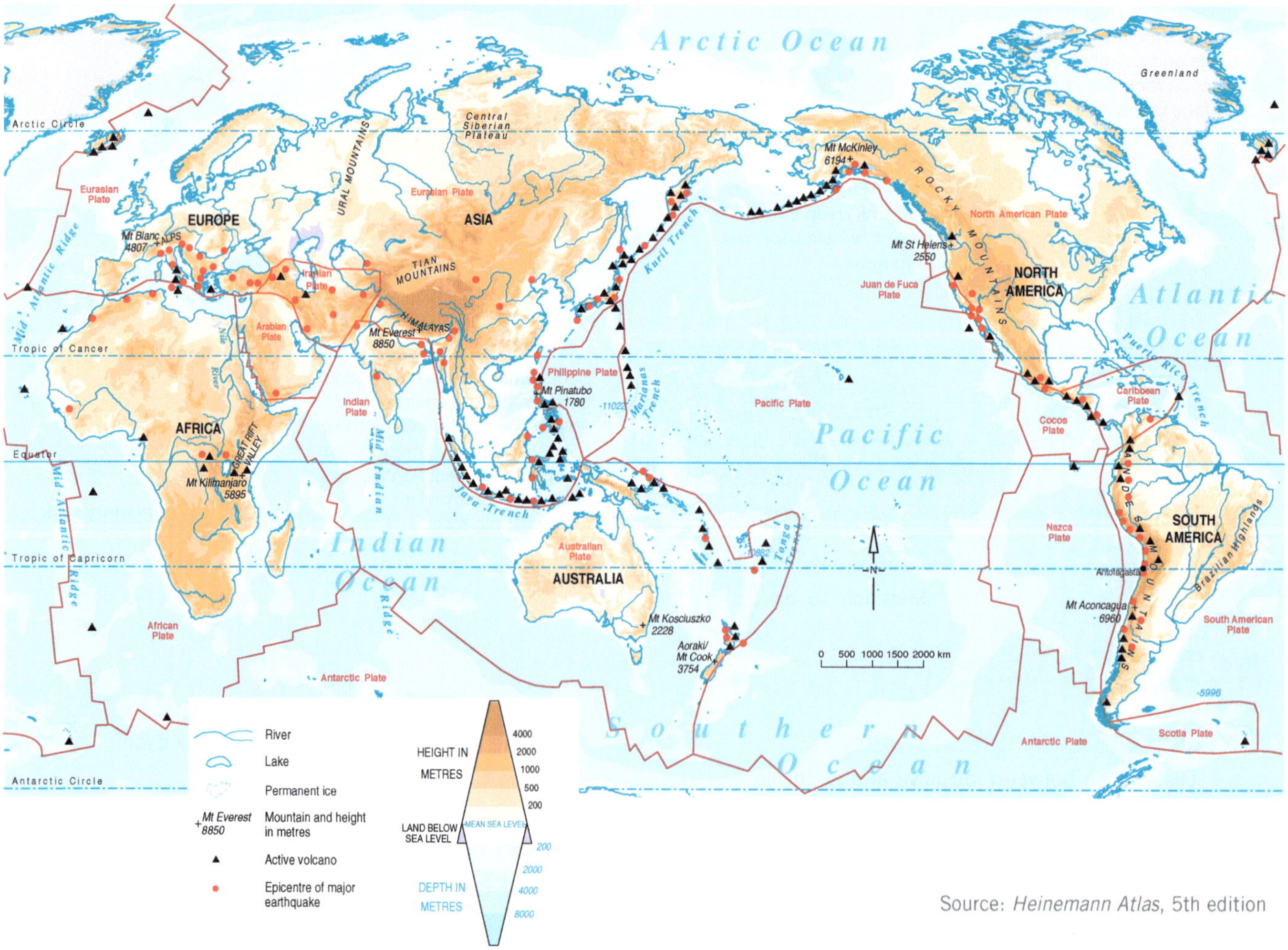

Source: *Heinemann Atlas*, 5th edition

Because the movement of the plates tends to be very slow, the mountain chains they produce are dominated by folded layers of rock. These mountain chains can, however, include landform features that are the result of faulting and volcanic activity. **Folding** results in wave-like patterns in the earth's crust (see Figure 4.12). **Faulting** occurs when there are fractures in the rock structure. Figure 4.13 shows both these features. **Rift valleys** and block mountains are examples of large-scale landform features associated with faulting.

As oceanic plates move apart (a process known as **seafloor spreading**), molten material fills the gap, forming a mid-ocean ridge. These ridges extend for 65 000 kilometres through all the earth's oceans. **Ocean trenches** are formed when oceanic plates are drawn down into the earth's mantle, where they melt.

4.12 Folded layers of rock

4.13 Folding and faulting of rock layers

A rift valley (graben) is a long valley area formed by the sunken land between two or more parallel faults. A series of sinkings at different rates may produce a series of step-like landform features.

A block mountain (horst) is an elevated area of land that has been uplifted between two or more parallel normal faults. The edges of many block mountains may be distinguished by the presence of an escarpment.

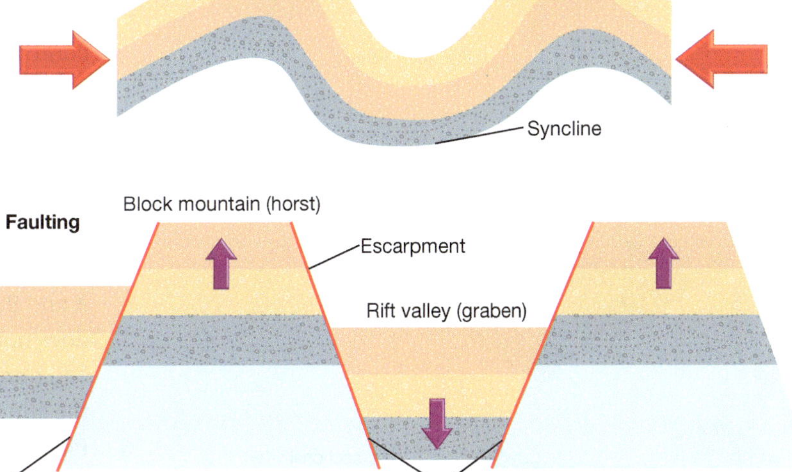

ACTIVITIES

Knowledge and understanding

1. Explain the difference between folding and faulting.
2. Identify the processes associated with the formation of rift valleys and block mountains.
3. Outline the conditions under which ocean trenches develop.

Geographical skills

4. Study Figure 4.11. Write a paragraph outlining the global distribution of earthquakes and volcanoes.
5. Create a pie graph of each continent's share of the world's mountainous areas using the following data:
   Asia (43.65%)
   Europe (6.7%)
   South America (8.4%)
   North America (15%)
   Australia and Oceania (1%)
   Africa (8.25%)
   Antarctica (17%).

4.4 Geoskills: Drawing cross-sections

A cross-section provides a side view, or profile, of a landscape. This view enables us to see how the shape of the land influences landuses such as settlement, drainage and vegetation.

Cross-sections

A cross-section is shown in Figure 4.14. Cross-sections are drawn from topographic maps, such as in Figure 4.15, and show the shape of the land.

4.14 A cross-section shows a side view, or profile.

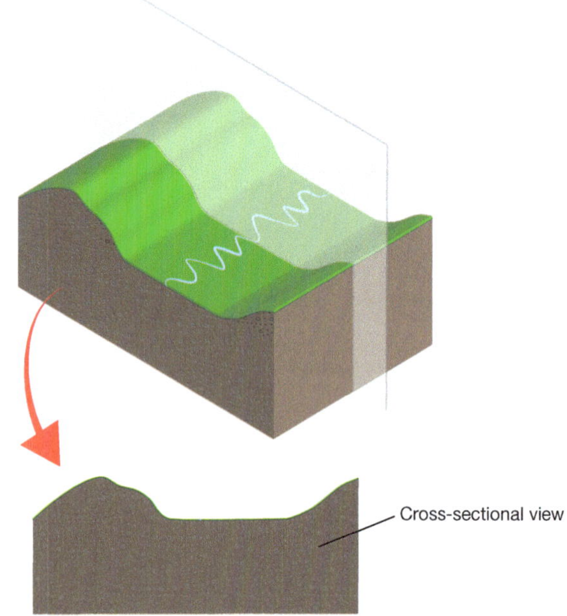

4.15 Twin peaks and a saddle

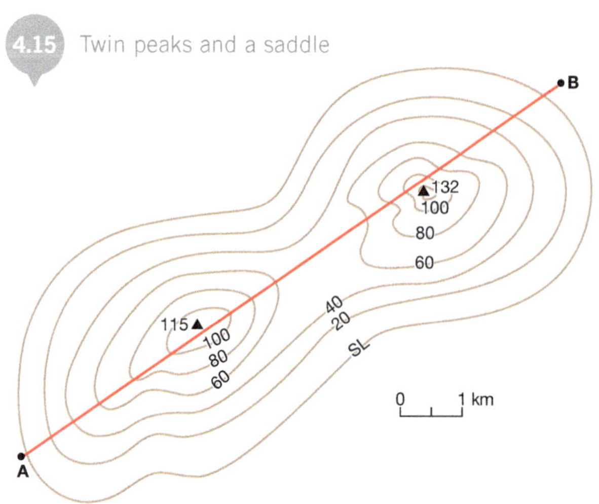

Constructing cross-sections

To draw a cross-section, follow the steps below and refer to Figure 4.16.

1 Locate the two points on the map between which the cross-section is to be made. Label these points 'A' and 'B' (see drawing *i*).
2 Place the straight edge of a piece of paper along an imaginary line joining points A and B. Mark points A and B on your paper (see drawing *ii*).
3 Mark the position where your paper crosses each contour line. Write the value of each contour line on your piece of paper (see drawing *ii*). You may have to estimate the height of your starting and finishing points.
4 On graph or squared paper draw the horizontal and vertical axes for your cross-section. The length of the horizontal axis should equal the distance between A and B. The vertical axis should use a scale that does not exaggerate your vertical scale too much. You don't want a range of low hills looking like the Himalaya!
5 Place your piece of paper along your horizontal axis. Lightly plot, in pencil, the contour points and heights as if you were drawing a line graph (see drawing *iii*).
6 Join the dots with a single, smooth curved line.
7 Label any features intersected by your cross-section, such as rivers and major roads.
8 Finish off your cross-section by:
   a shading in the area below the landform you have drawn
   b labelling the scale on the horizontal and vertical axes
   c giving it a title.

i

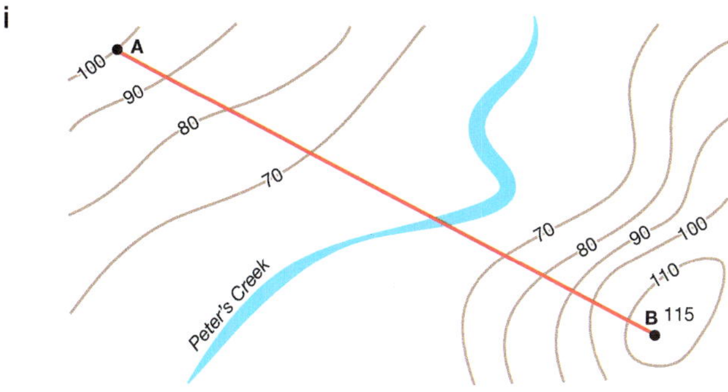

ii

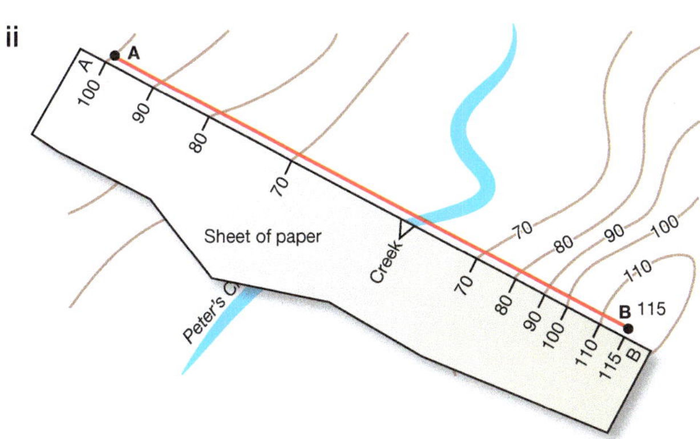

iii

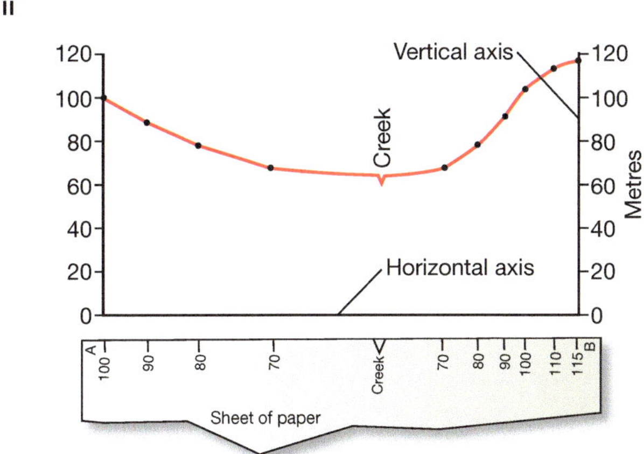

iv

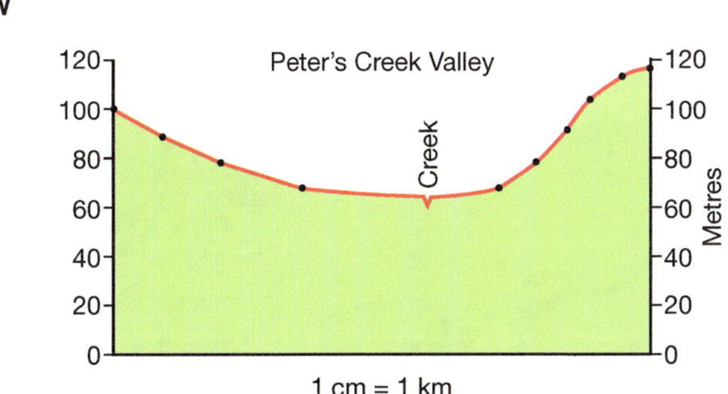

Skillsbuilder

Calculating vertical exaggeration

Calculating vertical exaggeration (VE) shows how much a cross-section has been exaggerated vertically.

The vertical exaggeration of a cross-section is given as a number. For example, 5× means that the vertical scale is 5 times greater than the horizontal scale. A value of 1× indicates that horizontal and vertical scales are identical. This means that the cross-section has no vertical exaggeration.

To calculate the vertical exaggeration (VE) of a cross-section we use the following formula:

$$VE = \frac{\text{Vertical scale (VS)}}{\text{Horizontal scale (HS)}}$$

Vertical scale (VS): The vertical scale used on the vertical axis of the cross-section

Horizontal scale (HS): The scale of the map from which the cross-section was drawn

Answers must be expressed as a single number. Vertical exaggeration has no units of measurement, nor is it expressed as a fraction.

For example, to calculate the vertical exaggeration of the cross-section shown in Figure 4.16:

$$VE = \frac{VS}{HS} = \frac{\text{1 cm represents 20 m}}{\text{1 cm represents 1 km}}$$

Make sure that the same unit of measurement is used on the top (numerator) and bottom (denominator) of the formula.

$$VE = \frac{VS}{HS} = \frac{\text{1 cm represents 20 m}}{\text{1 cm represents 1000 m (1 km)}}$$

$$VE = \frac{\frac{1}{20}}{\frac{1}{1000}}$$

Invert the denominator and then multiply:

$$= \frac{1}{20} \times \frac{1000}{1}$$

$$= \frac{1000}{20}$$

$$= 50 \text{ times}$$

ACTIVITIES

Geographical skills

Study Figure 4.15. Construct the cross-section A–B. Calculate the vertical exaggeration of the cross-section you have drawn.

4.5 Weathering, erosion and deposition

The shape and appearance of the earth's biophysical landscapes are constantly changing. The processes responsible for these changes are known as weathering, erosion and deposition.

Weathering

Weathering involves the physical or chemical breakdown into smaller pieces of rocks that do not undergo transportation from their original position.

Physical weathering

There are three main types of physical (or mechanical) weathering.

- ***Temperature change*** As rocks are heated by the sun they expand. At night, as temperatures fall, the rocks contract. Over time, this expansion and contraction causes cracks to appear and the rock begins to break down. This process is called **exfoliation** and is common in deserts in which there are large daily differences in temperature (see Figure 4.17).

4.17 Changes in temperature have caused the outer layer of this rock to crack and break away.

- ***Freeze–thaw action*** Water collects in the cracks in rocks, and when the water freezes it expands. This places a pressure on the rock and causes the cracks to widen and deepen. The pressure may eventually cause parts of rock to break off, as is illustrated in Figure 4.18. The small broken-down rocks are called scree. This type of weathering is common in high mountainous areas where the water freezes.

4.18 Freeze–thaw weathering

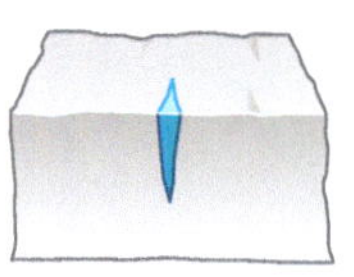

A crack in the rock fills with water

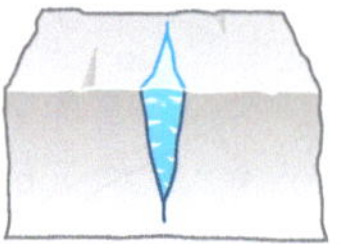

The water expands when it freezes and makes the crack wider

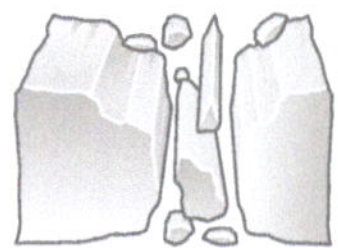

Eventually the crack gets so wide that the rock splits

- ***Organic action*** The growing roots of plants can exert a force that causes cracks in rocks to widen (see Figure 4.19). Gradually, the rock breaks down into smaller pieces.

4.19 Organic weathering

4.20 Jenolan Caves in New South Wales

Chemical weathering

When rainwater mixes with carbon dioxide (which is present in the atmosphere), it forms a weak acid that attacks many of the minerals contained in rock. This acid is especially effective in dissolving limestone. Limestone caves (such as the Jenolan Caves in New South Wales, shown in Figure 4.20) are a result of this type of weathering.

Rocks that contain iron are affected by a process known as oxidation. When oxygen and water come into contact with the iron in rocks they change the chemical make-up of the rocks, which take on a rusty appearance. The surface of the rock gradually decays and is eroded away.

Organic acids are produced when water mixes with decaying vegetation. These acids help to break down minerals in rocks.

Erosion

Erosion is the transportation of material that has been weathered (worn away) from one place to another by water, wind and ice. Erosion occurs when rivers carve out deep canyons, glaciers grind out massive valleys, and water drips through limestone to create caves and sinkholes.

The processes of erosion are as follows.

- ***Attrition*** Rocks collide as they are transported and are worn into smoother, rounded stones.
- ***Abrasion/corrasion*** Material rubs against riverbanks or valley sides as it is transported, or is thrown against cliffs by the sea. This acts like sandpaper on the landscape, gradually eroding it.
- ***Corrosion*** Some rocks, especially limestone, are dissolved by the natural acids in water and are carried away by the water.
- ***Hydraulic action*** Riverbanks or cliffs can be worn away by the sheer force of water hitting them, or can be blasted apart as air is forced into cracks.

Agents of erosion

The agents of erosion are water, wind and ice.

Water

Running water is the most powerful agent of erosion, transportation and deposition. It is especially effective in very dry areas and where humans have damaged the protective cover of vegetation that binds the soil together. When rain falls on exposed earth it causes deep channels called gullies to form as shown in Figure 4.21.

4.21 A landscape deeply scarred by running water—Death Valley, United States of America

Rivers

Rivers shape the land by eroding, transporting and depositing material. In a mountainous area, the river erodes downwards, creating narrow, V-shaped valleys. Away from the mountains, valleys become wider and some of the river's load of sediment is deposited. Closer to the sea, the river weaves, or meanders, across a wide, flat plain, depositing fine particles of soil called alluvium. These alluvial soils are usually very fertile and are often used for agriculture.

Waves

Coastlines are constantly changing. Some are eroded by storm waves and are dominated by landform features formed by the processes of erosion, such as headlands, cliffs, rock platforms and arches. Other coastlines advance towards the sea as waves deposit large amounts of sand. Such coastlines are dominated by landform features formed by the process of deposition, such as sand dunes, sand bars and spits.

Wind

Wind is a very effective agent of erosion in areas with little or no vegetation, in deserts and in areas where the land has been damaged.

Wind can pick up weathered rock material and, with it, effectively 'sandblast' larger rock features. This process is known as abrasion. It results in sculpted rock formations such as that shown in Figure 4.22. On a larger scale, rock-strewn desert surfaces form when strong winds sweep away the finer surface materials. This process is known as deflation.

Ice

Glaciers are slow-moving rivers of compacted snow (glacial ice). They form when compacted snow that has gathered over many years gradually moves downhill under the influence of gravity. Glaciers erode land by transporting rock (see Figure 4.23).

Deposition

Deposition is the process by which eroded material is added to a landscape. Water, wind and glacial ice transport weathered and eroded rock. As the speed of water slows, the strength of a wind declines and as a glacier melts and retreats, the load it carries is deposited, building up layers of sediment. Depositional landforms include beaches, sand bars and dunes; natural (river) levees and desert sand dunes.

4.22 Wind-borne sand grains have worn away the base of a desert rock, leaving this pedestal-shaped feature.

Findel Glacier, Zermatt, Switzerland

Changing landscapes

Weathering is sometimes called the passive, or inactive, agent of erosion because weathered material usually remains in place. Erosion and deposition (the removal and laying down of transported material) are known as active processes because the material is moved from its original position or location. A summary of this process is outlined in Table 4.24.

Weathering and erosion in summary

Weathering	
Physical weathering	**Chemical weathering**
Temperature change Freeze–thaw action Organic action	Weak acids (when rainwater and carbon dioxide mix) Organic acids
Erosion	
Erosional processes	**Agents of erosion**
Attrition Abrasion/corrasion Corrosion Hydraulic action Deposition	Running water, rivers and waves Wind Ice

ACTIVITIES

Knowledge and understanding

1. Explain the difference between physical and chemical weathering.
2. Outline the three processes of physical weathering.
3. Explain how chemical weathering helps to weaken and break up rock.
4. Explain why weathering is described as a passive, or inactive, agent of landform development.
5. State what erosion is.
6. Explain the following terms:
   - attrition
   - corrosion
   - abrasion
   - hydraulic action.
7. Explain what is meant by the term 'deposition'.

Applying and analysing

8. Identify the conditions under which water is the most effective agent of erosion.

Geographical skills

9. Construct a series of diagrams like those in Figure 4.18 to explain how organic weathering occurs.

4.6 Geoskills: Landscape photos

The world's landscapes have been developed by wind, water and plate movements. These landscapes are constantly undergoing change.

World facts

- The world's largest gorge is the Grand Canyon in the United States of America. Carved by the Colorado River, it is 20 kilometres wide at its widest point, 349 kilometres long, and more than 1.5 kilometres deep.
- The world's deepest canyon is El Cañon de Colca, in Peru, South America. It is 3 kilometres deep.
- With a length of 515 kilometres, the Lambert-Fisher Ice Passage in Antarctica is the world's longest glacier.
- The world's fastest-moving glacier is Greenland's Quarayaq Glacier. It moves 20 to 24 metres per day.

1

2

3

4

ACTIVITIES

Applying and analysing

1. Match the following captions with the photographs:
   a. Mt Fuji, Japan
   b. Californian coast, United States
   c. fjord, New Zealand
   d. the Sahara, Africa
   e. Grand Canyon, United States
   f. glacial landscape, Alaska, United States
   g. Victoria Falls, Africa.
2. Identify the photographs that best match each of the following descriptions:
   a. landscape shaped by wind
   b. landform shaped by ice
   c. landform shaped by wave action
   d. landform feature produced by volcanic action.
3. Write a sentence describing the type of landform shown in each photograph.
4. Select one of the photographs and write a paragraph explaining the processes responsible for the landscape's landform features.

CHAPTER 5 LANDSCAPE HAZARDS

In 2004, the Indian Ocean earthquake and tsunami cost the lives of at least 230 000 people. In 2011, the Tohoku earthquake and tsunami killed almost 16 000 people. Such natural disasters are, unfortunately, all too common and reflect the dynamic nature of the earth's crust.

The earth is a restless planet, continually shifting under our feet. Powerful processes deep inside the earth's crust create volcanoes and earthquakes, push up mountains and pull continents apart. The earth's rocks provide clues to how these forces have shaped and changed our planet over billions of years.

The surface of the earth is made up of massive plates that rest on a molten core. The heat beneath the plates drives their movement. This process is called plate tectonics.

In this chapter we focus on a range of landscape hazards, including earthquakes, volcanoes, coastal erosion, mass movement, mudslides and avalanches.

KEY IDEAS

- To investigate the causes and consequences of selected landscape hazards
- To understand the challenges of living in areas of risk
- To develop skills in interpreting satellite images and digital terrain models

5.0 Mudslide in southern Seoul, South Korea, 2011

GLOSSARY

caldera	a volcanic landform feature formed when a violent eruption blasts away the top of an existing volcanic cone or shield
dormant volcano	a volcano that is said to be 'sleeping', as it has not erupted for a period of time but may still erupt
epicentre	a point on the earth's surface that is directly above the centre (focus) of an earthquake
extinct volcano	a volcano that has not erupted for a long period and will not erupt again
focus	a point under the earth's surface at which a sudden movement in the earth's crust occurs; the origin of an earthquake
hot spot	a location where hot molten magma from deep within the earth rises up through the crust to reach the surface
lahar	a mudflow formed when volcanic material mixes with water
lava	molten rock at the earth's surface
magma	molten rock below the earth's surface
magnitude	the energy released by an earthquake
maar	a broad, roughly circular, flat-floored volcanic crater with steep inner walls and a low surrounding rim made up of fragments of rock material blown out of the crater during eruptions
pumice	a very light and porous volcanic rock
reclaimed land	land that has been gained from the sea or wetlands, such as waterlogged land that has been drained of water, or that has been restored for human use, such as a former quarry that has been filled in
seamount	a volcanic topographic feature rising from the seafloor
volcanic cone	a cone-shaped volcanic landform made up of layers of lava and volcanic ash
volcanism	processes associated with volcanoes and volcanic activity
volcanologist	person who studies volcanoes

5.1 Earthquakes

Earthquakes occur when energy that has built up in the earth's crust is released. While earthquakes are most commonly associated with the movement of the earth's plates, smaller, often less destructive earthquakes can occur well away from plate margins. These earthquakes are usually associated with fault lines (lines of weakness) in rock.

Shock waves

When the built-up energy within the earth's crust is released, shock waves go out in all directions. The point underground at which the sudden movement occurs is called the **focus**. The place at which the shock waves first reach the earth's surface is called the **epicentre**, as shown in Figure 5.1.

Measuring earthquakes

A seismograph measures the duration, magnitude and direction of an earthquake. The magnitude (or amount of energy released by an earthquake) is given a measurement on the **Magnitude scale** outlined in Table 5.2. The Magnitude scale replaced the Richter scale in 2002. Each point on the scale is ten times greater than the point below. This means, for example, that the Mexico City earthquake of 1985 was ten times stronger than the earthquake that struck San Francisco in 1989.

5.1 The focus, epicentre and shock waves of an earthquake

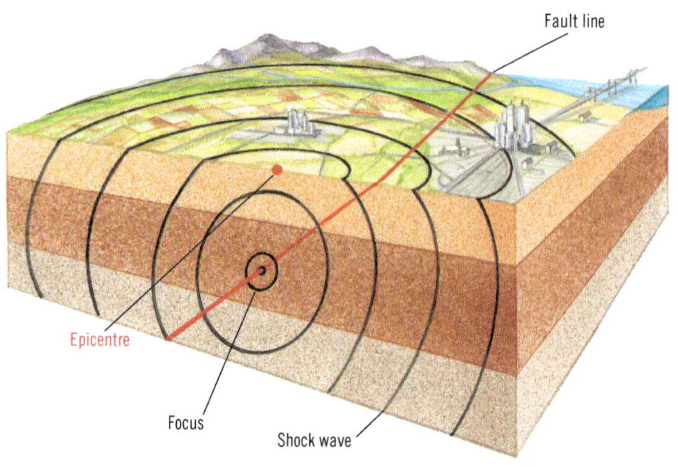

5.2 The Magnitude scale

Magnitude	Effects	Average number per year worldwide
2.5 or less	Usually not felt, but can be recorded by seismograph	900 000
2.5 to 5.4	Often felt, but only causes minor damage	30 000
5.5 to 6.0	Slight damage to buildings and other structures	500
6.1 to 6.9	May cause a lot of damage in very populated areas	100
7.0 to 7.9	Major earthquake. Serious damage	20
8.0 or greater	Great earthquake. Can totally destroy communities near the epicentre	One every 5 to 10 years

Effects of earthquakes

Primary effects

Primary effects are effects that are directly related to the earthquake, such as collapsed buildings (see Figure 5.3) and open fissures.

Secondary effects

Secondary effects are the short-term effects of the earthquake, such as fires caused by broken gas pipes and electrocution caused by downed power lines.

Tertiary effects

Tertiary effects are the long-term effects of an earthquake, such as homelessness, business failure, disease outbreaks and famine.

Place and effect

While the strength of an earthquake is important, other factors can also influence the extent of damage and loss of life. These include:

- distance from the epicentre—energy waves lose their intensity as they spread out
- level of preparation—building structures designed to withstand earthquakes means that the number of deaths caused by collapsing buildings is greatly reduced
- population density—damage and destruction in densely settled urban centres are more severe than in rural areas
- time of day—whether people are at home or out and about makes a difference to the number of deaths and injuries
- season—the secondary effects of earthquakes are much worse in winter
- type of land that cities are built on—silt and **reclaimed land** tend to magnify the effects of earthquakes.

5.3 The primary devastation caused by the 2010 earthquake in Christchurch, New Zealand

Tsunamis

A tsunami is a series of ocean waves caused by underwater land movement. Movement can be caused by a landslide, a meteor strike or an earthquake, as shown in Figure 5.4. Tsunami waves resemble surges of water tens of metres high. The length of time between waves can range from minutes to hours. Although the impact of tsunamis is limited to coastal areas, their destructive power can be enormous.

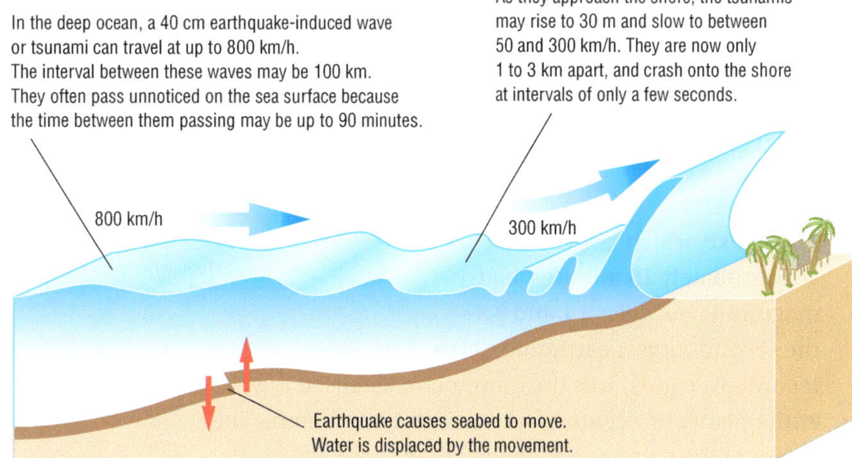

5.4 Formation of a tsunami

ACTIVITIES

5.1

Knowledge and understanding

1. State what an earthquake is. Explain why and where earthquakes occur.
2. Define the terms 'focus' and 'epicentre'.
3. Identify the instrument used to measure the magnitude of earthquakes.
4. State the difference between the primary, secondary and tertiary effects of earthquakes.
5. Outline the reasons why some places are more seriously affected by earthquakes than others.

Applying and analysing

6. Decide whether an earthquake measuring 7.0 on the Magnitude scale in a remote rural area will cause less or more damage than a 6.0 earthquake in a big city during winter. Explain your answer.

5.2 Case study: Asia's tsunamis

In recent years, Asia has experienced a number of massive natural disasters. Two of the biggest involved earthquake-induced tsunamis. One of these affected the countries surrounding the Indian Ocean. The other devastated low-lying coastal areas of eastern Japan.

Indian Ocean

On 26 December 2004, an undersea earthquake with an epicentre off the west coast of Sumatra triggered a series of devastating tsunamis that killed more than 225 000 people in eleven countries surrounding the Indian Ocean, and inundated coastal communities with waves up to 30 metres in height. Officially known as the Great Sumatra–Andaman earthquake, the event is more commonly referred to as the Asian tsunami or the Boxing Day tsunami. Indonesia, Sri Lanka, India, and Thailand were hardest hit (see Figure 5.5).

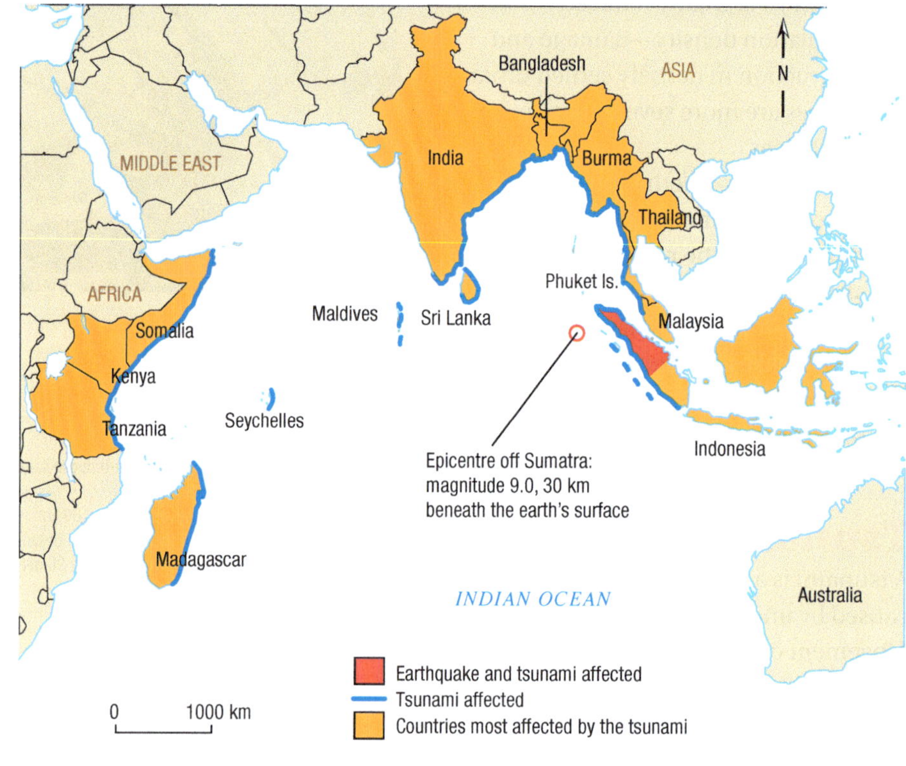

5.5 The source and extent of the Asian tsunami, 26 December 2004

The quake, which lasted approximately 10 minutes, had a magnitude between 9.1 and 9.3—the second-largest earthquake ever recorded. So great was the movement, it caused the entire planet to vibrate by as much as 1 centimetre. The extent of the damage and the plight of those affected shocked the world and led to a worldwide humanitarian response. More than US$14 billion of humanitarian aid was given to the countries affected.

East Japan

On the afternoon of 11 March 2011, a 9.0 magnitude undersea earthquake occurred at a depth of 32 kilometres, 70 kilometres off the east coast of Japan. It was one of the most powerful earthquakes ever recorded, moving the island of Honshu, Japan's largest, 2.4 metres to the east.

The earthquake caused a 180-kilometre-wide section of the seabed to lift by 5 to 8 metres. The resulting displacement of water triggered a massive tsunami that reached heights in excess of 40 metres. The surging waters travelled up to 10 kilometres inland, sweeping aside all that lay in their path (see Figure 5.6). More than 15 800 people died, 27 000 were injured and over 3000 remain missing, feared dead. Over a million buildings were destroyed or partially damaged. More than 4.4 million households were left without electricity and 1.5 million without water. The tsunami also initiated a meltdown at the Fukushima Daiichi Nuclear Power Plant. Residents within a 20-kilometre radius of the plant were evacuated and still cannot return to their homes. Figure 5.7 shows the travel times of the waves of the tsunami.

5.6 The estimate for the cost of rebuilding in Japan after the earthquake and tsunami damage is US$300 billion.

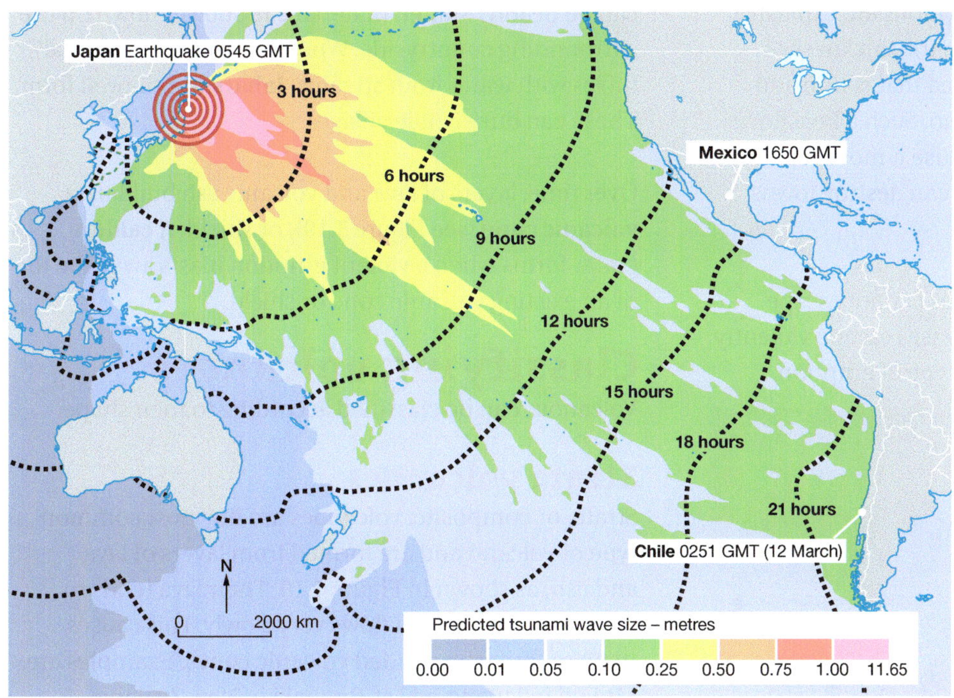

5.7 Estimated travel times of the waves generated by the 2011 Great East Japan earthquake

ACTIVITIES

Knowledge and understanding

1 Copy and complete the following table.

	Earthquake and tsunami	
	Indian Ocean	Great East Japan
Earthquake details		
Damage		

Applying and analysing

2 Study Figure 5.6. Comment on the scale of the disaster shown in the photograph. As a class, discuss the descriptive terms used in your individual responses.

Geographical skills

3 Study Figure 5.5. Describe the extent of the 2004 tsunami's impact. How far did the wave of destruction travel?

4 Study Figure 5.7. How long did it take for the tsunami wave to travel to:

a Mexico

b Papua New Guinea

c the North Island of New Zealand

d South America.

5.3 Volcanoes

Volcanic eruptions occur when molten material (magma) forces its way to the earth's surface through cracks or faults in the earth's crust. The molten material that flows from a volcano is called lava. The most active volcanoes lie in lines that coincide with the collision zones of the earth's plates.

Volcanic landform features

Volcanic eruptions can produce landforms in a rapid and spectacular manner. Some eruptions are explosive, while others are not—it depends on the thickness of the **magma**. Figure 5.8 shows magma that is thin and runny, flowing easily out of a volcano. Such a **lava** flow is rarely a threat to human life because it moves slowly enough for people to escape, but it can destroy towns and villages in its path.

If the magma is viscous (thick), gases cannot escape easily. Pressure builds up until the gases escape violently and explode. Figure 5.9 shows this type of eruption, as magma blasts into the air and breaks into pieces called volcanic bombs. Explosive volcanic eruptions can be deadly. Molten material can shower down on the surrounding countryside. When hot volcanic material mixes with water from streams, **lahars** (mudflows) form. These can bury whole villages.

Over time, layers of lava and volcanic ash build up a **volcanic cone** (see Figure 5.10), or shield. A **caldera** crater forms when a violent eruption blasts away the top of an existing volcanic cone or shield.

5.8 A lava flow

Classifying volcanoes

Volcanoes can be classified according to their shape.

Strato volcanoes

Strato, or composite, volcanoes are the most common type of volcano and are formed from layers of lava and ash, as shown in Figure 5.10. Their lava tends to be viscous (thick) and flows very slowly. This causes them to form steep-sided volcanic cones. Examples are Mt Fuji in Japan and Mt Taranaki in New Zealand.

5.9 Volcanic bombs

5.10 A cross-section through a strato volcano formed of layers of lava and ash

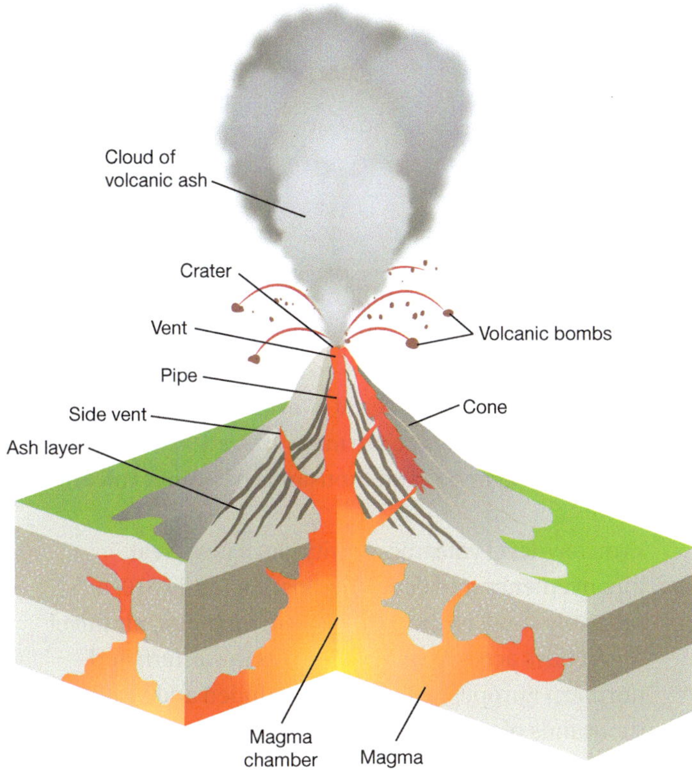

Shield volcanoes

Shield volcanoes are among the world's largest volcanoes. They develop a broad base because the very fluid lava from which they are made runs downhill rather than piling up. Examples include the Likauea and Mauna Loa volcanoes of Hawaii.

Calderas

While calderas are the least common form of volcano, they are often huge and potentially very explosive. When they erupt they tend to collapse in on themselves and so they don't build up a cone. Famous calderas are Yellowstone in the United States of America and Greece's Santorini caldera in the Aegean Sea, shown in Figure 5.11.

Recent volcanic activity

Some of the world's great mountains and many of the islands that dot the world's oceans are the result of past volcanic activity. There are 550 known active (that is, still capable of erupting) volcanoes on earth. Large eruptions may result in short-term climatic change as millions of tonnes of volcanic ash and smoke are released into the atmosphere, reducing the amount of sunlight reaching the earth's surface. Table 5.12 lists some volcanic eruptions that have caused great loss of life.

5.11 Santorini clings to the rim of a collapsed volcanic caldera.

5.12 Volcanic eruptions that have caused great loss of life

Year	Place	Impacts
1815	Tambora, Indonesia	92 000 people died, mostly in Indonesia—because of starvation caused by the loss of crops and livestock
1883	Krakatau, the Sunda Strait	The resulting tsunami between Java and Sumatra killed 36 000
1902	Pelee, Martinique	Poisonous volcanic gases and pyroclastic flows killed 36 000
1985	Nevado del Ruiz, Colombia	A wave of mud smothered 23 000
1991	Mt Pinatubo, the Philippines	900 people were killed, mostly from collapsing roofs

ACTIVITIES

Knowledge and understanding

1 Explain why some volcanic eruptions are violent while others are not.
2 Explain what volcanic bombs and lahars are.
3 Describe how volcanic eruptions can bring about short-term climatic change.

Applying and analysing

4 Study Figure 5.12.
  a List the main causes of loss of life.
  b Which cause killed the most people? Explain why.

5.4 Volcanoes transforming landscapes

Volcanic activity is always occurring somewhere in the world. However, sometimes there is an eruption on a vast scale—an eruption so massive that it transforms landscapes.

Active volcanism

Current areas of active **volcanism** include the Bagana and Manam volcanoes in Papua New Guinea; Dukono and Semeru in Indonesia; Karymsky and Shiveluch in Russia; Kilauea in Hawaii; Mount St Helens in the United States of America, shown in Figure 5.13; Colima in Mexico; Fuego and Santa Maria in Guatemala; Masaya in Nicaragua; Arenal in Costa Rica; Sakura and Suwanose in Japan; Sangay and Tungurahua in Ecuador; Soufriere Hills in Montserrat, West Indies; and Stromboli and Etna in Italy. There are many more inactive, or dormant, volcanoes and thousands of **extinct volcanoes**.

Even low-level volcanic activity can cause great inconvenience and economic loss. In April 2010, a relatively small eruption in Iceland disrupted air travel for six days across twenty countries in western and northern Europe. Flights were grounded and hundreds of thousands of travellers had their travel plans disrupted.

5.13 The Mt St Helens eruption, United States of America, 1980

Mt St Helens

One of the most spectacular volcanic eruptions of recent times was the Mt St Helens eruption of 18 May 1980. The volcano, located in the Cascade Mountains of the United States of America (see Figure 5.14) had been dormant for over a century. Figure 5.15 shows the causes of the eruption. At 3.62 a.m. an earthquake of magnitude 8.1 rocked the earth directly beneath the mountain. The earthquake was caused by the collision of two plates, the Juan de Fuca Plate colliding with and descending under the lighter North American Plate. The earthquake started an avalanche, which was followed by a massive blast of gas, rock, ash and ice. The devastating effects are outlined in Figure 5.16.

- One hundred and twenty metres of the summit vanished, and in its place was a crater 2 kilometres wide, 4 kilometres long and 1.5 kilometres deep.
- Three hundred and eighty square kilometres of land to the north of the mountain was stripped of its vegetation.
- Volcanic ash, carried by the wind, spread 1500 kilometres to the west.
- Sixty-two people died.

5.14 Location of Mt St Helens

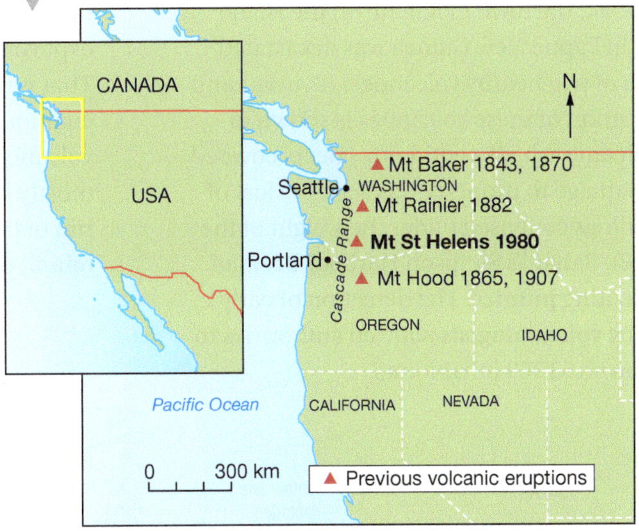

5.15 Causes of the Mt St Helens eruption

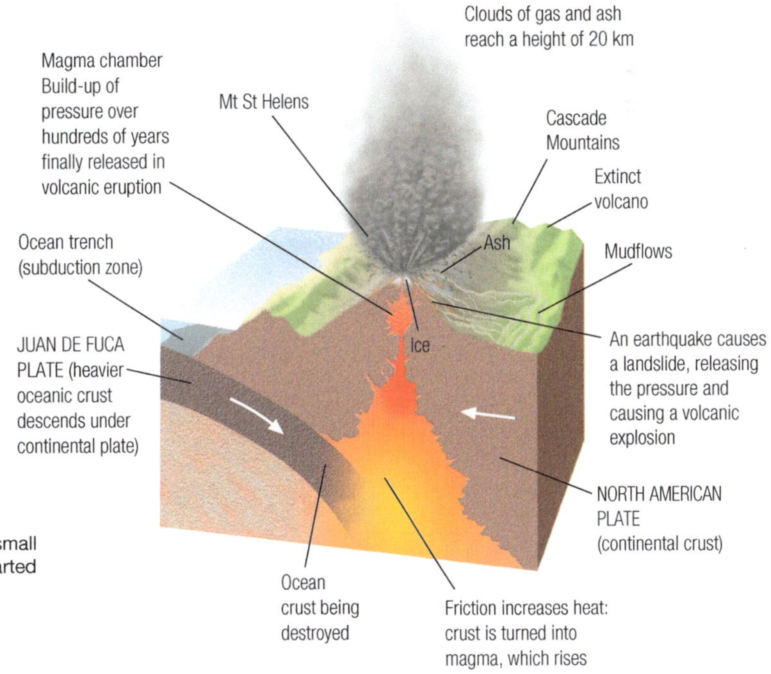

5.16 Damaged caused by the Mt St Helens eruption, 1980

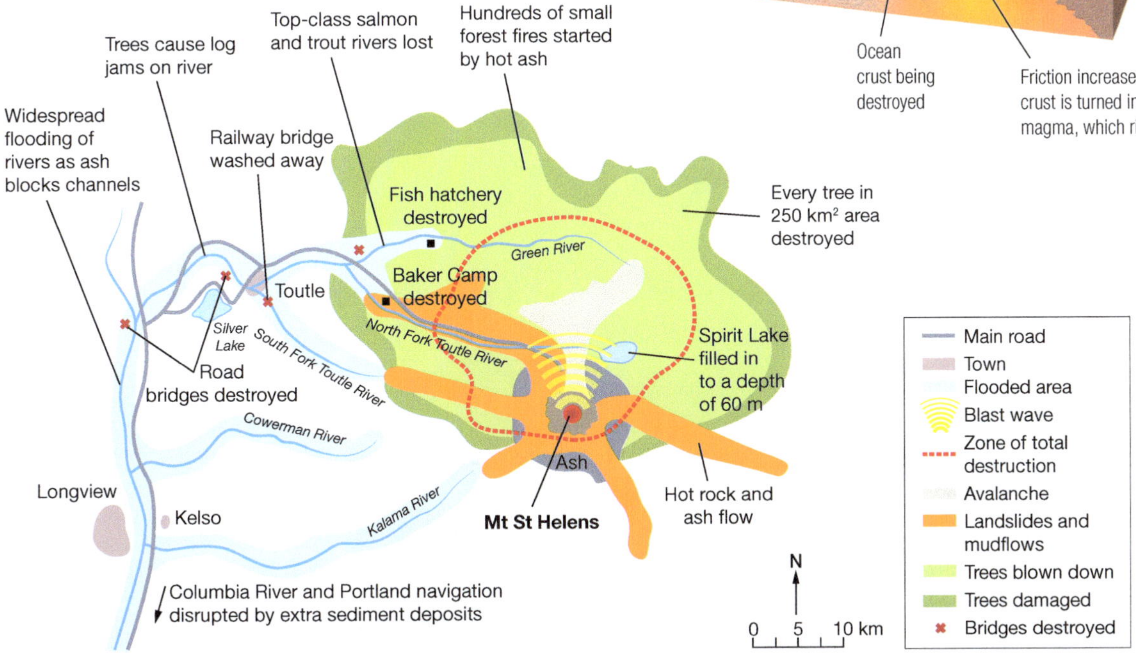

Rabaul

In September 1994, the town of Rabaul on the island of New Britain in Papua New Guinea was devastated by the eruption of two nearby volcanoes: Tavurvur and Vulcan. The location of these volcanoes is shown in Figure 5.17. Volcanic ash, 75 centimetres deep, covered the town and damage to property was severe. Most of the town's buildings collapsed under the weight of the ash (Figure 5.18). Rabaul's Simpson Harbour became clogged with floating **pumice**. The detection of early warning signs by **volcanologists** allowed authorities to evacuate the town's 53 000 inhabitants.

Rabaul's harbour is a large caldera, about 10 kilometres in diameter. Its collapsed floor is the result of a large, explosive eruption that took place some 1400 years ago. That eruption partly emptied the underlying magma chamber (the reservoir of molten rock feeding the volcano). Many smaller eruptions have occurred since, mostly from minor cones that have developed near the rim of the caldera. After the 1994 eruption, Rabaul was rebuilt on a new site.

5.17 The Rabaul caldera and the Tavurvur and Vulcan volcanoes

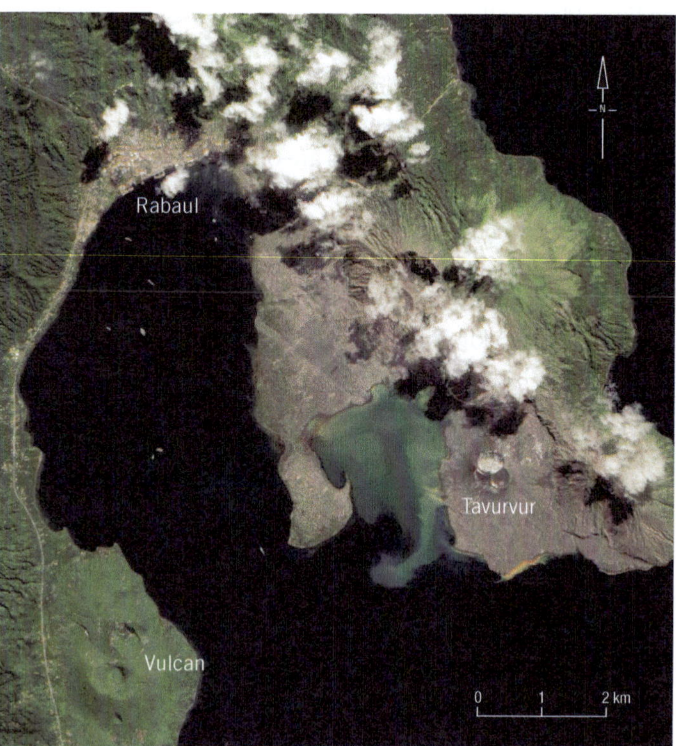

5.18 Rabaul was buried with volcanic ash to a depth of 75 centimetres.

5.19 The Hawaiian islands formed as the north-westerly moving plate moved over a hotspot in the earth's mantle.

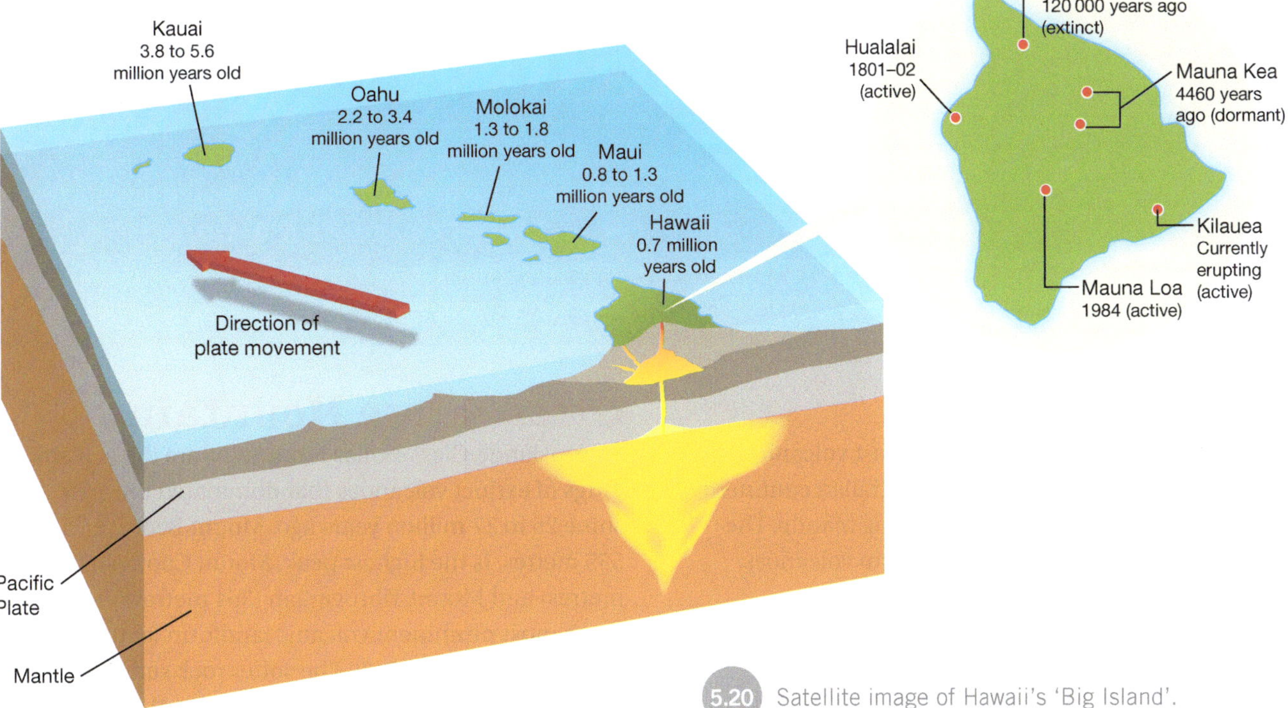

Hawaii

Hawaii is a chain of volcanic islands in the north Pacific Ocean. It consists of hundreds of islands spread over a 2400-kilometre-long archipelago. The five largest of these islands—Hawaii ('Big Island'), Maui, Oahu, Kauai and Molokai—are located at the south-east end of the archipelago. The islands are the exposed peaks of a great undersea mountain range known as the Hawaiian-Emperor Seamount Chain.

Figure 5.19 shows the formation of the Hawaiian archipelago. The volcanic islands formed over millions of years as the north-westerly moving plate moved over a **hot spot** in the earth's mantle. As the Pacific Plate moved, a new volcano or island was formed over the hot spot. Currently, the hot spot is under Hawaii's 'Big Island', shown in Figure 5.20.

5.20 Satellite image of Hawaii's 'Big Island'.

ACTIVITIES

Knowledge and understanding

1. Study Figure 5.15. Explain the cause of the Mt St Helens eruption.
2. Study Figures 5.15 to 5.18. List the damage caused by the Mt St Helens and Rabaul eruptions.
3. Study Figure 5.17. Draw a sketch map of the Rabaul caldera. Locate the now abandoned Rabaul township and the volcanic vents Tavurvur and Vulcan.
4. Study Figure 5.19. Explain how hot spots have created the Hawaiian archipelago.

5.5 Volcanic landscapes in Australia

There are no active volcanoes on the Australian mainland or in Tasmania. There are, however, a number of extinct volcanoes and numerous landform features created by volcanic activity.

Volcanic history

Scientists believe that Australia's east coast volcanic landform features developed as the Australian continent moved across a hot spot in the underlying mantle. The hotspot 'melted' through the plate to form volcanoes. Figure 5.21 outlines this process. Today, the only active volcanoes on Australian territory are located 4000 kilometres to the south-west of Perth on the Australian territories of Heard Island and the nearby McDonald Islands.

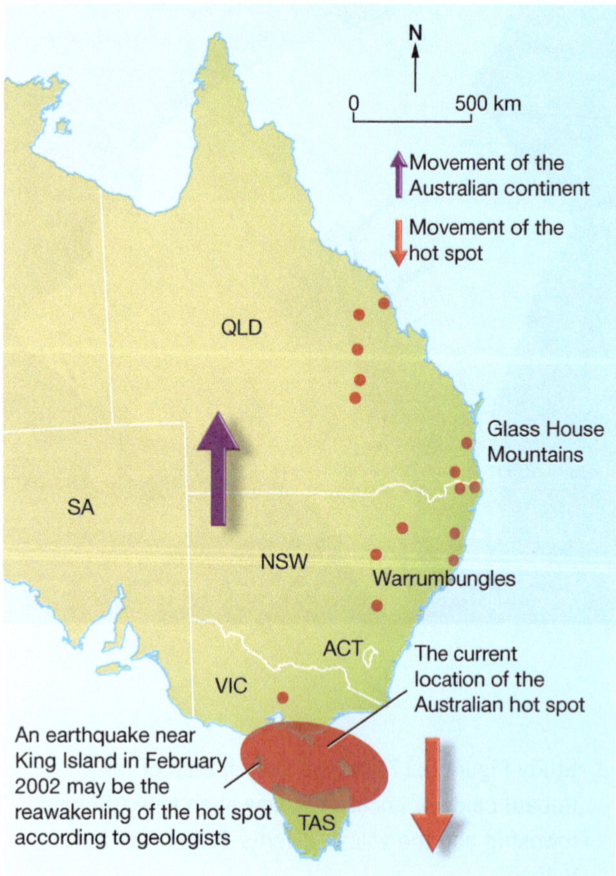

5.21 Volcanoes and the predicted position of the hot spot on Australia's east coast

Glass House Mountains

Queensland's Glass House Mountains are the volcanic plugs of extinct volcanoes that dominated the area some 26 to 27 million years ago. Mount Beerwah, at 555 metres, is the highest peak. Mount Coonowrin (377 metres) and Mount Tibrogargan (364 metres) are the area's most prominent volcanic landform features. They are shown in Figure 5.22. The softer rock surrounding the plugs has eroded over time, exposing the harder volcanic material.

Warrumbungles

The Warrumbungle National Park's landscape, shown in Figure 5.23, originated 180 million years ago when the region was covered by large shallow lakes. At the bottom of these lakes, accumulated layers of sediment were slowly compressed to form sandstone. About 18 to 15 million years ago hot magma exploded through the layers of sandstone. Over millions of years of volcanic eruptions, a huge shield-shaped volcano developed. Weathering and erosion have since removed large amounts of the softer sandstone and exposed the inside of the original volcano.

The exposed volcanic landform features include Crater Bluff and Belougery Spire (vents from which magma once erupted) and the Belougery, Split Rock and Bluff Mountain lava domes (where molten rock has bubbled to the surface and clogged up the source vent). Perhaps the most striking landform feature is the 90-metre tall rock wall called the Breadknife. Known in geological terms as a dyke, the Breadknife formed when magma was forced through a fault, or crack, in the earth's surface. Because the rock surrounding the Breadknife was softer than the solidified magma, erosion has left a long sliver of stone exposed today.

5.22 The distinctive volcanic landform features Mt Beerwah and Mt Coonowrin of Queensland's Glass House Mountains.

5.23 This distinctive volcanic landform feature of the Warrumbungle National Park, the Breadknife, is the remnants of a large heavily eroded shield volcano that was active from 18 to 15 million years ago.

Newer Volcanic Province

Compared with Australia's other volcanic landscapes, the Newer Volcanic Province shown in Figure 5.24 is relatively young, having been formed in the last 20 000 years. Figure 5.25 shows Mount Noorat, which last erupted between 5000 and 20 000 years ago. Mount Noorat is one of the major volcanoes of the Newer Volcanic Province in Victoria. It is the deepest enclosed crater in Victoria. Other volcanic landform features of the province include maars. A **maar** is relatively uncommon volcanic landform feature comprising a broad, roughly circular, flat-floored volcanic crater with steep inner walls and a low surrounding rim built up of fragments of rock material that were blown out of the crater during eruptions. The Leura Maar is a shallow, oval-shaped depression about 2.5 kilometres long, 1.7 kilometres wide and up to 50 metres deep. It was formed by a series of major volcanic explosions. Some of the material forced from the vent was thrown high into the air, but most surged rapidly across the ground surface as a dense cloud of steam, gas and rock fragments. Mount Leura and Mount Sugarloaf began as lava vents on the rim of what is now Leura Maar.

5.24 East coast volcanic activity and Victoria's volcanic plain

5.25 Mount Noorat is a dormant volcano.

5.26 Mt Gambier's Blue Lake

Blue Lake

Figure 5.26 shows South Australia's famous Blue Lake, in Mt Gambier. The lake is located in one of four lake-filled maars (extinct volcanic craters) found in the area. The others are Valley Lake, Leg of Mutton Lake and Brownes Lake (the latter two lakes are now dry as a result of a lowered water table).

Blue Lake's crystal-clear water has filtered through the limestone rock that lies beneath the city. Each year in November the lake starts its colour change from a winter steel-blue to brilliant turquoise-blue, changing back to steel-blue the following March.

Other places in Australia

Elsewhere in Australia there are only a limited number of volcanic landform features. Tasmania has a number of offshore **seamounts** while in Western Australia there are no active or dormant volcanoes, although there are a number of extinct ones. For example, there are a significant number of small extinct volcanoes in the valley of the Fitzroy River in the Kimberley region.

DID YOU KNOW?

The most recent volcanic eruption on the Australia mainland was about 4000 to 5000 years ago at Mt Gambier in South Australia.

ACTIVITIES

Knowledge and understanding

1 Explain why Australia has no active volcanoes.
2 Outline the origins of Australia's volcanic landforms.
3 State where Australia's only active volcanoes are located.

Geographical skills

4 Study Figure 5.21. Identify the principal areas of past volcanic activity in Australia.
5 Select one of the photographs in this unit and construct a photo sketch. Label the volcanic features shown in the image.

5.6 Geoskills: Satellite images

Satellites orbiting the earth take satellite images. The processes involved are referred to as 'remote sensing'. Geographers use remote sensing to study the spatial distribution of natural, managed and constructed environments. Remotely sensed images are especially useful for investigating change over time.

Collecting data

Remotely sensed images are produced from data gathered by satellite-mounted sensors. These sensors are so sensitive that they can record the radiation given off by features on the earth's surface. This data is then converted into images. Often these images are referred to as false-coloured images and the observer needs to know what each colour represents to be able to interpret them (see Table 5.27).

As satellites became more sophisticated they were able to capture the data necessary to produce true-coloured images. These images feature colours as they appear to human eyes. However, we still need to know what each colour represents (see Table 5.28).

5.27 Colour guide for false-coloured images

Colour	Feature
Dark blue to black	Deep water in oceans, lakes and dams
Mauve to steely blue	Urban and industrial areas
Blue to light blue	Arid scrubland; very shallow water
Dark green	Deep muddy floodwaters, clear shallow water
Light green	Moist, ploughed, bare soils; light grass cover
Brown	Drier vegetation such as eucalypts and arid woodlands; bare rock
Red	Healthy growing vegetation; rainforest (deep red); growing crops and pastures; mangroves (deep red)
Pink to red	Early growth of crops and grasslands; suburban gardens, lawns and parks
Yellow	Areas with little vegetation cover, heavily grazed areas, deserts and sand dunes
White to cream	Bare ground; dry sand and salt areas, dunes and beaches; clouds

5.28 Colour guide for true-coloured images

Colour	Feature
Dark blue to black	Deep clear water in oceans, lakes and dams
Light blue	Shallow water
Mauve to steely blue	Urban and industrial areas
Brown to light brown	Drier vegetation such as eucalypt and arid woodlands; bare rock
Bright light green	Grassland, growing crops and pastures; suburban parks and gardens
Bright green	Healthy growing green vegetation; rainforest and mangroves
Light pink to orange to brown	Cleared farming land; early growth in crops and grasslands
White to cream	Bare ground; dry sand and salt areas; dunes and beaches; clouds

Mt St Helens

In the three decades since the 1980 eruption of Mt St Helens in Washington State, United States of America, geographers have been given a unique opportunity to observe the steps through which a devastated landscape is reclaimed. The scale of the eruption and the beginning of recovery in the Mt St Helens blast zone are documented in this series of images captured by NASA's Landsat satellites between 1979 and 2011 (Figures 5.29 and 5.30).

The 1980 eruption began with an earthquake that caused the northern face of the mountain to collapse, producing the largest landslide in recorded history. The avalanche buried 23 kilometres of the North Fork Toutle River valley with an average 46 metres—but in places up to 180 metres—of rocks, dirt and trees. In the years since the eruption, the river has recarved a shallow path through the buried valley.

When the mountain collapsed, hot rocks, ash, gas and steam exploded upward and outward to the north. The outward blast spread volcanic debris (grey in the images) over 600 square kilometres. All around the southern half of the mountain, volcanic mudflows (lahars) poured down rivers and gullies.

The first plant to grow in the devastated landscape was the prairie lupine. This plant attracted insects and herbivores, and captured leaves and other organic matter. Dead plants and insects, wind-blown organic matter and the droppings of herbivores (plant-eating animals) slowly created pockets of soil on the volcanic deposits. Gradually other plants began to grow.

5.29 A sequence of false-coloured satellite images of Mt St Helens before and after the 1980 eruption

29 August 1979

24 September 1980

10 September 2009

5.30 True-coloured image of Mt St Helens, 1984 and 2011

ACTIVITIES

Knowledge and understanding

1 Explain what is meant by the term 'remote sensing'.

2 Outline why geographers find remotely sensed images useful.

3 Explain the difference between false-coloured and true-coloured images.

Geographical skills

4 Study Figures 5.29 and 5.30. Outline the changes you can observe in the sequence of satellite images.

5.7 Geoskills: Digital terrain models

Digital terrain (or elevation) models (DTMs) are computer-generated 3-D representations of part of the earth's surface. The models are generated using elevation data collected by satellites orbiting the earth. DTMs are an excellent way of investigating landform features. They are also used in the advertising, publishing and engineering industries for reports, advertisements and tourist promotions.

Block DTMs

Block DTMs are 3-D representations of the earth's surface presented in block form (see Figure 5.31). While such models are usually used to illustrate the mountainous regions of the earth, they are also used to show the topography of the world's oceans.

5.31 Mt St Helens, United States of America, an example of a block terrain model

Digital terrain fly-throughs

DTMs can be used to construct digital fly-through simulations. These are constructed by superimposing topographic data on satellite or aerial images. They are typically developed along a predetermined flight path (see Figure 5.32).

5.32 A still image from a Swiss digital terrain flight

SPOTLIGHT

Southern California DTM

Figure 5.33 shows a 3-D DTM produced by placing a Landsat satellite image over a topographic map. This image covers an area stretching from the Mojave Desert to the ocean and shows a variety of landscapes and environments. The topography shown in the model plays a critical role in southern California's climate. Winds usually bring moisture to this area from the west, moving from the Pacific Ocean, across the coastal plains, to the mountains, and then to the deserts. Most rainfall occurs as the air masses rise over the mountains and cool with altitude. Moving further east, the air, drained of its moisture, drops in altitude and warms as it spreads across the desert. The mountain rainfall becomes groundwater and stream flow that supports citrus, avocado, strawberry and other crops, as well as a large and growing population on the coastal plains.

5.33 3-D digital terrain model of southern California, United States of America

ACTIVITIES

Knowledge and understanding

1 Explain what a DTM is. How is it produced?
2 State what block DTMs are used to illustrate.

Investigating

3 Use the internet to find examples of DTMs and create an annotated visual display about the image.
   a Select one of the following landforms or landscapes: desert landscape, mountain, volcano, coral reef, ocean trench, coastal region, glacier.
   b Annotate the DTM image with the following:
   - a map of the location of the landform feature
   - notes on the type of landform feature and the processes responsible for its formation
   - an overview of the vegetation
   - details of the population.

5.8 Coastal erosion

Coastal erosion is the wearing away of land and the removal of beach and dune sediments by wave action. Wind-generated waves attack coastal landforms (cliffs, beaches and sand dunes) and then deposit the eroded materials elsewhere. Where the coastline has been settled by people, coastal erosion can destroy people's homes and damage infrastructure.

Australian coast

Fifty per cent of the Australian coastline is made from sand. These types of shorelines created by the movement of sand are open to change as a result of storm waves and rising sea levels. While few parts of Australia's long coastline need protection, there are some that require human intervention.

Gold Coast

The main attraction of the Gold Coast of south-east Queensland is its 42-kilometre stretch of beach, from Coolangatta in the south to South Stradbroke Island in the north. It is shown in Figure 5.34. Under natural conditions sand is transported north from New South Wales by the movement of water, which is powered by south-easterly winds. The original beach zone also extended several hundred metres inland and included extensive wetland and dune areas.

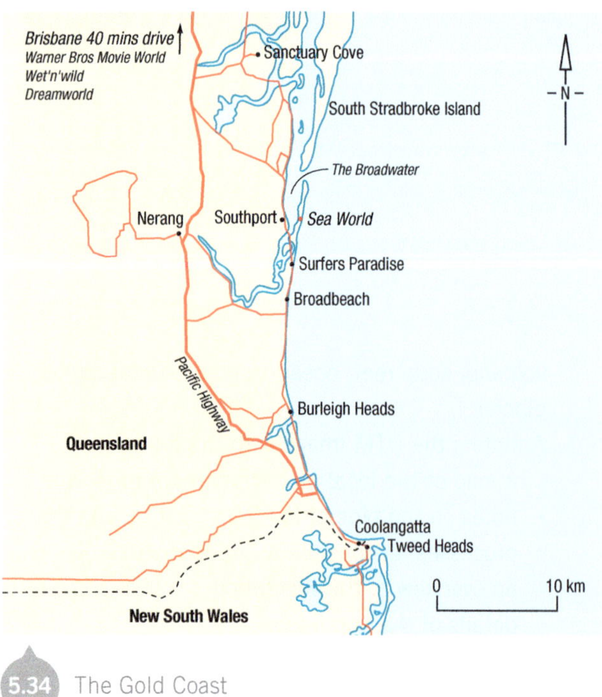

5.34 The Gold Coast

Upsetting the balance

The Tweed River in the north of New South Wales has a history of instability due to a large spit, which used to cause the river mouth to relocate periodically. In 1962, large 'training walls' were constructed on either side of the river mouth to stabilise it. These can be seen in Figure 5.35. In the following years, millions of tonnes of sand were trapped by these training walls, disrupting the natural process of longshore drift. As a result, the beaches of the Gold Coast did not get the sand needed to replace what was eroded by waves. The construction of residential towers and protective barriers for the towers on what was once a protective dune system added to the problem.

Storm attack

In the late 1960s, a series of severe storms caused havoc in the Gold Coast region. Roads and dwellings were washed into the sea. Since this time, the Queensland Government and local councils have spent millions of dollars restoring the beach zone. This has included a large-scale beach replenishment program that involved pumping sand (mixed with water) from south of the Tweed River, and a beach and dune stabilisation program. In 2001, an artificial sand reef was constructed off the northern end of Surfers Paradise beach in an attempt to reduce erosion. Its aim was to copy the effect of a sandbar by protecting the beach from destructive waves.

Bringing back the sand

The beaches are central to the economic wellbeing of the Gold Coast's tourism-based economy. Sand retention is a critical issue. A key response has been the $25 million Tweed River Entrance Sand Bypassing Project (TRESBP). This major engineering solution aims to re-establish the northward (longshore) movement of sand that was interrupted by the construction of the stone training walls at the entrance to the Tweed River.

The TRESBP involves pumping sand from where it accumulates (south of the Tweed River training walls) via a system of pipes across the Tweed River to one of four outlets, where sand is released onto beaches (see Figure 5.35). The impact of sand pumping onto Coolangatta's beaches is shown in Figure 5.36. The Bypass Project is a joint initiative of the New South Wales and Queensland governments.

5.35 The Tweed River Entrance Sand Bypassing Project sand redistribution system

5.36 Oblique aerial photograph showing the replenished beaches of Coolangatta

Rising sea levels

Rising sea levels, a result of climate change, could have a major impact on coastal communities. In Australia, the Department of Climate Change and Energy Efficiency has been modelling low (50 centimetres), medium (80 centimetres) and high (110 centimetres) sea-level rise along Australia's coasts. A medium sea-level rise along Surfers Paradise is shown in Figure 5.37.

5.37 Sea-level rise and inundation, Surfers Paradise

Source: Commonwealth Department of Climate Change and Energy Efficiency

ACTIVITIES

Knowledge and understanding

1 Define the term 'coastal erosion'.
2 List the reasons why the Tweed River Entrance Sand Bypassing Project was established.

Applying and analysing

3 Assess the success of the Tweed River Entrance Sand Bypassing Project.
4 Study Figure 5.37. List the areas (suburbs) that will be affected by a medium (80 centimetres) sea-level rise.

5.9 Mass movements

Mass movements occur when rock material moves downhill under the influence of gravity. These movements may be rapid, as in the case of rock falls, landslides, earthflows, mudflows and slumps, or slow, as in the case of soil creep.

Role of water

Water plays an important role in most types of mass movement. It acts like a lubricant, weakening the binding properties of the soil and rock material. The amount of water is important. Think about building sand castles at the beach. Water is needed to bind the sand together, but too much water results in the castle collapsing.

Slow movements

Soil creep

Soil creep is a long-term process. Over time, soil moves downslope under the influence of gravity. As a general rule, the steeper the slope, the faster the creep. Leaning fences and power poles, and the curved trunks of trees, are all evidence of creep, as illustrated in Figure 5.38.

5.38 Evidence of soil creep

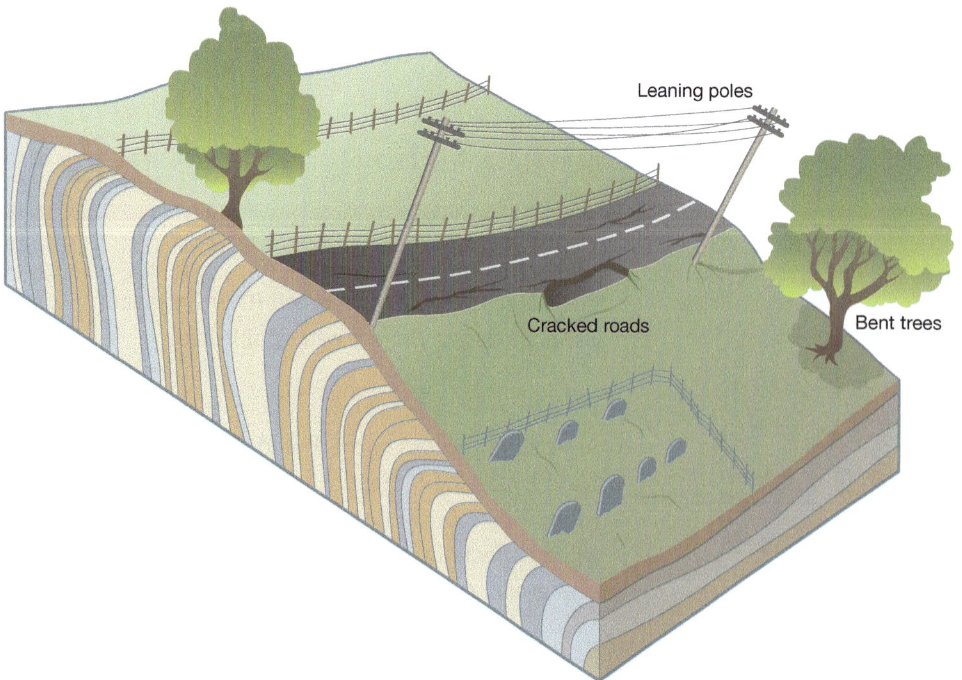

Rapid movements

Rock falls

A rock fall occurs when weathered rock material moves downslope but is not enough to cause a landslide. Rock falls usually occur on very steep slopes such as cliff faces. Earthquakes, expanding ice, plant roots, the undercutting action of waves and water penetration may all dislodge rock material. The rock accumulation at the base of the cliff is called talus. A rockfall and talus can be seen in Figure 5.39.

Earthflows

Figure 5.40 shows an earthflow, which is the downslope flow of saturated fine-grained materials. The materials most commonly involved are small rocks, clay, fine sand and silt. When the earth cannot hold any more water, the flow will begin. The rate at which the earth flows depends on the amount of water present: the higher the water content, the greater the speed, or velocity.

5.39 Rock fall and associated talus slope

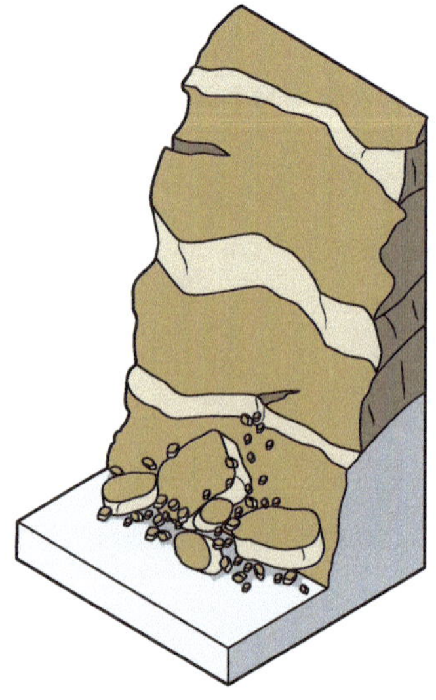

5.40 When soil and rock cannot hold any more water they will start flowing.

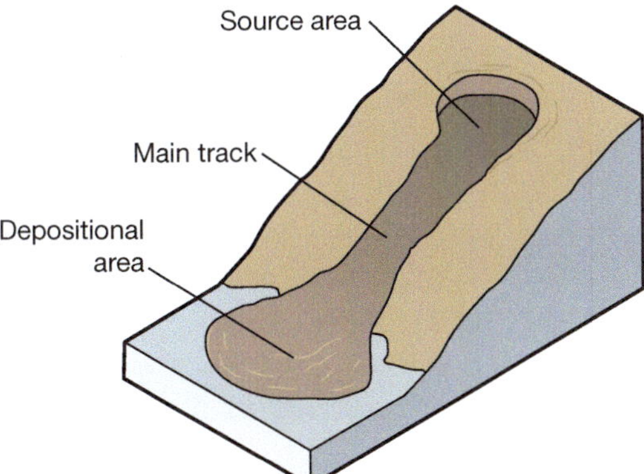

5.41 Mudslide in El Salvador, Central America, 2001. Because they occur suddenly and without warning, mudflows can prove deadly.

Mudflows

A mudflow (or mudslide) is the downhill movement of saturated earth and debris, made fluid by rain or melted snow. Mudflows are often able to build up great speeds (up to 80 kilometres per hour). Mudflows are most likely to occur in areas that have been stripped of their vegetation, on slopes that have been altered for the construction of buildings, and on roads and other areas to which surface run-off has been redirected Damage caused by a mudflow can be seen in Figure 5.41.

Slumps

A slump occurs when a well-defined mass of soil or rock layers moves downslope. Slumps can be initiated by earthquakes, water saturation and the freezing and thawing of a mass of soil or rock layers. Figure 5.42 shows an earthflow with slump features in the upper section.

5.42 An earthflow with well-developed slump features in the upper section

ACTIVITIES

Knowledge and understanding

1 Define the term ‘mass movements’.
2 Outline the role of water in mass movements.
3 Explain the difference between landslides and earth- and mudflows.

Applying and analysing

4 Study Figure 5.38. List the damage caused by soil creep.
5 Study Figure 5.41. Describe the damage you observe in the image.

5.10 Mudslides

Mudslides and landslides often occur with little, if any, notice and the results can be deadly. The following hazard 'snapshots' illustrate the deadly consequences of such disasters.

La Paz, Bolivia

In February 2011, at least 400 homes were destroyed after torrential rains caused a massive slump and mudslide in La Paz, the capital of Bolivia. The slump and slide caused widespread damage in the poor neighbourhood of Callapa. Roads cracked and houses were buried under mud and debris, as shown in Figure 5.43. Fortunately, no one was killed. Residents had started evacuating the neighbourhood a day or so earlier when the hillside, saturated after prolonged rainfall, began to slide and cracks appeared in streets and homes.

5.43 Four hundred homes were destroyed after torrential rains sparked a massive slump and mudslide in La Paz, the capital of Bolivia.

Rio de Janeiro, Brazil

In early 2011, a series of floods, mudslides and landslides in the Brazilian state of Rio de Janeiro killed at least 903 people. It was one of the worst natural disasters in Brazilian history. In a 24-hour period between 11 and 12 January 2011, the local weather service registered more rainfall than would be normally expected for the entire month. Flooding of many areas in the region immediately followed. The soil on the region's steep hillside quickly became saturated and in many places collapsed, with devastating effects. Rescuers can be seen among the damage in Figure 5.44.

5.44 Officials and volunteers search for the dead after the floods and mudslides that killed more than 900 people in the Brazilian state of Rio de Janeiro on 16 January 2011.

Gansu, China

The Gansu mudslide of 8 August 2010 killed more than 1600 people when a massive torrent of mud engulfed houses and demolished multistorey blocks of flats. The disaster followed a period of heavy rain during which water built up behind a dam of debris blocking a small river near the city of Zhouqu. When the debris dam collapsed, 1.8 million cubic metres of mud and rocks swept through the city. The surge levelled an area 5 kilometres long and 300 metres wide, the mud measuring up to 5 metres deep. More than 300 low-rise homes and more than a dozen multistorey buildings were destroyed. One village was entirely buried. Figure 5.45 shows rescuers at work at the site of the disaster.

5.45 Rescuers work at the scene of a mudslide in Zhouqu County, Gansu Province, China, in August 2010.

ACTIVITIES

Knowledge and understanding

1 Copy and complete the following table.

Location	Type and cause/s of landslide	Damage
La Paz		
Rio de Janeiro		
Zhouqu		

Discuss the similarities and differences between these landslides.

5.11 Avalanches

An avalanche is a sudden, downslope movement of a mass of snow. Avalanches are triggered when the snow pack (the extent of the snow) becomes overloaded due to natural factors such as earthquakes or heavy snowfalls, or human agents such as skiers, snow-boarders and snowmobile riders.

Avalanche damage

Avalanches can reach speeds of 130 kilometres per hour within just 5 seconds. The size and impact of an avalanche is influenced by the snow type and the rate at which it accumulates, the temperature, the nature of the sliding surface, the trigger, and the angle and aspect of the slope. Figure 5.46 shows an avalanche in action. Note the steep slope of the mountain.

Powerful avalanches can cause large rocks, trees, and other debris to move downslope, adding to their destructive power. Avalanches cause loss of life, destroy settlements and disrupt transport when they block roads and railway lines. Because of their ability to move enormous amounts of snow rapidly over large distances, avalanches are a major hazard to life and property in mountainous regions.

Avalanches kill more than 150 people each year. Ninety per cent of avalanche deaths are the result of snow slides triggered by the victim or someone in the victim's party. Skiers, snowboarders and riders of snowmobiles are the most frequent triggers of avalanches.

Avalanche control

In populated alpine regions, avalanche control is an important management issue. Avalanche control begins with an assessment of the level of risk. This involves studying the topography, vegetation pattern and seasonal distribution of snowfall to determine the areas likely to be affected by avalanches. Once the avalanche risk areas have been identified, threatened elements of

5.46 Two mountaineers were swept away by this avalanche on Europe's Mont Blanc. Fortunately, they escaped.

the constructed environment such as roads, railways, settlements and ski resorts can be protected. Avalanche prevention plans involve monitoring the snow pack and constructing structures to protect people and property.

Avalanche control includes the following measures.

- Management of the snow pack, by setting off small detonations, prevents large and dangerous snow packs from forming.
- Snow racks (see Figure 5.47) and avalanche bridges (see Figure 5.48) can be built to protect towns, roads and railway lines.
- Dams, ditches and earth mounds can be built to deflect and slow an avalanche.
- Planting trees, as forested areas will slow an avalanche and may prevent one from occurring.
- Avalanche awareness programs educate and protect people engaged in recreational activities in alpine regions.

5.47 Snow racks are used to retain snow, thereby reducing the risk of avalanches.

5.48 Avalanche bridges are used to protect vital transport infrastructure.

Surviving avalanches

If you are ever caught in an avalanche, try to get out of the avalanche pathway as quickly as possible. Figure 5.49 outlines safety tips for skiers and snowboarders riding in avalanche-prone areas.

5.49 Avalanche safety checklist

Avoiding avalanches

- Be aware when an avalanche is likely to occur, especially on steep slopes after heavy snowfalls.
- Always follow avalanche warnings and never ski or board alone in avalanche-prone areas.
- Avoid crossing steep slopes.
- If skiing or boarding in a group, always spread out. That way, if some of you are trapped in an avalanche, the others can dig you out.
- Avalanches can happen without notice. Be prepared and know what to do.

If caught in an avalanche

- Do not attempt to out-ski or board the avalanche. It will be travelling faster than you can ski. Try skiing towards to edge of the avalanche, where the mass of snow is thinner and less powerful.
- Try to stay on the surface, grab onto a tree or rock, or thrust yourself upward by kicking.

If you are knocked over

- Use a swimming-like stroke to avoid being covered in snow.
- When you come to a stop, curl up into a ball and use your hands to cover your face. Rotate your head to make an air pocket.
- To see which way is up, spit into your hands and feel which way the saliva runs. Remember you may be in complete darkness unless you are close to the surface.
- If you can see a hint of daylight, try to push one hand to the surface to attract attention.
- Breathe steadily to preserve energy and oxygen.

ACTIVITIES

Knowledge and understanding

1. Explain what an avalanche is and the conditions under which avalanches take place.
2. List the factors that determine the nature of the avalanche.
3. Outline what can be done to manage, or control, the avalanche threat.

Applying and analysing

4. Study Figure 5.49. Produce a podcast to educate people engaged in alpine recreational activities about avalanche safety.

5.12 Living in areas of risk

Scientists study the earth's crust in order to understand the processes involved in plate movements. As a result, they are able to assess the risk of earthquakes and volcanic eruptions. With this knowledge, authorities are able to plan for, and respond to, natural hazards such as earthquakes and volcanic eruptions.

Predicting hazards

Earthquake prediction

Scientists study the past frequency of large earthquakes in order to determine the future possibility of similar large shocks. Where this possibility is high, special building laws can be made. Authorities can also plan emergency responses to earthquakes.

Tsunami warning systems

Tsunami warning systems are used to detect tsunamis and issue warnings to minimise the loss of life and the damage to property. A tsunami warning system consists of two equally important parts: a network of sensors to detect tsunamis, shown in Figure 5.50, and a communications infrastructure to issue alarms in time for people to evacuate coastal areas.

5.50 A tsunami warning system

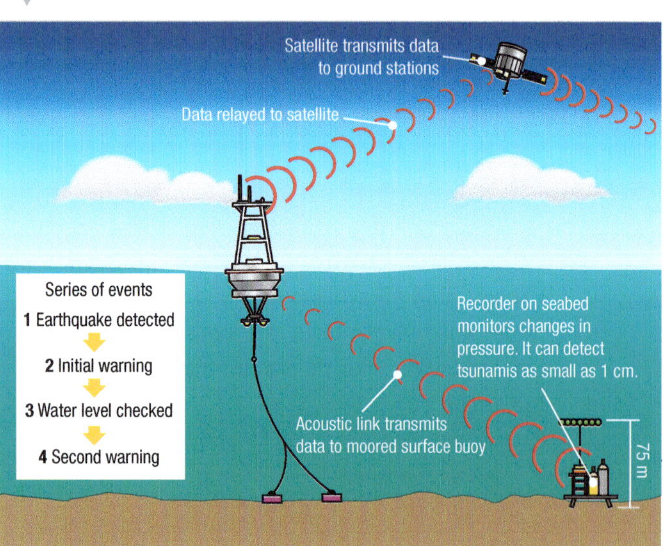

Volcano monitoring

There are currently about 500 active volcanoes in the world. About fifty of these volcanoes erupt each year. Volcanologists monitor high-risk volcanoes so that they can determine the level of risk and forecast eruptions. Experts now issue alerts using a classification system. In the United States of America, volcanoes are now classified according to the following stages: normal, advisory, watch or warning.

Adapting to threats

The forces involved in earthquakes and volcanic eruptions are so great that they are beyond the control of humans. All we can hope to do is minimise their impact and make sure we are ready to respond in a natural disaster.

Earthquakes

When an earthquake strikes, land may be moved up and down, and from side to side. Modern engineering techniques are used to construct buildings that can withstand such movements.

In earthquakes, the biggest danger is not the earthquake, but falling debris and collapsing buildings. Broken glass becomes deadly and falling concrete can crush people. Engineers have designed windows and walls that will not break when shaken by the intense motion caused by earthquakes, and specially designed foundations absorb the energy of the earthquake and allow buildings to flex rather than shake violently.

Tsunamis

Natural barriers such as sand dune systems, mangroves, plantations and native forests can reduce the impact of tsunamis. During the 2004 Indian Ocean tsunami, some coastal communities escaped with little loss of life and damage to property because trees such as coconut palms and mangroves absorbed the tsunami's energy.

For example, while the entire Nagapattinam district in the state of Tamil Nadu, India, was severely affected by the tsunami, the town of Naluvedapathy emerged virtually unscathed. This was due to the presence of a very large windbreak of 80 000 casuarina trees planted in 24 hours in 2002 as part of a Guinness World Record.

Tree planting along tsunami-prone coastlines has been recommended as a cheap and long-lasting way of reducing the effects of tsunamis and as an alternative to the building of expensive barriers. In India's worst tsunami-hit state, Tamil Nadu, tree planting has begun along the coastline to act as a barrier against future tsunamis.

In coastal areas where it is not possible to use forest barriers, for example in existing urban areas, concrete walls are constructed and seawater barriers that deploy automatically when tsunami waves approach shorelines are built, such as the barriers shown in Figure 5.51.

5.51 A tsunami barrier and gate, Numazu, Japan

Volcanoes

Worldwide, an estimated 500 million people live near active volcanoes. Fortunately, the loss of life from volcanic eruptions has been relatively low. A notable exception was the Mt Pinatubo eruption (in the Philippines) of 1991. Despite the large-scale evacuation of people in the days leading up to the eruption—itself a demonstration of the effectiveness of eruption prediction by volcanologists—847 people were killed. Most died when roofs collapsed under the weight of accumulated ash. Mt Pinatubo's dense forest was home to several thousand indigenous people, the Aetas, many of whom were displaced by the disruption.

SPOTLIGHT

Surviving earthquakes

If you are indoors, remain there and:

- take shelter under a sturdy table or desk, stand or crouch in a strong doorway in a load-bearing wall, or brace yourself in an inside corner of the room
- shield your head with a blanket, doona or cushion
- stay clear of windows, mirrors, or other glass that might shatter
- keep clear of bookcases, cabinets and other pieces of heavy furniture that might topple or spill their contents
- stay away from fireplaces, and any area where bricks might fall from the chimney.

If you are outside:

- quickly move into an open space where nothing can fall on you
- stay clear of power lines and poles, trees or branches, building facades, chimneys, or anything else that might fall.

If you are in a city centre:

- avoid falling glass and masonry by taking shelter in a strong doorway or under a large vehicle
- don't use elevators or stairs during the quake.

ACTIVITIES

5.4

Knowledge and understanding

1 Explain why scientists study the earth's dynamic crust.

2 Outline the ways in which we can respond to a high-level earthquake risk assessment.

3 Identify the ways in which coastal communities can be protected from the impacts of a tsunami.

Applying and analysing

4 Study Figure 5.50. Explain how a tsunami warning system operates.

Investigating

5 Undertake internet research. Investigate the ways in which buildings can be made earthquake-resistant.

6 Investigate a recent volcanic eruption. Identify the ways in which people were informed about the impending eruption and the actions they were able to take to minimise the impacts of the eruption.

7 Research the different ways to respond to hazards and develop a poster highlighting how people should respond in the case of a natural disaster such as an earthquake, a tsunami or a volcanic eruption.

CHAPTER 6 BUSHFIRES

Bushland that has been affected by bushfire experiences an explosion of new plant growth. The explanation for this is that many Australian native plants are dependent on fire for their survival. Many seeds will lie dormant in the soil for years. Only when fire moves through the area will these seeds germinate. This is the case for plants such as banksias, hakeas and most wattles. Bushfires, therefore, play an important role in shaping the Australian landscape.

Fire is both feared and harnessed by humans. Aboriginal and Torres Strait Islander people have long used fire as a land management tool and the European settlers used it to clear land for agriculture. Today we use it to protect properties from intense, uncontrolled fires in a process known as 'hazard reduction'.

Bushfires are a costly natural hazard. The loss of life and injury can be high. According to the Australian Institute of Criminology, in an average year insurable losses due to bushfires are $80–100 million.

KEY IDEAS

- To understand the causes and impacts of the bushfire hazard in Australia
- To examine how landforms affect bushfire behaviour
- To investigate Australia's deadliest and most costly bushfire disasters
- To discuss how individuals, communities and governments can assist in minimising the impacts of bushfires

6.0 A raging bushfire, Kimberley region, Western Australia

GLOSSARY

bushfire	fire burning out of control; sometimes referred to as a wildfire
cold front	the leading edge of an advancing mass of cold air
combustion	the process of burning; involves the production of heat and light
drought	a prolonged period of below-average rainfall
firestorm	an intense fire capable of generating strong convection currents and winds ahead of the fire front
fire front	the leading edge of a fire
flammable	easily set on fire
fuel	any material that can be ignited and can sustain a fire
fuel load	combustible material that builds up in the bush over time
heatwave	a relatively short period (usually no longer than a few days) of well above-average temperatures
ignition	the beginning of flame production or smouldering combustion; the starting of a fire
incinerate	to reduce to ashes
radiant heat	heat that travels through the air directly from the flames of a fire and can contain enough energy to cause burns to the flesh and materials
running crown fire	a fire that burns in the crowns, or tops, of trees; usually hot and destructive
spot fire	a fire that occurs well ahead of the fire front, caused when glowing embers are caught up in the convection column and carried ahead of the main fire by the wind

6.1 Bushfires in Australia

'Bushfire' is an Australian word that is used to refer to any fire burning out of control in bushland areas. In other countries, and increasingly in Australia, the term 'wildfire' is used to describe such events. Bushfires are common in Australia and are associated with the explosive burning of the 'bush', the eucalypt forests and woodlands. Grassfires are also common.

How bushfires start

A fire requires material that can burn (**fuel**), a source of ignition (heat) and a supply of oxygen. Fire is the **combustion** of fuel as it reacts with oxygen in the air. This process is known as the fire triangle. It is shown in Figure 6.1.

6.1 The fire triangle

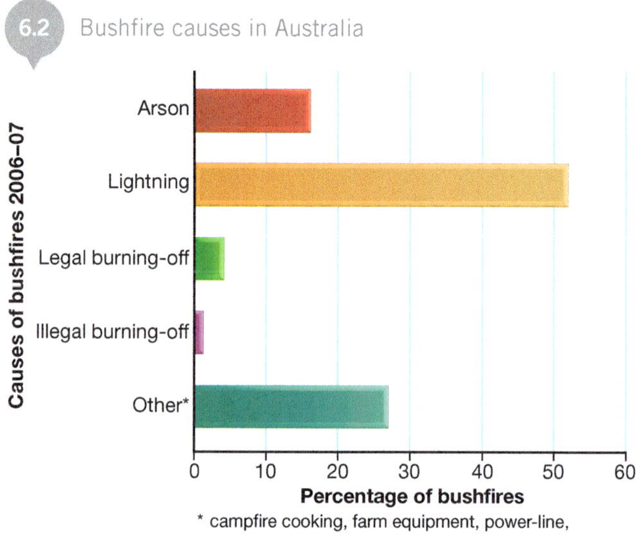

6.2 Bushfire causes in Australia

Source: Emergency Management Australia

Fuel

The greater the **fuel load** (the amount of combustible material), the hotter and more intense a fire is. The main fuel for **bushfires** comes from trees, debris (fallen leaves and bark) and undergrowth. During **droughts** and in very hot, windy weather, the leaves and small branches of trees can become very dry and **flammable**. Dead or dried-out grass can also fuel grassfires, and some grasses can burn even when they are green.

Ignition

Bushfires have long occurred naturally in Australia, but they can also originate from human activity. Bushfire causes are shown in Figure 6.2. Lightning strikes account for over half of all bushfire **ignitions**. Sometimes controlled fires are initiated in an effort to reduce fuel loads. Occasionally, however, changes in the weather can cause these fuel-reduction fires to get out of control and become bushfires. Some very serious fires have been lit by arsonists (people who deliberately light fires).

Fires can also be the result of accidents, such as when a mower blade hits a rock and produces sparks, or power lines are brought down during high winds.

Oxygen

Combustion requires oxygen, and once lit, a fire will only continue to burn if it has a supply of oxygen. Loose litter on the ground is vulnerable to fire in open spaces because there is plenty of oxygen available for it to burn.

How bushfires attack

Bushfires are a threat to life and property in various ways.

The fire front

Flames at the **fire front** will ignite anything flammable with which they come into contact. The most dangerous of all fronts is that of a **running crown fire**. In eucalypt forests, bushfires can advance at alarming speeds through the upper layers of the forest, as outlined in Figure 6.4.

Ember attack

Spot fires can break out when hot embers (burning debris, such as twigs, bark and small branches) from the fire front are picked up by winds and deposited elsewhere. Embers can be carried as far as 30 kilometres downwind from the fire front.

Wind

Winds drive a fire by delivering a continuous supply of oxygen. They also blow the flames towards fresh fuel. Even a slight increase in wind speed results in a significant increase in the rate at which a fire can advance. Changes in wind direction also affect fire behaviour and can quickly cause the size of a fire front to increase and the direction in which it moves to change. Winds also enable fires to spread by causing spot fires, as they lift the burning embers into the sky and carry the embers ahead of the fire front.

6.4 How a bushfire attacks

Grassfires and bushfires

Grassfires move quickly, passing in 5 to 10 seconds, and smoulder for only minutes after the fire has passed. They are low to moderately intense fires, but can do a lot of damage to crops, livestock, fences and farm buildings (see Figure 6.3).

Bushfires usually move more slowly than grassfires, but are much hotter and more intense. They pass in 2 to 5 minutes, but logs and tree stumps can smoulder for days afterwards. Bushfires can **incinerate** vast tracts of country and destroy whole townships.

6.3 A grassfire

ACTIVITIES

Knowledge and understanding

1 Define the term 'bushfire'.
2 Identify the three things that must be present for a fire to occur.
3 Distinguish between a bushfire and a grassfire.
4 Explain the danger of ember attack.
5 Explain how winds influence bushfire behaviour.

Applying and analysing

6 Study Figure 6.2 and complete the following table.

Natural causes	Human causes

a Write the fire ignition type in the correct column.
b Is the most common type of ignition due to humans or nature?
c Considering your answer to b, what measures would you suggest to reduce the incidence of bushfires?

7 Design an annotated visual display illustrating how bushfires start, progress and attack.

6.2 Severity of bushfire hazard

Bushfires have long been part of the Australian landscape and vegetation has evolved with them. Indigenous and European occupation of the continent have had a significant impact on the biophysical environment and the incidence and intensity of bushfires.

Drying Australian climate

When Australia was part of Gondwanaland, the climate was warm and wet. The separation of the Australian continent, and its northward drift, carried it into the latitudes characterised by dry air. The dominance of high-pressure systems over the continent produced fine, sunny conditions, resulting in low, unreliable rainfall and prolonged dry periods. The Australian continent today is generally hot, dry and has frequent droughts. Such conditions dry out the country and increase the bushfire threat.

Evolution of Australian flora

The drying of the climate resulted in vegetation that has adapted to the dry climate and poor soils. Plants such as the eucalypt (gum) trees developed characteristics that meant they require less water and fewer nutrients—small, thick, leathery leaves are tough and need to be replaced less often.

During times of drought, Australian eucalypts survive by reducing their water loss through shedding their leaves. The dead leaves build up on the forest floor with discarded bark, creating fuel. In dry conditions, the rate of decay slows significantly and so the fuel load builds up even further.

Eucalypt leaves also contain aromatic oils and waxes to discourage insects from eating them. The presence of these substances and the low water content in the leaves explains why they burn explosively. On hot days the oils in eucalypt leaves pass into the atmosphere as a vapour. This vapour is quite combustible given the right conditions, and when a fire passes through, tree crowns appear to explode when the oil-rich vapour given off by the leaves ignites in a fireball.

Evolution with fire

The Australian environment evolved in the presence of fire. Lightning strikes have ignited the country for millions of years. Many plant species developed a range of adaptations to survive bushfires and others even came to rely on regular burning to trigger the release or germination of seeds for the next generation of plants. Figure 6.5 shows regeneration after a bushfire.

6.5 The bush regenerating after a bushfire

Australia's eucalypt forests and woodlands

Eucalypts dominate the less dry regions of the continent in the south-east and south-west. The forests occupy areas with relatively high rainfall nearest the coast, while the woodlands of smaller, more sparsely planted trees are further inland, where it is drier. These coastal and inland areas have the highest bushfire risk in Australia, as shown in Figure 6.6.

Human impact

While bushfires have always been part of Australia's natural environment, human attitudes and management of fire have had a significant impact on the severity of bushfires.

Aboriginal use of fire

When the first Aboriginal people arrived more than 60 000 years ago, they adapted their lifestyle to the conditions. They also learnt to manage the land in ways that met their needs. The main tool they used to do this was low-intensity fire, and their patterns of burning were quite sophisticated.

Aboriginal people made frequent and purposeful use of fire:

- to keep the country open and therefore easy to move through
- to promote the growth of fresh green grass and herbs that would attract the animals they hunted
- as a means of signalling and hunting
- for the purposes of warmth and cooking.

6.6 The east coast of Australia has the highest risk of bushfire.

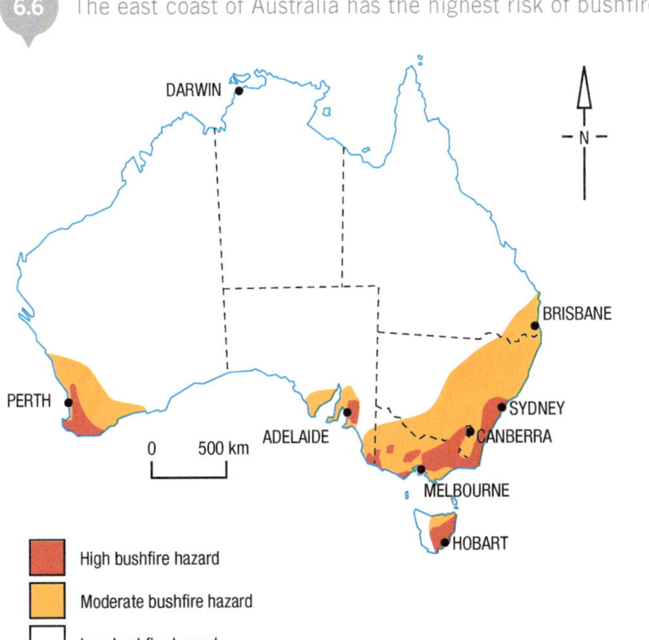

European settlement

European settlers considered the Australian environment to be difficult and hostile, and they were especially fearful of bushfires. The explosive infernos in the Australian bush were unlike anything they had seen before. In Europe, fires had to be put out because they destroyed forests. There was no understanding that the vegetation in Australia actually benefited from burning. Bushfires were terrifying. They killed people and stock, and destroyed homes and properties. The response was to try to extinguish bushfires or to prevent them from occurring.

As the Europeans displaced the Aboriginal and Torres Strait Islander people, the regular burnings stopped and, as a result, there was a build-up of fuel loads in the bush. Fires, being part of the natural cycle in Australia, were inevitable, and with more fuel to burn they became more intense and destructive.

The pattern of European settlement also placed people in areas where the bushfire hazard was greatest. The majority of Australians live on or near the south-east coast of the continent, where the eucalypt forests are also located.

When the Europeans first arrived they settled on the flat land near the sea or along ridges where it was easier to build. They avoided the valleys, as these were too steep and, being moister, were covered by thick forest. This pattern of settlement can still be seen where the valley bush remains and homes are located on ridges. As fires race uphill, these properties are at severe risk of fire.

Factors affecting bushfires

The occurrence of bushfires and the rate at which bushfires spread are determined by a number of factors. These are shown and discussed in Table 6.7. A combination of these factors can result in 'blow-up' days when severe fires blaze out of control and spread rapidly.

DID YOU KNOW?

The oil-rich vapour from eucalypt leaves gives the Australian bush its bluish appearance when it is viewed from a distance.

6.7 Factors that can affect the ignition and progress of a bushfire

Factor	Explanation
Amount of fuel	High fuel loads hold the potential for disastrous fires should they be ignited. Also of importance is the 'vertical structure' of the fuel. Dry shrubs and grasses, for example, enable fires to build in intensity and support the development of more advanced running crown fires.
High air temperatures	The higher the temperature, the more oils are released from trees and so the more likely the fuel will catch alight because the air temperature is closer to its ignition point. Hot air can lower the moisture content of forests and grasslands to about 5 per cent and, in extreme cases, as low as 2 to 3 per cent. This greatly increases the speed at which a fire moves and the heat of the fire.
Prolonged drought	Prolonged periods of below-average rainfall dry out the fuel load, creating what is commonly referred to as 'tinderbox' conditions.
Low humidity Humidity (%) 100 25 0 Time LOW	Hot, dry air with humidity below 25 per cent creates critically dangerous bushfire conditions, as it causes fuels to dry out and become more flammable. Dry air creates more intense fires.
Strong winds	Air movement provides the oxygen to keep fires burning. Strong winds mean extra oxygen and more intense fires. They fan the fire and accelerate the speed at which the fire spreads. In fact, a doubling of the wind speed increases the rate of spread by a factor of four. An added danger of strong winds is that they can carry burning embers great distances ahead of the fire front. These ignite spot fires, thus enabling the fire to leap ahead of itself.
Terrain Most severe fire risk Preheats the fuel in front of it Fires spread faster uphill	Fires tend to spread more quickly up hillsides and slopes because the fires preheat their fuel source. As hot air rises up hills, fires accelerate when travelling uphill. For every 10° of slope, a fire doubles its forward speed. (So for a 10° slope the fire is twice the speed it would be on level ground, for 20° it is four times the speed and for 30° it is eight times the speed.) This is reversed when the fire is travelling downhill. (On a 10° slope, for example, the speed of a downward fire is halved.)

SPOTLIGHT

Eucalypts are messy trees

Every year, eucalypt trees grow another layer of bark. This increases the thickness of the trunk and branches. In some species the outermost layer of bark dies, cracks and falls off each year. Large strips and flakes hang from the tree until they fall to the ground to become part of the fuel load, as shown in Figure 6.8. When a bushfire occurs, the old bark burns like an oily rag, especially when there are strong winds to fan it.

6.8 Dead bark peeling off a eucalypt to reveal smooth, new bark underneath. The discarded bark is very dry and catches alight easily.

ACTIVITIES

Knowledge and understanding

1 Explain how the climate of Australia changed as the continent drifted northwards.
2 Describe the adaptations of eucalypts to the difficult Australian environment.
3 Explain why eucalypts are so fire-prone.
4 Compare the attitudes to fire of the Aboriginal and Torres Strait Islander people and the European settlers.

Applying and analysing

5 Discuss how the actions of Europeans have increased the severity of the bushfire hazard in Australia.

Geographical skills

6 Study Figure 6.6. List the states that have a high bushfire hazard.

Investigating

7 Find a eucalypt tree and observe and photograph its features and surrounds.
   a Collect a leaf and paste it in the middle of a page. Add annotations around the page to identify the adaptations of the eucalypt to a difficult environment. (Note features of the leaves and how they hang on the tree.)
   b Using your photographs to support your findings, assess the potential of a eucalypt tree to burn easily. (Check for debris from the tree that would increase the fuel load on the ground.)

6.3 Severe bushfire days

Severe bushfires often occur after long dry spells and when high temperatures, low humidity and strong winds combine to produce ideal conditions for the rapid spread of fire. Some of the most dangerous fires have been a result of a sudden change in wind direction that has altered the fire's behaviour.

High fire-danger weather

Weather systems can produce a combination of conditions that increase the severity of the bushfire risk.

High temperatures

High temperatures are common in Australia because of Australia's position in or near the tropics. Summer temperatures can reach the high 30s and even exceed 40°C. When this excessive heat continues for days it is known as a **heatwave**. In such extreme conditions the risk of a fire starting, and quickly becoming an inferno, increases.

Thunderstorms are also common in the hot summers, and the associated lightning strikes often start bushfires. Many such storms are dry storms with lots of lightning, but no rain to put out subsequent flames.

Low humidity

A lack of moisture in the air is called a moisture deficit and is often used as an indicator of extreme bushfire weather. This is the time that has lapsed since rain last fell in an area and is important because prolonged dry spells lower humidity. Such weather conditions can become severe when winds blow in from the arid interior of the continent, as these winds are extremely dry.

6.9 Wind direction ahead of a cold front

Strong winds

Strong winds develop with weather systems such as lows and cold fronts moving across southern Australia. The steep pressure gradients associated with low-pressure systems cause strong winds that can drastically increase the intensity and spread of fires.

Cold fronts that pass across southern Australia are probably the most powerful influence on weather because of the strong winds they generate. Ahead of a cold front, winds are generally warm to hot and blow from the north-west. Fires driven by these steady strong winds are usually long and narrow. Figure 6.9 illustrates the wind direction before a cold front.

As the cold front passes, the wind can shift 90° to a south-westerly direction. While these south-westerly winds are cooler, they are very dangerous, as the flank (or former side of the fire) becomes the much wider fire front, moving off in a different direction. Figure 6.10 illustrates what happens to the fire front after a wind change. This change in direction makes such fires very erratic and unpredictable, and often catches people unawares. Those towns and people not previously under threat suddenly find themselves in the new path of the fire and under attack from the fire front.

6.10 When the wind direction shifts, the flank of the fire becomes the main front.

High-risk weather patterns

South-east Australia

By mid-summer, the grasslands, woodlands and forests of south-east Australia have usually dried out. The most dangerous fire conditions occur when a strong cold front approaches a slow-moving high in the Tasman Sea. The weather conditions associated with the Black Saturday fires of 2009 are shown in Figure 6.11.

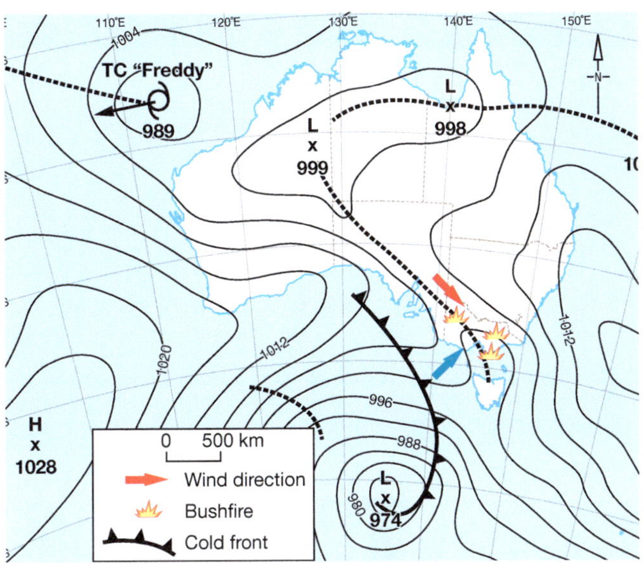

Source: Australian Bureau of Meteorology

6.11 Weather conditions on 7 February 2009 in south-eastern Australia

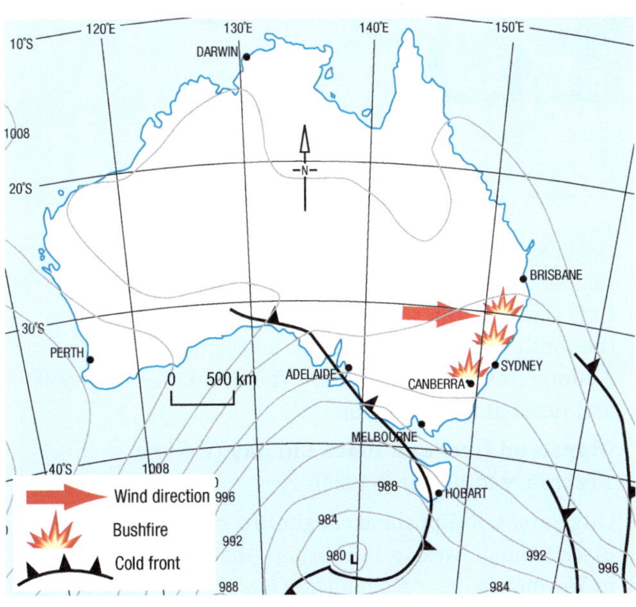

Source: Australian Bureau of Meteorology

6.12 Weather conditions on 7 January 1994 on the east coast of Australia

New South Wales and southern Queensland

The worst conditions occur when deep low-pressure systems near Tasmania bring strong, dry, westerly winds to the coast. This can be seen observed in Figure 6.12, which shows the weather pattern when disastrous fires struck New South Wales in January 1994.

Southern Western Australia

The hot dry winds from the interior are easterly winds in this part of the continent. The fires in the Perth Hills in February 2011 were brought on by such winds (see Figure 6.13).

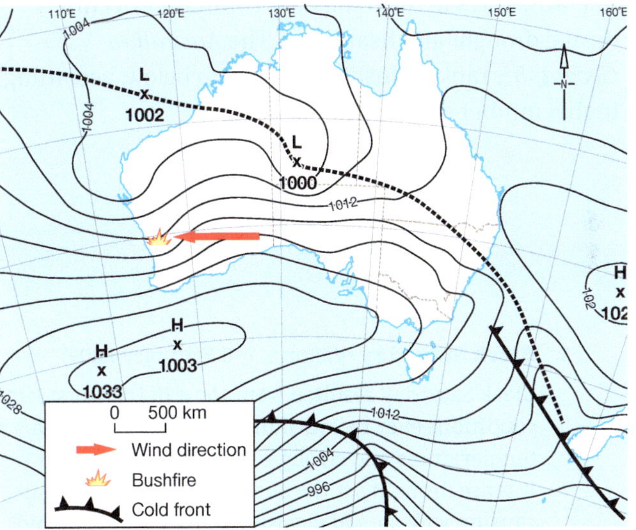

Source: Australian Bureau of Meteorology

6.13 Weather conditions on 6 February 2011 in the Perth Hills

ACTIVITIES

Knowledge and understanding

1. Identify the main conditions that lead to extreme fire danger days.
2. Describe the temperatures experienced in Australia.
3. Explain how winds from the interior of the continent influence weather.

Geographical skills

4. Study Figures 6.9 and 6.10. Write two paragraphs explaining how the passage of a cold front affects bushfire behaviour.
5. Study Figures 6.11, 6.12 and 6.13. Assess the role of wind direction in creating extreme fire danger on each of the days shown.

6.4 The worst bushfires

Australian communities are vulnerable to a number of natural hazards. Bushfires account for an annual average cost of about $77 million and for the most fatalities. From 1900 to 2008, Australian bushfires caused 552 deaths. Then, on 7 February 2009, Black Saturday, several large fires broke out across Victoria, killing 173 people. These fires were precipitated by a heatwave across southern Australia.

Deadliest bushfires

The worst fires in Australia have commonly occurred during drought and heatwaves. The *Australian Geographic* ranked bushfires as shown below, according to the number of fatalities.

1. **Black Saturday** (Victoria, 7–8 February 2009)

   Black Saturday resulted from some of the worst fire conditions ever recorded in Victoria. Record-high temperatures and strong winds after a season of intense drought set the bush alight across the state, causing widespread devastation, 173 fatalities and the destruction of 2029 homes.

2. **Ash Wednesday** (Victoria and South Australia, 16–18 February 1983)

   Widespread drought, gale-force winds, high temperatures and low humidity set the scene for a series of fires across Victoria and south-eastern South Australia. Accidents and arsonists started most of the fires, which spread rapidly through residential regions near Melbourne and Adelaide, resulting in the deaths of 75 people and the destruction of nearly 1900 homes (see Figure 6.14).

3. **Black Friday** (Victoria, 13–20 January 1939)

   Drought conditions and water shortages also preceded Black Friday, but the usual combination of high temperatures, strong winds and low humidity finally triggered fires throughout bush communities near Melbourne. Well-intentioned locals and graziers tried to use controlled burns to protect themselves from disaster, only to help spread the flames. In all, 71 people were killed and 650 houses were destroyed.

4. **Black Tuesday** (Tasmania, 7 February 1967)

   An unusually abundant spring covered Tasmanian forest floors with debris, providing excess fuel for the bushfire season. Strong northerly winds and high temperatures coupled to help fuel at least 80 different fires across southern Tasmania. The fires swept over the south-east coast of the state and came within 2 kilometres of central Hobart. The fires killed 62 people and razed almost 1300 homes.

6.14 The destruction caused by the Ash Wednesday bushfires in 1983

5. **Gippsland fires and Black Sunday** (Victoria, 1 February to 10 March 1926)

   Large areas of Gippsland caught fire, culminating in the Black Sunday fires on 14 February that killed 31 people in Warburton, near Melbourne. Over the two-month period, a total of 60 people were killed.

Source: Adapted from 'The worst bushfires in Australia's history', by Liz T Williams, *Australian Geographic*, 3 November 2011

SPOTLIGHT

The Canberra bushfires, 2003

The firestorm that swept through the Australian Capital Territory on 18 January 2003 brought unparallelled scenes of chaos and destruction to the streets of the nation's capital. Driven by fierce winds, fires ripped through Canberra, leaving four people dead, dozens injured and 506 homes destroyed in nineteen suburbs (see Figures 6.15 and 6.16). The city's sewerage, power supply and telecommunications infrastructures were extensively damaged. Ninety-five per cent (or 95 000 hectares) of Namadgi National Park was affected and, tragically, 90–95 per cent of the fauna of Tidbinbilla Nature Reserve were incinerated.

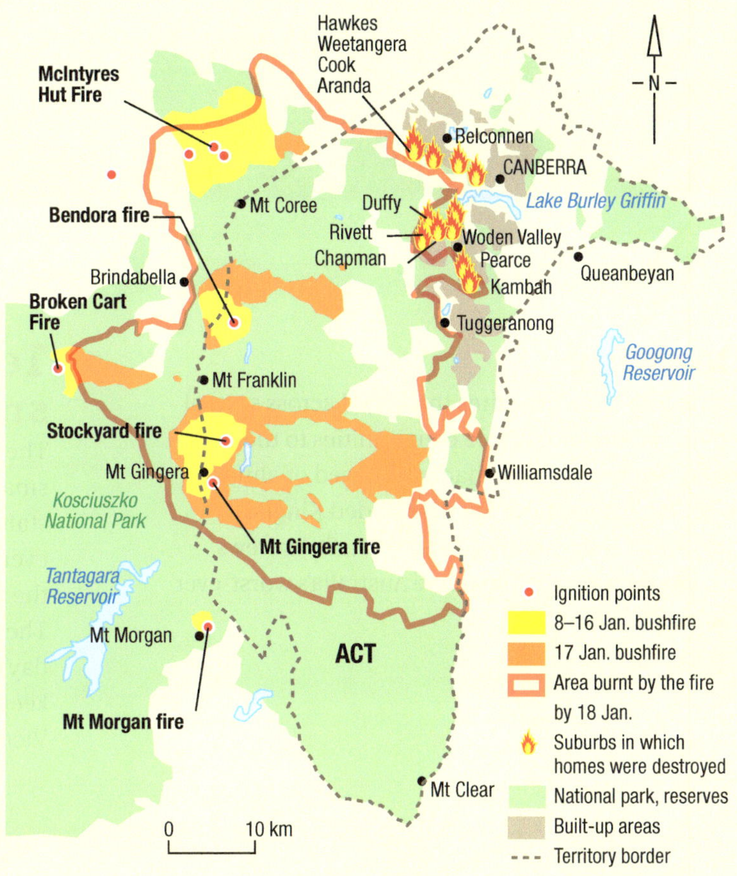

6.15 On 18 January 2003, a lightning strike caused a fire in Brindabella and Namadgi national parks.

6.16 Fire burns a house in Duffy, Canberra, one of thirty-six homes destroyed in the suburb. This picture was taken at approximately 7 p.m. on 18 January 2003.

ACTIVITIES

Knowledge and understanding

1 Outline when the deadliest bushfires are likely to occur.

2 List the annual average financial cost of bushfires in Australia.

Geographical skills

3 Study Figure 6.15.
 a List the locations of ignition points.
 b Describe the area not burnt by the bushfires in the Australian Capital Territory.

Investigating

4 Conduct research into a wildfire that has occurred overseas. Prepare a report and include the following:
 a a map of the area, showing the terrain, vegetation and size of population
 b information about the climatic conditions that existed before the ignition of the fire
 c a list of the factors that affected its progress
 d a summary of the damage caused by the fire.

6.5 Victorian bushfires

On 7 February 2009, a fierce firestorm destroyed entire towns in central Victoria and the state's west Gippsland district. One hundred and seventy-three people lost their lives, 2029 homes were destroyed, up to 7000 people were made homeless and 4500 square kilometres of forests and pastures were burnt.

Damage

Figure 6.17 shows the fires that raged across several regions of Victoria. Whole communities to the north-east of Melbourne were badly damaged or almost completely destroyed. These included Kinglake, Marysville, Narbethong, Strathewen, St Andrews and Flowerdale. The fires were Australia's worst-ever natural disaster.

The Black Saturday fires were, on average, only 12 kilometres from the outlying suburbs of Melbourne.

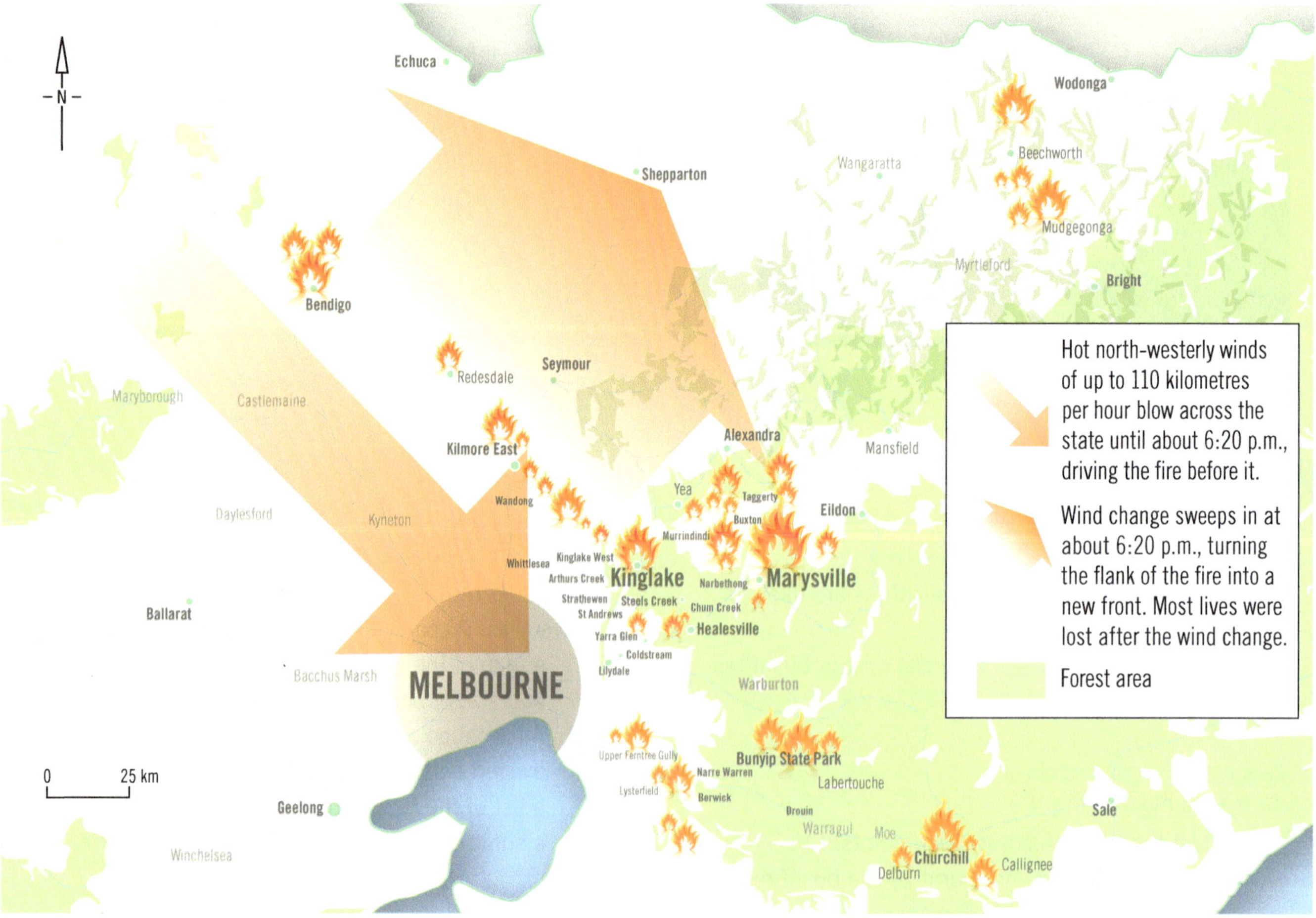

Temperature records smashed

The heatwave that accompanied the 7 February blazes smashed temperature records. Nearly 90 per cent of the state experienced the highest February temperatures ever recorded. The temperatures across Australia for the week ending 10 February are shown in Figure 6.18. The temperature in Melbourne reached 46.48°C on the day of the disaster, the highest in 154 years of record keeping. Figure 6.19 shows the record temperatures in Victoria in the weeks before the disaster.

N

0 25 km

Hot north-westerly winds of up to 110 kilometres per hour blow across the state until about 6:20 p.m., driving the fire before it.

Wind change sweeps in at about 6:20 p.m., turning the flank of the fire into a new front. Most lives were lost after the wind change.

Forest area

Source: *Firestorm: Black Saturday's Tragedy*, by Glenvale School, Lilydale

The soaring temperatures were accompanied by strong, gusty winds and very low humidity. The bush was already tinder-dry following weeks of high temperatures and little, if any, rainfall. Victoria had suffered 12 years of drought. The rainfall for the period March 2006 to February 2009 is shown in Figure 6.20. The fuel load, which helped feed the fire, was, according to experts, at record levels.

The passage of a cold front in the late afternoon of 7 February brought more disaster, rather than relief. The 90° change in wind direction transformed relatively small localised fingers of fire into wide fronts and led to a sudden increase in the intensity and speed of the fires. The mountainous terrain of the region also had an impact on the behaviour of the fire, as fires are able to race up hills at great speeds.

Human cost

People perished as their homes exploded in the face of the raging inferno, while others were burned in their cars as they tried to flee the flames. Witnesses described seeing trees exploding and skies raining ash in the furnace-like conditions.

The fire roared through Marysville at 6 p.m. on 7 February. Once known as Melbourne's honeymooning centre, the town, with its pretty guesthouses and century-old oaks, was quickly reduced to a smouldering wasteland.

DID YOU KNOW?

If the power could have been harnessed from the Black Saturday fires it would have met Victoria's complete energy needs for two years.

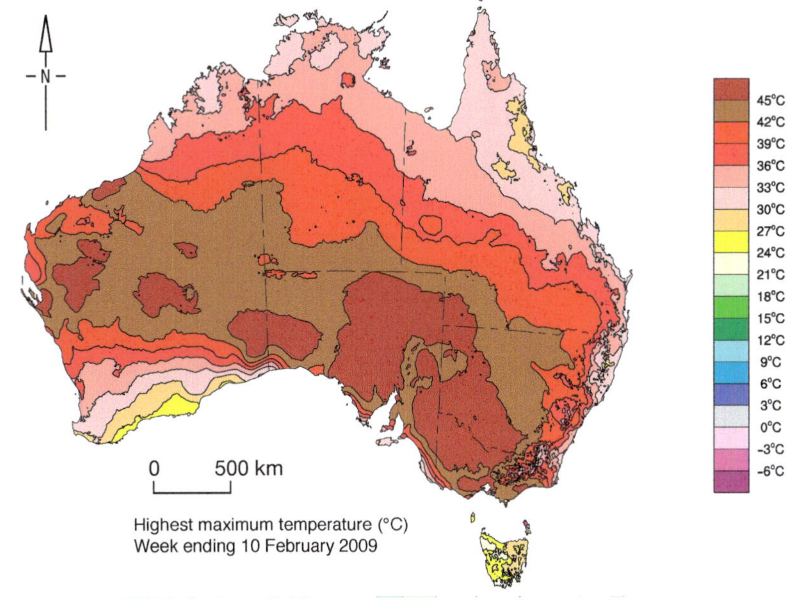

Source: Australian Bureau of Meteorology

6.18 Southern Australia experienced a record heatwave in February 2009.

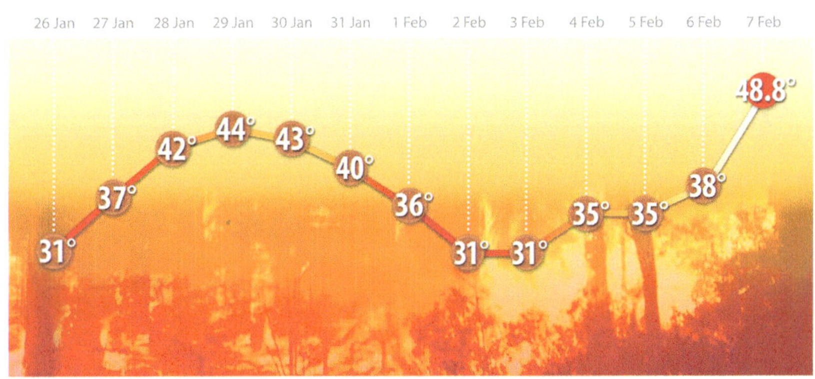

Source: *Firestorm: Black Saturday's Tragedy*, by Glenvale School, Lilydale

6.19 Victoria's daily maximum temperatures leading up to Black Saturday and the highest temperature reached on that day

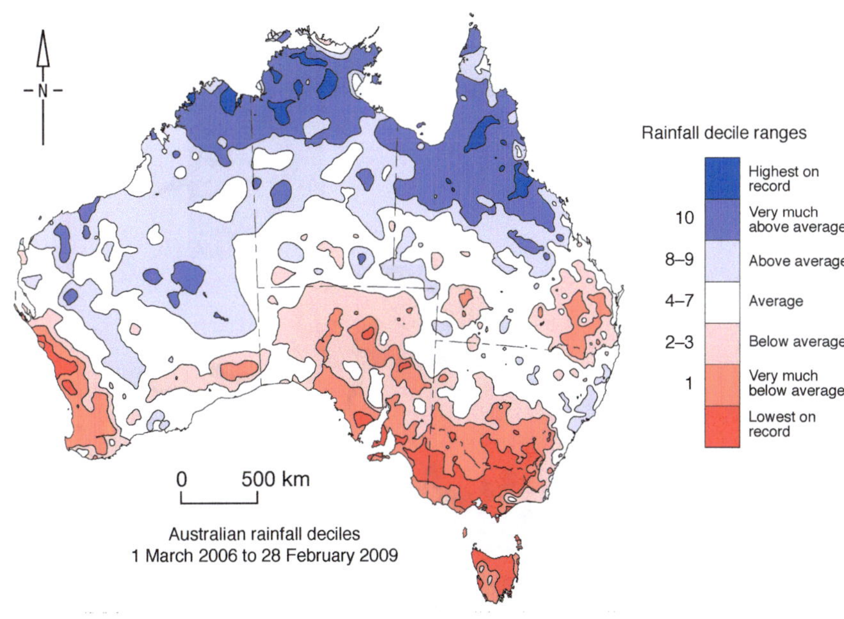

Source: Australian Bureau of Meteorology

6.20 Rainfall deciles for March 2006 to February 2009 throughout Australia

SPOTLIGHT

Megafire

The Black Saturday fires were fuelled by wild winds of up to 120 kilometres per hour, which snapped mature trees in half. The fires developed a towering smoke plume and created their own weather, triggering storm clouds and lightning strikes that started even more fires (see Figures 6.21 and 6.22) Flames leapt over 100 metres in the air over tree tops, and fireballs barrelled ahead of the fire front, landing with destructive embers in bushland, townships and paddocks. As they fed on their own heat and intensity, these supercharged megafires became even fiercer, and more destructive, multiplying the heat, wind and energy that they had already created.

6.22 The towering smoke plume above an advancing fire

6.21 The Black Saturday megafire was an exploding fireball that spiralled upwards and generated its own thunderstorm.

Prevailing wind

CONVECTION COLUMN
Smoke rotates as it would do at the start of a fire tornado. No known tornado developed on Black Saturday, but it did in the fires that burnt into Canberra in 2003.

On Black Saturday the smoke column developed into a white, pyrocumulus cloud which produced a thunderstorm and lightning which then spawned new fires in Melbourne's water catchments.

EMBER ATTACK
The fires send up burning bits of bark leaves or twig that fly in front of it. These embers or 'firebands' spark spot fires up to 15 km away from the centre of the storm and are often what burn people's homes.

EXPLODING FIREBALL
The fuel on the day was so dry and the temperatures were so high that burning plants gave off volatile gases quickly. The gases moved through the air faster and then ignited into huge balls when they reached clearer air at the edge of the fire.

Spotfires the size of normal bushfires that are within this 'fire area' can move against the prevailing winds when caught in the intense local weather created by a firestorm.

Graphic: Matt Davidson

Television footage of the town revealed a scene of utter devastation: house after house was a smoking ruin, with charred wooden beams, piles of blackened bricks and twisted iron roofing sheets, and burnt-out cars littered the streets (see Figure 6.23).

A family confronted by the fire in the Valley of Ewen survived by fleeing to a dam. They spent two hours under an old woollen blanket, going under the water every 30 seconds or so to clear the smoke that had formed and put out any hot embers that had stuck to the wool.

At the height of the crisis, thousands of volunteer fire fighters battled to contain about eighty blazes. While some volunteers fought to save the homes of others their own homes burnt to the ground. Some also lost family members.

The death toll would have been greater but for the bravery of many individuals and groups. At Yarra Glen's Grand Hotel, four friends, armed with garden hoses, drenched the building with water as flames approached on three fronts. Inside, up to 400 people who had sought refuge from burnt-out homes waited for the fire front to pass.

Response

In the days following the blazes, a massive recovery effort was launched by authorities and community-based groups such as the Red Cross. The emergency services first had to secure the area, treat the injured, and locate and identify the dead. Relief centres provided essential items for survivors and a nation-wide appeal raised millions to support those made homeless by the disaster.

A Royal Commission appointed by the Victorian Government investigated the causes of the fires and the policies and procedures used to protect life and property. The recommendations included the use of improved warning systems and the identification and advertising of neighbourhood safe places and community refuges to provide shelter to residents and visitors in high-risk towns who are unable to escape from bushfires.

Policy change

The Royal Commission examined the 'stay or go' policy, which advised people to 'Prepare, stay and defend, or leave early' in the event of a bushfire. This policy was heavily criticised after the Black Saturday disaster. The Royal Commission recommended a new policy, which advises residents to 'Prepare', 'Act' and 'Survive' (see Fig 6.24). This policy has been adopted by every state and territory in Australia.

6.23 Aerial view of Marysville the day after the disaster

6.24 Prepare. Act. Survive

ACTIVITIES

Knowledge and understanding

1 Outline the conditions that led to the Black Saturday fires.
2 Describe how the fires developed into a megafire.
3 Explain why the death toll was so great.
4 Outline the actions of emergency services in response to the disaster.

Applying and analysing

5 Discuss how effective you think the Royal Commission's recommendations would be in preventing future loss of life and property from bushfire.

Geographical skills

6 Study Figures 6.18, 6.19 and 6.20. In what way did the conditions on 7 February cause the fire to become an inferno?
7 Study Figure 6.20. Australia has been described as a country with 'drought and flooding rains'. How does this map of rainfall deciles support this statement?

6.6 Surviving bushfires

In a country such as Australia, where the bushfire hazard is so extreme, individuals, communities and governments must recognise their responsibilities and undertake measures to reduce the impact of the hazard.

Knowing the risk

Australians will continue to live in places that are at risk from bushfires. It is therefore important that people are aware of what they can do to prepare their properties and improve their chances of survival should they get caught by a bushfire.

Protecting property

Figure 6.25 outlines the measures that you can take to protect your property.

Increasing your chance of survival

The decision to stay and actively defend your property or leave the area must be made well before a fire threatens. It is also important to follow advice to evacuate. Waiting until the last minute to leave can have deadly consequences. It is hard to see through the smoke, and fallen trees and power lines may block the road. You are also at risk of being exposed to **radiant heat**. This scorching, invisible heat surrounds the flames and kills anything caught in its path. It is not always the flames but the inhalation of smoke that presents the greatest danger in a bushfire.

Some of the best ways to increase your chances of survival if you are in the path of a rapidly approaching bushfire are listed in Table 6.26.

6.25 Protecting your property from bushfires

Preparing your property

- Mature trees can help shield against radiant heat and embers. They must be strategically located and well managed.
- Remove flammable items from around your home.
- Keep grass short.
- Keep woodpiles away from the house.
- Store flammable liquids away from the house.
- Get rid of dry grass, leaves, twigs and loose bark.
- Prune shrubs well away from branches of mature trees.
- Cut back overhanging tree branches close to property—no branches within 10 m.
- Do not have large shrubs next to or under windows.
- Use pebbles and rocks in your garden (not flammable mulch).
- Keep gutters and roof areas clear of leaf litter.

Note: A new code red fire rating has been created for use throughout Australia. What this means is:

- These are the worst conditions for a bush or grass fire
- Homes are not designed or constructed to withstand fires in these conditions
- The safest place to be is away from high-risk bushfire areas

Home improvements

- Roof: seal gaps and install sarking (reflective non-combustible sheeting).
- Windows: install shutters, seal gaps and maintain window sills.
- Doors: install metal screen doors, seal gaps and install non-combustible door sill.
- Decks: use non-combustible deck materials, separate decking from dwelling and ensure no fuel under decking
- Ensure access to at least 10 000 litres of water independent of mains water.
- Install water pipes underground.
- Install a fire-fighting pump with separate power generator, including firefighting hoses.

Defending your property

- Block downpipes and fill gutters with water. Hose down the house and surrounding areas.
- Place a ladder in the manhole so that you can inspect the roof cavity at regular intervals.
- Put wet towels against spaces under doors. Close curtains and blinds.
- Wear personal protective clothing (wide-brimmed hat; eye protection such as goggles; mask/scarf to filter smoke; long-sleeved shirt and pants made from a cotton fibre; tough leather gloves and sturdy boots with woollen socks).
- Ensure you have fire-fighting equipment: torch, shovel, bucket, mop, radio and water sprayer.
- Install a sprinkler on the roof.
- Fill buckets, basins, baths and sinks with water to put out spot fires.

Source: CFA Victoria

6.26 Bushfire survival

Personal survival indoors
Wear as much cotton or woollen clothing as possible to cover the whole body (jeans are good); no synthetic fibre garments as they will melt onto the skin.
Sturdy leather boots with thick soles are best.
Crouch or lie down on the floor of a room that is away from the approaching fire; the air close to the ground contains less smoke
The fire front and associated radiant heat usually pass in 2 to 4 minutes. Even if the house catches fire, it is safer to stay indoors until the fire front has passed.
Personal survival outdoors
If you see smoke ahead go in the opposite direction.
Find the clearest or most open area. Never run uphill through unburnt country or shelter at the bottom of a valley, as the wind will suck the fire across or along these.
If possible, lie down in a depression, pond or dam, or cover yourself with loose earth or rocks. Thick, woollen clothing or woollen blankets offer some protection from the radiant heat.
If in a car, park by the roadside in the clearest area possible. Stay in the car, wind up the windows and put on the headlights. Crouch down and shelter under a rug, floor mat or anything similar that is available.

Government responsibility

Government authorities are responsible for advising the community of the current level of fire danger and fire bans that are in force. In the event of a bushfire occurring, the authorities coordinate evacuations and ensure that residents are updated on the status of the fire. Laws also require local governments to develop strategies to protect homes, farms, water supplies and roads from fire. These may include preventive slashing of undergrowth or prescribed burning (controlled burning in advance of fires so as to reduce the availability of fuel) and keeping watch from fire towers.

Community support

In Australia there are 250 000 volunteer fire fighters who willingly answer the call for assistance. Leaving their jobs and families, they travel great distances to fight fires. State Emergency Services volunteers assist by staffing road blocks, refuelling helicopters and assisting the police with evacuations.

Fighting a bushfire is often likened to a military campaign, with highly trained fire fighters deployed strategically on the ground and air support striking from above. See Figure 6.27 for strategies commonly used to fight bushfires.

6.27 Strategies commonly used to fight bushfires

ACTIVITIES

Knowledge and understanding

1 Outline the steps that individuals can take to protect themselves, their families and their property.

Applying and analysing

2 Assess and analyse the readiness of your home or school for a bushfire event. Prepare a sketch with annotations indicating changes required.

CHAPTER 7

ALPINE LANDFORMS

The world's alpine landscapes are not defined by their landforms. Rather, it is the distinctive climate found at high altitudes that determines their location and extent. Many alpine areas are, however, home to landforms shaped by the processes of glaciation, in either the past or the present. These harsh but spectacular places, and the related landforms, plant and animal life, are among the most culturally important of all environments. In addition, they are increasingly important as an economic resource. Tourists are attracted to alpine areas in all seasons and winter-based snow sports are becoming increasingly popular.

KEY IDEAS

- To investigate the main types of erosion shaping alpine landform features
- To understand the distinctive landform features associated with glaciation and how they are formed
- To investigate Australia's alpine environments
- To examine why alpine landscapes are worth protecting

7.0 Mt Assiniboine, British Columbia, Canada

GLOSSARY

abrasion	the wearing down or wearing away of rock by friction
accumulation zone	an area where snow and ice accumulate at a faster rate than they melt
arête	the sharp-sided ridge separating two cirques
cirque	a small valley that is shaped like an amphitheatre
endemic	unique to a defined geographic location
firn	compacted layer of new snow that survives the summer melt
fjord	a long, narrow flooded valley with steep sides, carved by glaciation
glacier	slow-moving rivers of compacted snow and ice
ground moraine	rock debris deposited across a valley floor as a glacier retreats
hanging valley	valley formed at the point where a smaller (tributary) glacier once joined a larger, deeper glacier
horn (pyramidal peak)	the pyramid-shaped peak of a mountain formed by three or more cirques
lapse rate	the rate at which the temperature of a body of air declines with elevation
lateral moraine	rock that accumulates along the edge of a glacier
medial moraine	the rock debris found in the centre of a valley, resulting from the merging of two lateral moraines when two glaciers merge
melting zone	an area where snow and ice melt at a rate faster than they accumulate—the end of a glacier
moraine	rock material deposited by a glacier
tarn	a lake at the base of a cirque
terminal moraine	rock material exposed as a glacier melts and retreats
treeline	the edge of the habitat in which trees are able to grow
U-shaped valley	a valley with a flat floor and steep sides, formed by the abrasive power of material embedded in a glacier
V-shaped valley	a valley eroded by a river

7.1 Alpine landscapes

The earth's highest mountains and plateaus are home to its alpine landscapes. The extent of such landscapes is determined by the distribution of alpine climates. An alpine climate is defined by the conditions that exist above the treeline—the elevation at which it is too cold for trees to grow.

Alpine climates

The climate is the average weather conditions experienced over a long period of time. One of the most important elements of weather is temperature. Alpine temperatures range from –18°C to 10°C. Precipitation is low and often falls as snow. Night temperatures are almost always below 0°C. The growing season of plants in alpine climates is usually less than 180 days per year.

The treeline in an alpine environment appears well-defined from a distance. On closer inspection it is actually a zone of slow change. The growing season of trees becomes shorter with increased altitude and more extreme climate conditions, and eventually there is a point where trees stop growing. In Australia's Snowy Mountains the treeline is approximately 1800 metres above sea level. In the Swiss Alps the treeline is approximately 2200 metres above sea level and in the Canadian Rockies it is 2400 metres.

Temperatures decline with elevation

Due to air expanding as it rises, temperatures drop as elevation increases. It is important to remember that the atmosphere is heated by the earth's surface, not from above. As a dry mass of air rises, it cools at a rate of –10°C per 1000 metres.

Shaping alpine landforms

When mountains are formed, weathering and erosion begin to wear them away. Exposed rock surfaces are broken down by physical and chemical weathering. The material is then removed by wind and water, and deposited (by rivers and glaciers) at lower elevations as sediment.

Work of rivers

Running water is the most powerful form of erosion in alpine landscapes that are not permanently covered with snow and ice, transporting and depositing rock and soil. In mountainous areas, rivers erode downwards, creating narrow, **V-shaped valleys**. Away from the mountains, valleys become wider and some of the river's load of sediment is deposited. Eventually, this material finds its way to the sea, where it is deposited in layers.

Glaciers: rivers of ice

At elevations where it is cold enough for snow and ice to accumulate over a long period of time, glaciers are the main type of erosion. **Glaciers** are slow-moving rivers of compacted snow and ice. Figure 7.1 shows a river of ice at the Gorner Glacier in Switzerland. Glaciers form when compacted snow, which has accumulated over many years, gradually moves downhill under the force of gravity.

Figure 7.2 shows the formation of a glacier. As the glacier moves, the surface of the land is scratched and worn down by rock fragments that have been picked up from the ground and frozen into the base of the glacier. This process is known as **abrasion**. The most spectacular landform features of glacial landscapes are **U-shaped valleys**. During periods of glacial activity, glaciers cause the valleys once occupied by rivers to deepen and widen. **Hanging valleys** form where smaller (tributary) glaciers join larger glaciers. Other erosional features include **cirques, arêtes** and **horns (pyramidal peaks)**. **Fjords** are formed when rising sea levels flood the valleys once occupied by glaciers.

The rocks that are picked up and transported by the ice can be carried long distances before they are deposited, forming features known as **moraines**.

7.1 A U-shaped valley at Gorner Glacier, Switzerland

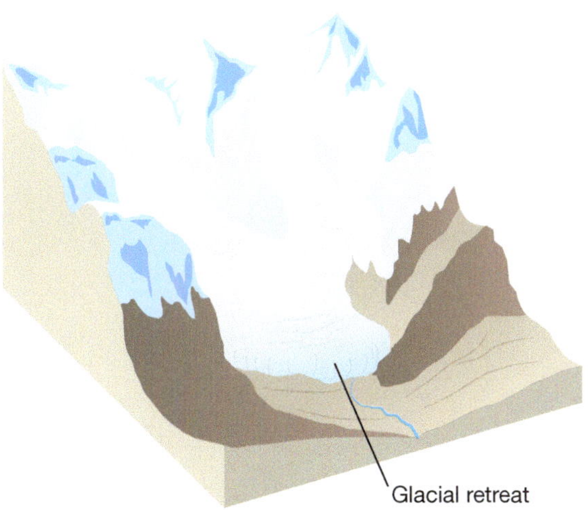

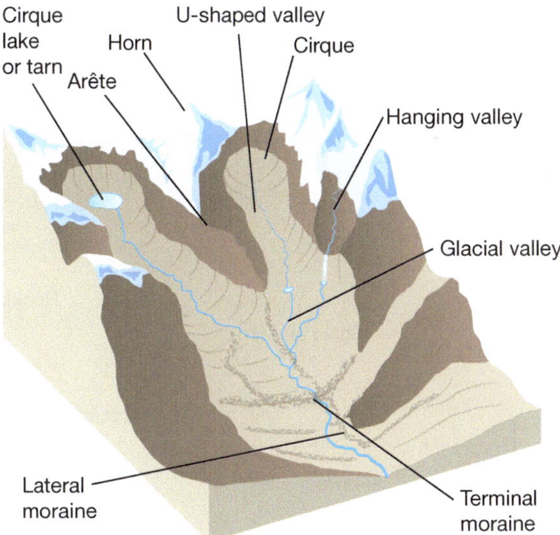

7.2 Glacial landforms during and after glaciation

ACTIVITIES

Knowledge and understanding

1. Explain how the extent of an alpine landscape is defined.
2. State what the term 'treeline' refers to.
3. Explain in your own words why the weather gets colder with altitude.
4. Identify the main types of erosion that shape alpine landscapes.
5. Explain what a glacier is.
6. Outline how glaciers shape the land.
7. List the landform features associated with the movement of glaciers (glaciation).

7.2 Glacial landforms and processes

A glacier is a slow-moving river of ice. Glaciers are found in alpine landscapes where temperatures are below zero for most or all of the year. Glaciers create a range of distinctive landform features that give alpine landscapes their spectacular appearance.

Glacier formation

When snow falls to the ground it forms a light, feathery layer that traps a lot of air. Where snow collects in a low area of land, or depression, it becomes compressed by more snowfalls and gradually develops into a more compact, thick layer of snow. When this compacted layer of snow experiences one winter's freezing and survives a summer's melting it is called **firn**. Air is progressively squeezed out of the firn and after 20 to 40 years it becomes solid ice. In parts of Antarctica and Greenland where there is no summer melting, the same process can take more than 200 years. Large masses of this ice may begin to 'flow' downhill, under the force of gravity, as a glacier. Figure 7.3 shows the main features of a glacier.

Glacier types

Glaciers are classified according to their size and shape. The five main types are:

- *niche glaciers*—very small accumulations of glacial ice that are found in depressions and gullies
- *cirque glaciers*—relatively small glaciers found in 'armchair-shaped' depressions in mountains
- *valley glaciers*—larger masses of ice that move down former river courses and are bounded by steep (almost vertical) valley sides
- *piedmont glaciers*—glaciers that form when valley glaciers extend onto lowland areas, spread out and merge.

7.3 Main features of a valley glacier

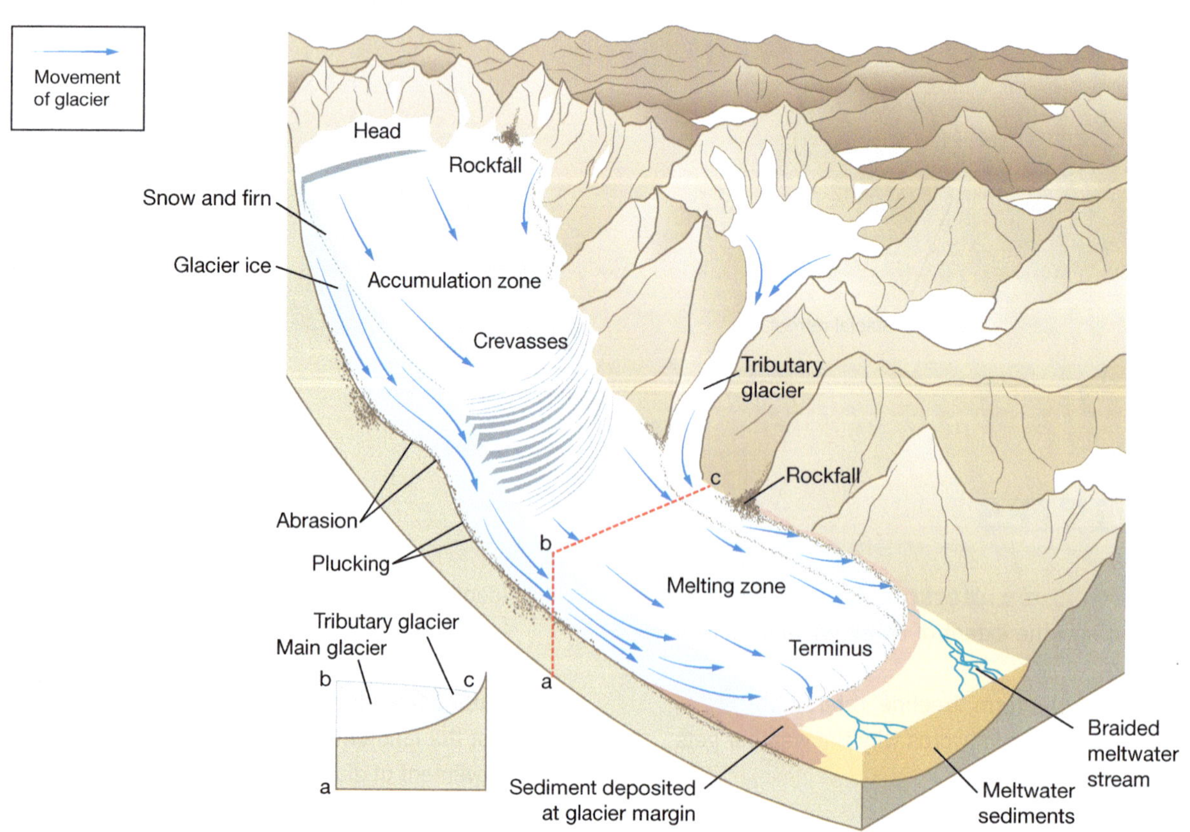

7.4 A U-shaped, former glacial valley with hanging valley and associated waterfall, Yosemite National Park, United States of America

- *ice caps and ice sheets*—huge areas of ice covering more than 50 000 square kilometres. Ice sheets once covered much of northern Europe and North America, but they are now found only in Antarctica and Greenland.

Glacial inputs and outputs

Water enters the glacial system mostly as snow and leaves as meltwater. At higher elevations, inputs (snow accumulation) exceed outputs (evaporation and meltwater). This area is called the **accumulation zone**. At lower elevations, outputs exceed inputs. This area is called the **melting zone**. It is shown in Figure 7.3.

Glacial erosion

Nearly all the glacial processes of erosion are physical, as the climate is too cold for chemical reactions to occur. Frost shattering is common and produces large amounts of loose rock. This rock material falls from valley walls to form **lateral moraine** along the edges of the glacier. Glacial **abrasion**, which is the scraping and grinding effect of rock material in the glacier, smooths and widens the U-shaped valley—a distinctive characteristic of areas shaped by glaciation.

At the top of glacial valleys large steep-walled bowls called cirques are located. When glaciers carve out valleys next to each other, a sharp-sided ridge, or arête, will separate them. If several of these valleys begin near the top of a mountain, a sharp peak, or horn, is formed. A glacial hanging valley forms where a smaller tributary glacier joins a larger valley glacier. After the glacial ice melts, the valley formed by the smaller glacier is left 'hanging' high above the valley floor. Hanging valleys often have spectacular waterfalls, as shown in Figure 7.4.

DID YOU KNOW?

Ten per cent of the earth's land area is covered with glacial ice, which includes glaciers, ice caps and ice sheets.

SPOTLIGHT

Formation of cirques

A cirque is a bowl-shaped landform feature often found at the beginning, or head, of a valley glacier. Carved by ice, this distinctive feature is usually surrounded on three sides by steep cliffs. The highest cliff is often called a headwall. The fourth side, known as the lip or sill, is the point at which the glacier flowed away from the cirque. Many glacial cirques are now occupied by **tarns**—lake-like features that form behind the till (rock debris) that collects at the lip (see Figures 7.5 and 7.6). The ridge between two adjacent cirques is called an arête.

A glacial horn, or pyramidal peak, is formed when three, or sometimes four, cirque headwalls and their arêtes join together to form a single peak that has a distinctive pyramid shape with very steep walls. The number of faces that make up the sides of a pyramidal peak depends on the number of cirques involved in its formation.

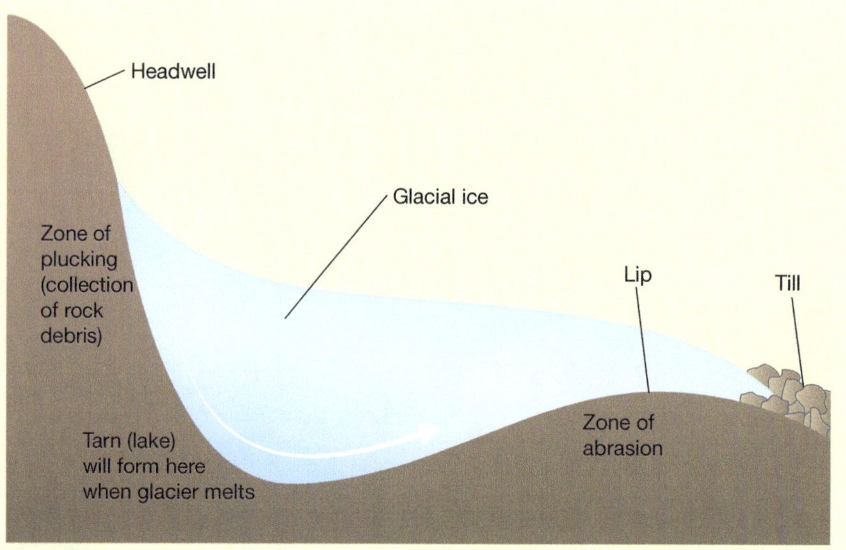

7.5 Formation of a cirque

7.6 Blue Lake in the Kosciuszko National Park, an example of a tarn

Aletsch Glacier, the largest glacier in Switzerland. The glacier's terminal, medial and lateral moraines are clearly visible.

1 Terminal moraine

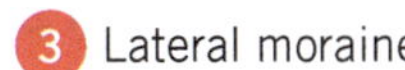

Glacial transportation

Glaciers are able to move large amounts of rock material, known as moraine. There are three main ways this material is moved downhill: on the surface of the glacier as **lateral** or **medial moraine**; as debris carried within the body of the glacier; or as material embedded in the base of the glacier, **ground moraine**.

Glacial deposition

The rock debris carried by glaciers is deposited at the point at which the glacier melts. When glaciers retreat (melt), these materials are deposited across the exposed landscape. A **terminal moraine** consists of a mound of material across the face of the glacier. The terminal moraine can be used to determine the maximum advance of a glacier. Ground moraine is deposited as rock debris across a valley floor as a glacier retreats (see Figure 7.7). Lateral moraine is rock debris deposited as an embankment along the valley sides as the glacier retreats. Medial moraine is found in the centre of a valley and results from the merging of two lateral moraines when two glaciers merge.

ACTIVITIES

Knowledge and understanding

1 Explain the process of glacier formation.
2 State the difference between valley and piedmont glaciers.
3 Explain why physical rather than chemical weathering processes dominate in alpine areas.
4 Outline what is meant by the term 'abrasion'.
5 State the conditions under which arêtes and horns develop.

Geographical skills

6 Study Figure 7.3. Sketch a glacier and label the main glacial landform features.
7 Study Figure 7.4. Write a paragraph explaining the circumstances in which hanging valleys develop.
8 Study Figure 7.4. Construct an annotated photo sketch of Yosemite National Park highlighting the key elements of the U-shaped and hanging valley.
9 Study Figure 7.6. Construct an annotated photo sketch of Blue Lake, highlighting the key elements of the cirque.

7.3 Alpine ecosystems

The factors that help to shape the alpine ecosystems are altitude, slope and aspect.

Altitude

With increasing elevation (or altitude), air temperatures decrease and wind speeds increase. These changes have a major impact on plant growth. As you go up a mountain, very distinct zones of vegetation can be seen. These zones are shown in Figure 7.8. At the highest zone there is no plant life, only ice and snow. Below this zone is the tundra. Figure 7.9 shows the flowering plants of the Rocky Mountains that grow in the tundra. Below the tundra is the coniferous forests zone. Figure 7.10 shows the forests on the slopes of Canada's Whistler Mountain. Beneath the forest is the temperate or deciduous forests zone.

7.8 Ecosystems change with altitude.

7.9 The tundra above the treeline, Rocky Mountains, United States of America. Flowering plants have a brief and intense life, with a growing season as short as thirty days. During this time, the plants must race through the processes of germination, growth and reproduction.

7.10 Coniferous forest on the slopes of Canada's Whistler Mountain

Slope

The degree of slope is important in alpine regions because it influences soil depth and moisture content. Rates of erosion are higher on steep slopes and moisture drains away quickly. This makes it difficult for steep slopes to support plant life.

Aspect

The aspect of a slope affects both light and temperature conditions. South-facing slopes in the Northern Hemisphere are more favourable for plant growth than those facing north (especially in winter). This is because north-facing slopes receive less direct sunlight, as illustrated in Figure 7.11. In the Southern Hemisphere, north-facing slopes receive more direct sunlight.

Alpine climates

Latitude, altitude and aspect interact to produce alpine climates. In general, the climate becomes colder with elevation—this characteristic is described by the **lapse rate** of air: air tends to get colder as it rises, because it expands. The rate at which dry air cools as it rises is 10°C per kilometre of altitude. Therefore, the temperature at the summit of a 4000-metre high mountain will be up to 40°C cooler than the sea-level temperature. This vertical variation in temperature determines which flora and fauna can survive at different altitudes.

DID YOU KNOW?

Alpine plants and animals are restricted to an area between the treeline and the mountain summit. In Australia, there are more than 250 species of alpine plants that grow only in this restricted habitat.

Impact of climate in a Northern Hemisphere valley

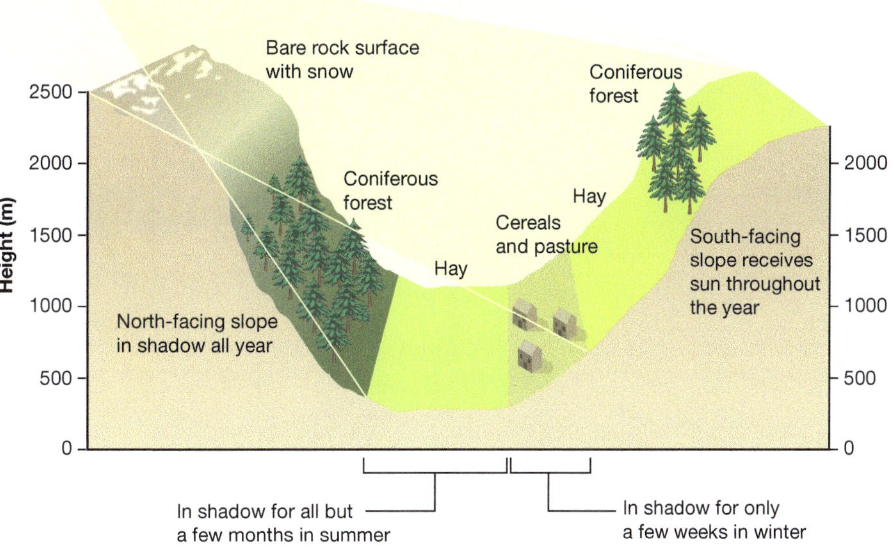

ACTIVITIES

Knowledge and understanding

1. Explain how altitude affects plant growth.
2. Outline how slope affects the potential for plant life in mountain environments.
3. Outline how aspect affects the pattern of vegetation in mountain environments.

Applying and analysing

4. Study Figures 7.9 and 7.10. Write down the vegetation differences you observe in the different alpine environments.
5. Study Figure 7.11. Explain the impact that aspect has on the valley shown in the diagram. If you were going to build a house, where would you build it? Justify your answer.

7.4 Managing Australia's alpine environments

Australia's alpine areas are found on both sides of the mountains and plateaus of the highest part of the Great Dividing Range. This is the only region on the mainland where snow accumulates on the ground, forming a snowpack that remains throughout winter. Australian alpine environments are unique and vulnerable.

Alpine environments

The word 'alpine' is used to describe all snow-covered areas. In the strictest sense it refers to the higher country where it is too cold for trees to survive, such as the Main Range of Kosciuszko National Park shown in Figure 7.12. On mainland Australia this occurs at 1800 metres (shown in the cross-section of the Kosciuszko National Park in Figure 7.13) and in Tasmania at 1200 metres.

7.12 The Main Range of Kosciuszko National Park, where it is too cold for trees to grow at the highest elevations

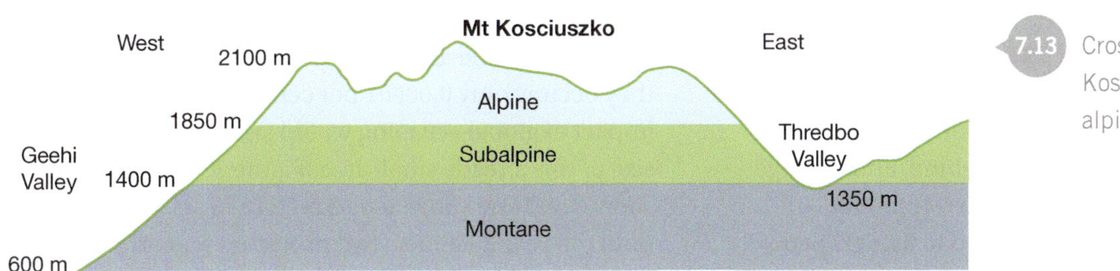

7.13 Cross-section of the Kosciuszko National Park alpine landscape

Environmentally significant features

Australia's alpine regions are an important part of Australia's natural and cultural heritage.

Natural values

Australia's alpine region includes:

- landforms such as cirque lakes (tarns) resulting from glacial activity during the last Ice Age
- unique flora and fauna communities adapted to the harsh conditions of low temperatures, frequent frosts and strong, cold winds. The alpine marsh marigold actually flowers under the snow in icy meltwater, enabling it to have enough time to flower and set seed in the brief summer
- rare cold climate animal species such as the mountain pygmy possum (see Figure 7.14) and the corroboree frog. Many of these species are **endemic**, as they are found naturally only in the alpine region. Several species are restricted in their distribution to within 5 or 6 kilometres of Mount Kosciuszko
- the only mountains in the world with such a substantial covering of soil. The humus soils are globally significant.

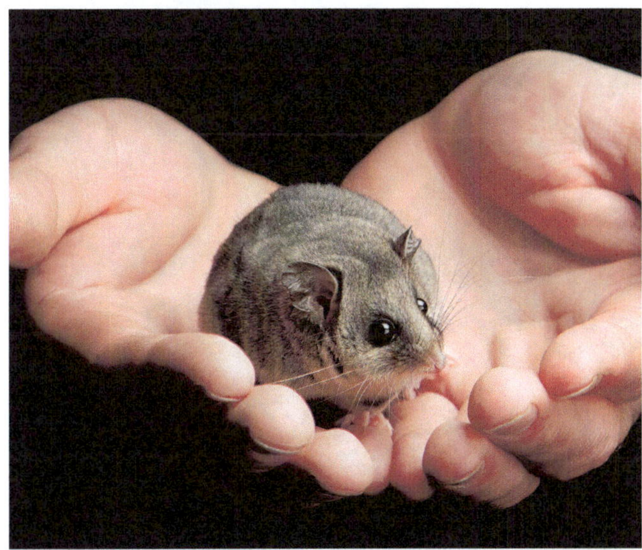

7.14 The mountain pygmy possum

Cultural values

The following aspects of the alpine region are an important part of Australia's cultural heritage.

- Aboriginal people have a living spiritual connection with the mountains. Researchers believe their summer visitation started 9000 years ago. Tribes travelled long distances to hold celebrations in the mountains and, while there, they feasted on the Bogong moths.
- There is a cultural history associated with European grazing of the alpine area that dates back to the 1830s. Legends of the high-country horsemen and relics such as stockyards and huts are valuable elements of Australia's colonial heritage.
- The mountains are significant water catchments and the Snowy Mountains Hydro-electric Scheme is one of the world's most complex engineering feats. The scheme enables water to be collected and diverted through the mountains to inland New South Wales and Victoria, where it is used for irrigation.
- People have long visited the mountains for recreation, to enjoy the spectacular scenery in the summer, and to ski or snowboard in the winter.

A fragile environment

While the species that survive in the climatic extremes of the alpine area are hardy, their survival is dependent on their environment remaining the same. Historically, considerable damage has resulted from human land-use practices in the mountains. Today, there are still a number of risk factors.

Damage from grazing

Records show that graziers started bringing large numbers of sheep and cattle up into Australia's alpine environments in the early 1820s in the search for pastures and water. The heavy, hard-hoofed animals caused considerable damage to the native alpine vegetation, resulting in severe soil erosion. Grazing continued for over 120 years.

The pressure of tourism

As numbers of tourists grow, increasing pressures are placed on the environment. Vegetation is trampled, rubbish and human waste are left behind, and damage is caused by people camping in the wrong areas. Victoria's five alpine resorts attract up to 900 000 people each winter and an increasing number visit outside the snow season. The New South Wales ski resorts, including Thredbo (see Figure 7.15) and Perisher Blue, attract 1.2 million visitors a year and are important sources of employment for those living in the region.

The threat of climate change

Climate change has the potential to lead to a reduction in snow cover, as milder temperatures will result in fewer snow days and a reduced depth of snow. If the change is too rapid, species will not have a chance to adapt. Australian alpine areas are particularly vulnerable, as they occupy only 0.00001 per cent of the continent; the impact of global warming would significantly reduce the size of this already small area. Figure 7.16 shows current snow cover and projected reduction in snow cover up until the year 2050. The two projected scenarios are:

- low scenario: the lowest projected level of global warming combined with the highest levels of precipitation
- high scenario: the highest projected level of global warming combined with the lowest levels of precipitation.

Environmental management

Australia's alpine national parks are featured on the Australian National Heritage List. With this recognition comes the responsibility to manage and protect them for the benefit of future generations.

7.15 Thredbo, Kosciuszko National Park. A range of sustainable practices have been implemented to minimise environmental impacts.

The management plans for the alpine national parks aim to:

- protect the unique mountain landscapes
- protect the region's natural and cultural values
- provide an appropriate range of outdoor recreation and tourism opportunities that encourage the enjoyment, education, understanding and conservation of the natural and cultural values and protect mountain catchments.

Management also includes safeguarding against the impacts of:

- introduced plants and feral animals
- soil erosion
- fragmentation or division of habitats
- pollution of waterways
- incompatible human activities.

Minimal Impact is one strategy used to protect the alpine environment. Under this strategy, all users of the parks are asked to reduce their environmental impact. Educational and information brochures on minimal impact are distributed to visitors. People are asked to take all rubbish out of the park, stay on tracks as shown in Figure 7.17, camp well away from water courses and fragile landscapes, use fuel stoves instead of wood fires for cooking, leave all Aboriginal and historic sites undisturbed, and avoid disturbing any plants, rocks, logs or animal nesting sites.

Paths and boardwalks are used to protect fragile landscapes at Mt Kosciuszko.

Current and projected decline in Australia's snow cover. The two predictions show the impact of 'high' and 'low' temperature rises due to climate change.

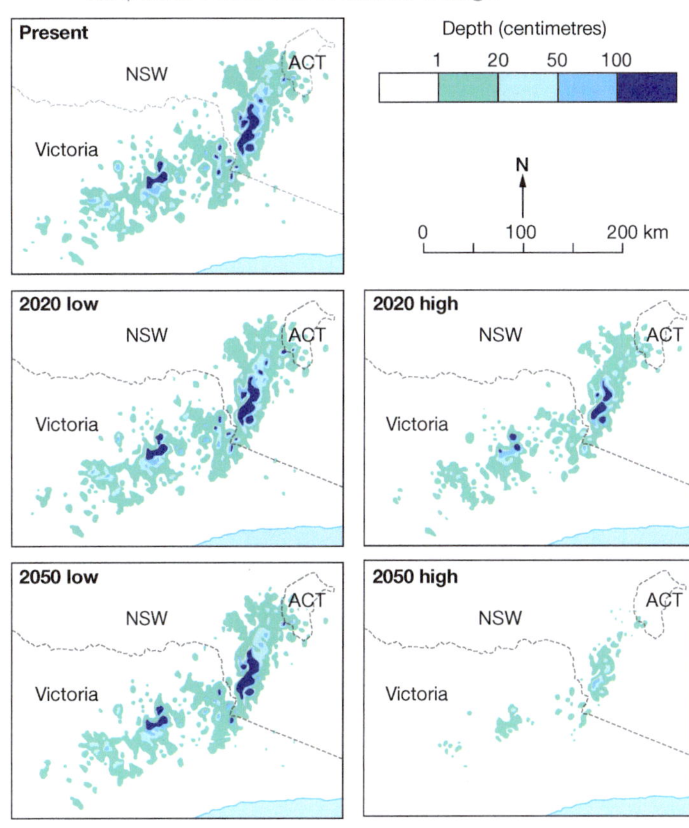

ACTIVITIES

Knowledge and understanding

1. Justify the protection of Australian alpine environments.
2. Define the term 'endemic species'.
3. Explain how the Australian Alps differ from mountainous areas elsewhere in the world.
4. Outline the threats to Australian alpine environments.

Applying and analysing

5. Use the 'think, pair, share' strategy to consider the following tasks.
   a. Consider why the treeline that marks the beginning of the alpine area is at a lower elevation in Tasmania than on the mainland.
   b. Assess the impact of grazing on soil formation in the alpine environment of Australia.
   c. Develop a poster designed to educate visitors to the Australian alpine parks on how to minimise their environmental impact.

CHAPTER 8 COASTAL LANDFORMS

Coastlines are dynamic environments, undergoing constant change. The main agents of change are waves, tides and currents. The formation and shape of a coast is also influenced by its geology—the harder the rock, the less likely it is to be affected by eroding forces. Variations in rock type create different-shaped coastlines.

Coastal landforms can be divided into two major categories—erosional and depositional. Erosional coasts are those dominated by cliffs and wave-cut platforms. Depositional coasts are those dominated by landforms made up of eroded materials, for example beaches and sand dunes.

Coastlines are especially vulnerable to the impacts of human-induced climate change. Global warming is predicted to accelerate the rate of coastal erosion. As sea levels rise, beaches and coastal dune systems will be eroded and coastal wetlands inundated. A global temperature increase of 4–5°C and the resulting rise in sea levels will mean that many of the world's coastal cities will have to be abandoned, as will the intensively settled and farmed coastal lowlands.

This chapter investigates the processes shaping coastlines and the erosional and depositional landforms that result. The chapter also examines how coastal landforms are managed and protected.

KEY IDEAS

- To investigate processes shaping coastlines
- To understand the main erosional and depositional coastal landforms and how they develop
- To explain how coastal landforms can be managed and protected.

8.0 Nash Point, Wales, United Kingdom

GLOSSARY

backwash	the return flow of water down a beach after a wave has broken
berm	a small wall of earth or sand at the back of the swash zone
blowout	a gap, or break, in the dune system formed by the inland movement of sand; often caused by the destruction of dune vegetation
clinometer	an instrument for measuring slope angles
corrasion	the process by which surging water bombards a cliff with rock fragments and drags other fragments backwards and forwards over rock surfaces, wearing them away; also called abrasion
depositional coast	a coastline dominated by landforms made up of eroded materials
erosional coast	a coastline dominated by landforms created by erosion
fetch	the length of water over which wind has blown
foredune	the sand dune closest to the beach
groyne	a wall of rocks or timber built to trap sand on a beach
hydraulic action	the process whereby waves breaking against a cliff cause air to be compressed in the cracks in the rock, resulting in the cracks widening and the rock loosening and breaking away over time
joint	a gap in layers of solid rock
longshore drift	the process whereby sand is moved along a beach shoreline as a result of waves approaching the shore at an angle; also called littoral drift
ocean current	a flow of water in the ocean with a different temperature or salinity from the water through which it is passing
riprap	a barrier built from large rocks or boulders to protect the shoreline
sand dune	a hill of sand shaped by the wind
sand bar	an area of sand deposited in shallow water
swash	the rush of water up a beach after a wave breaks
tide	the rise and fall of sea levels caused by the gravitational forces exerted by the moon and the sun, and the rotation of the earth
wave-cut platform	a flat rock surface found at the base of a cliff or headland

8.1 Processes shaping coastlines

Coasts are very complex environments that are constantly being changed by the forces of nature. Waves pound the coastline, sometimes bringing sediments such as sand onto the beach and at other times striking the coast with so much force that they erode its landforms.

Coastline types

Coastal landforms are categorised into two main types—erosional and depositional coastlines. Erosional coasts are formed by the powerful action of waves and other forces that weather and erode rock. The Victorian and South Australian coastlines are mostly erosional coastlines. Figure 8.1 shows an erosional coast. Depositional coasts are made up of eroded materials transported along the coast. The Queensland and northern Western Australian coastlines are mostly depositional coasts. Figure 8.2 shows a depositional coast.

Waves

Of all the processes shaping coastlines, waves are the most important. They play a role in both erosion and deposition. The sea's surface is constantly moving. When wind blows across the surface of the sea, energy is transferred from the wind to the water surface. A wave is energy travelling through the water.

In the open sea, far away from land, waves move as ocean swell in a circular motion. As the wave approaches the shallow shoreline it begins to interact with the seabed. The seabed disturbs the circular motion of the wave, causing it to rise up, and then gravity eventually causes it to break. This process is shown in Figure 8.3.

8.1 An erosional coastline

8.2 A depositional coastline, Fraser Island, Queensland

Swash and backwash

When a wave breaks it creates turbulence in the water. This turbulence is called **swash**. As the swash water surges up the beach it carries sediment it has picked up with it. As the energy in the wave is gradually lost, most of the water soaks into the sand. The remaining water, called **backwash**, flows back down the beach and into the sea.

Constructive and destructive waves

Waves are very powerful—the energy within them is strong enough to wear down solid rock and cause widespread destruction to coastal zones, especially during storms. However, waves are also builders of the coastal system. They are able to carry and deposit large amounts of sediments and play an important role in constructing the coastal system.

When a low-energy wave crashes on a beach, it has less backwash and most of its sediment is dropped during the swash, so it becomes a constructive wave. However, if the wave has considerable energy, much of this energy will return to the sea in the backwash. In this instance, the wave can continue to pick up sand in the backwash and return it to the sea, becoming a destructive wave.

In Australia, waves are generally constructive during summer in southern Australia and during the dry season in northern Australia. During this time the beach expands as sand is deposited onto the beach. This sand builds up depositional landforms such as sand dunes. Wave size is determined by speed of wind, duration of wind and fetch.

8.3 Waves increase in height when the wind blows strongly, for a long time and over a long distance (fetch). When they enter shallower water the circular movement of energy is disturbed. The forward movement of the wave causes it to break as surf.

In the winter months, there is an increase in the number of large storms and strong winds in southern Australia. This type of weather results in the creation of destructive waves. These are waves with huge amounts of energy that wear away coastal areas. Destructive waves erode beaches and headlands, and create considerable damage.

Weathering

Although waves are the most important agents of erosion, there are other agents that impact on the wearing down of a coast.

- Salt-spray weathering gradually breaks up the surface of rock because the salt particles expand and contract as they dry out in the sun.
- Plant weathering breaks up rock, as plants send tiny roots into **joints** (gaps within the rock).
- Animal weathering occurs when sea creatures attach themselves to rock. The animals produce chemicals that attack minerals in the rock, breaking it down.

Erosion

Erosion is the transporting of material that has been weathered (worn away) from one place to another. There are two types of erosion: wind and water. In the coastal environment, water erosion occurs mostly through waves and the coastal rivers that carry swash sediments from inland areas out onto the coast. Wind erosion is very important in the formation of sand-based landforms such as coastal dunes. The factors affecting the rate of erosion are shown in Figure 8.4.

There are six main types of coastal erosion.

- ***Wave pounding*** Large waves transport huge amounts of energy. When waves break against rocks they weaken the rock's structure, causing small pieces of rock to break away.
- ***Hydraulic pressure*** When a wave strikes against a rock surface a small amount of air is trapped in cracks within the rock face. As the air is compressed by the weight of the water, pressure is exerted on the rock. Over time this weakens the rock face.
- ***Corrosion/solution*** This process is similar to the formation of rust on metal. A chemical reaction between air, the salt in the seawater and the rock surface causes small particles of rock to disintegrate.
- ***Abrasion/corrasion*** As a wave breaks it throws sand, small rocks and sometimes even large boulders against rocky coastlines. This action gradually wears away the rock, much like sandpaper does to a piece of wood.
- ***Attrition*** When a large piece of rock falls away, the waves gradually wear it down into smaller rock particles and eventually sand.
- ***Land-based erosion*** Water running from the land can also erode the coastal environment.

8.4 Factors affecting the rate of erosion along a coastline

Coastal erosion

Beach width. If the beach is wide then much of the wave's energy will be lost as it surges up the beach. This means that there will be less damage to the landform at the back of the beach.

Rock hardness. The strength of the rock will determine the rate at which it is eroded. Some rock, such as sandstone, is easy to erode due to its softness.

Depth of sea, fetch and configuration of coastline. Beaches with steep gradients create higher, steeper waves than beaches with gentle gradients. The longer the **fetch**, the more time there is for waves to collect energy from the wind. When waves approach a headland, wave energy is concentrated onto the headland.

Supply of beach sediment. If there is a large amount of sediment on the beach then the power of the waves is reduced and they become less erosive. The sediments help to absorb energy.

Wave steepness. Steep, destructive waves formed locally have more energy than gentle, swell-induced waves.

Breaking point of the wave. If the wave breaks as it hits the cliff, it releases more energy than if it breaks earlier then rolls up the cliff face.

Deposition

Many of the features you see at the coast (such as beaches, sandbars and sand spits) are built up by deposition. Waves and currents move sand along the coast and deposit it in large piles or strips. Beaches are the best known of these features and are formed only where there is a good supply of sand and some form of protection to prevent the sand from being eroded away.

Tides and currents

Tides are caused by the gravitational pull of the moon on the oceans. Figure 8.5 shows this effect. They vary greatly in height from one place to another. While most places have an average of a few metres between high and low tides, there are places where the height of the tide rises by 20 or 30 metres.

Ocean currents play a very important role in distributing sediments and also in regulating temperatures. Ocean currents carry cool water away from the North and South poles towards the Equator and warm water from the Equator towards the poles. In this way they help to create milder climates.

Longshore drift

Along many coastlines a special type of current known as **longshore drift** occurs. This current plays a major role in the movement and deposition of sediment, and is very common along the eastern coast of Australia. Longshore drifts, shown in Figure 8.6, run parallel to the coastline, carrying the sediments that destructive waves have eroded from beaches. These sediments can be carried very long distances. The prevailing wind also affects longshore drift.

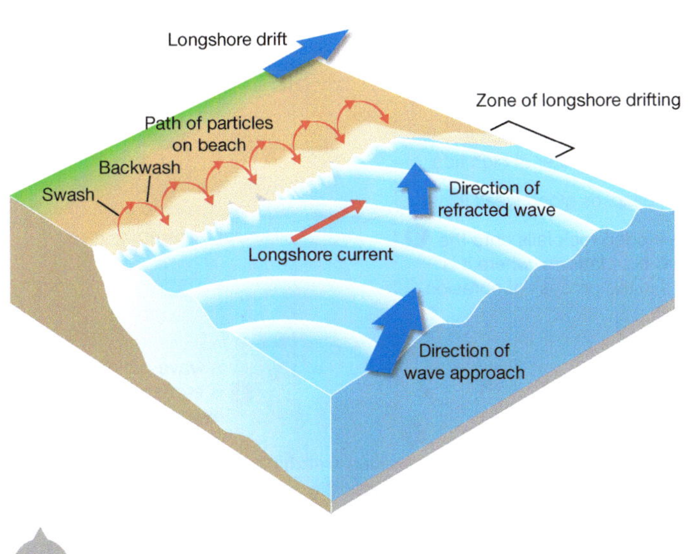

8.6 Longshore drift

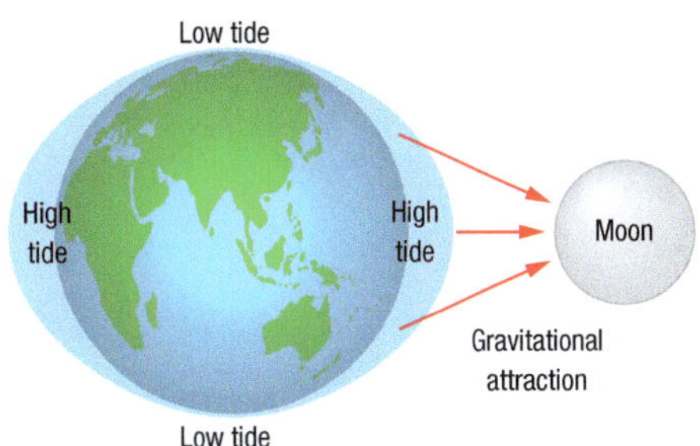

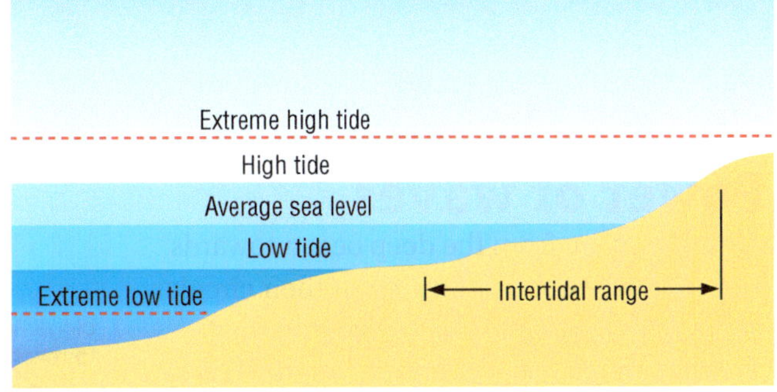

8.5 The effects of gravitational pull by the moon and the sun cause the daily and monthly pattern of tides.

ACTIVITIES

Knowledge and understanding

1 Distinguish between erosional and depositional coasts.
2 Explain how waves form.
3 Explain what weathering is.
4 Outline the various forms of weathering that shape the coastal environment.
5 Name three landform features associated with the process of deposition.

Geographical skills

6 Study Figures 8.1 and 8.2.
   a Construct a Venn diagram and list the similarities and differences between the coastal landscapes shown in the photographs. Include a definition explaining the differences between erosion and deposition.
   b Construct a mind map of all the different ways that the coast is changed by natural processes.
7 Study Figure 8.6. Explain, in your own words, the process of longshore drift.

8.2 Erosional landforms

Many coastlines are made up almost entirely of rock. Wave erosion dominates these coasts. Storm waves undermine cliffs and gradually wear away headlands. Sometimes the largest waves that break on coasts are caused by storms hundreds of kilometres away. The wind may be so strong that it builds up huge waves that travel uninterrupted across the sea.

Power of waves

As waves move from the deep ocean towards the shallower waters of the coast their circular motion becomes affected by the frictional drag of the ocean floor. This causes the bottom of the wave to move more slowly than the crest and the circular motion is increasingly distorted; the wave increases in height until it finally plunges forwards as a breaker. The energy of the wave is released and the work of erosion, transportation and deposition takes place. This process is outlined in Figure 8.7. On exposed coasts, these are the dominant processes in the area between low and high tides. Waves are most active during storms, when high winds create large, high-energy waves.

Coastal erosion

On coasts with cliffs, wave action is concentrated at the base of the cliff. This process is outlined in Figure 8.8 and illustrated in Figure 8.9. Coastal erosion occurs very quickly when the cliffs are formed from softer rock, such as sandstone.

Processes of wave erosion

At high tide, deep water allows bigger waves with greater energy to reach the cliffs, increasing erosion.

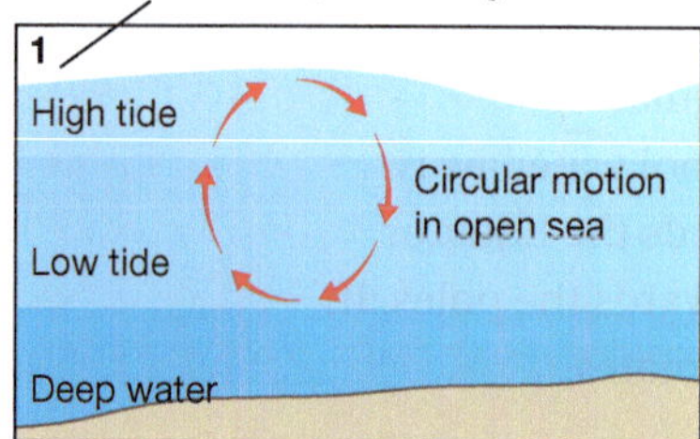

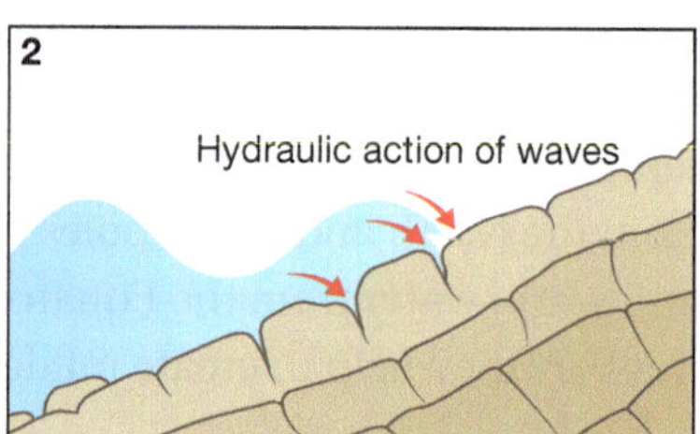

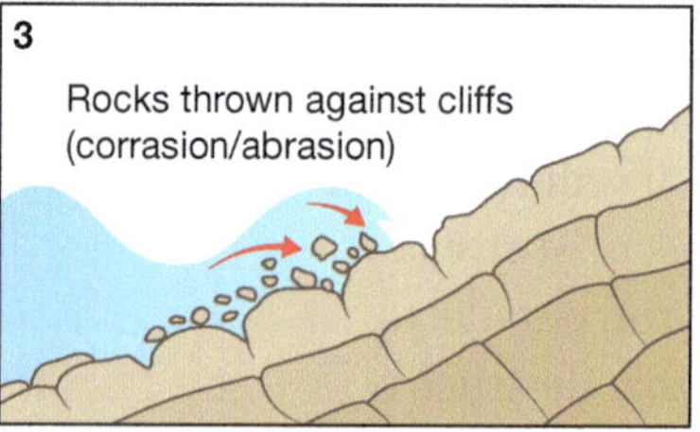

8.8 Coastal processes and landform features on a rocky coast

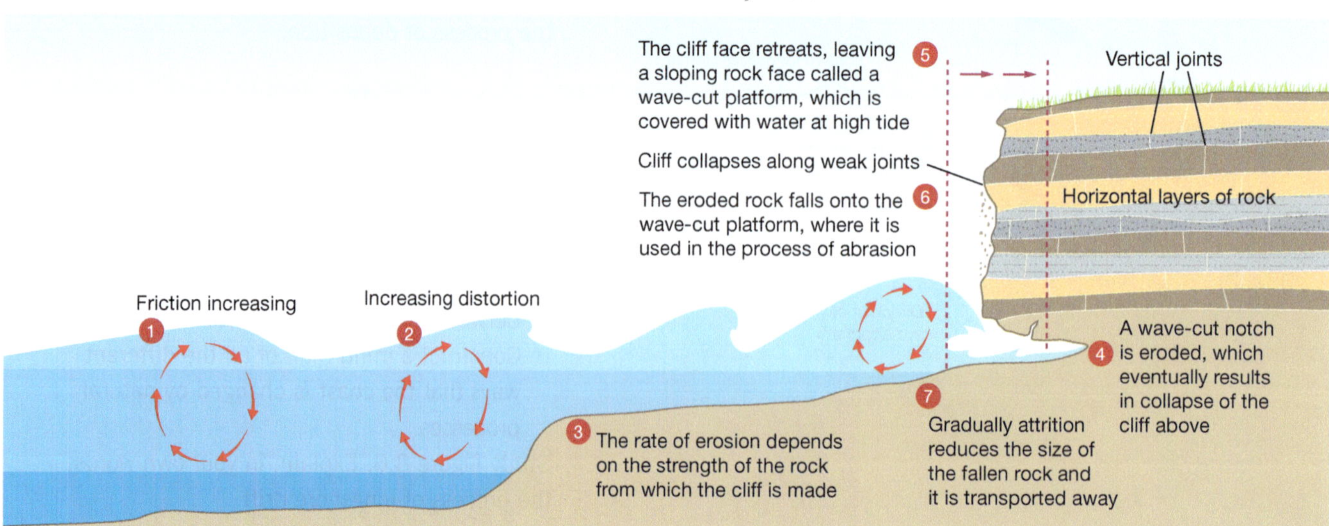

8.9 Cliff face, notch and wave-cut platform, Newfoundland, Canada

Stacks and arches

Some sections of rock within a cliff face may be more erosion-resistant than others. The softer rock around more resistant rock is eroded more quickly, leaving the harder rock as an outcrop along the coast. The sea may then eventually erode the less resistant rock on all sides of this outcrop. The resistant rock is left as a small island or stack, shown in Figure 8.10.

Occasionally, the stack may be left linked to the headland by an arch. This arch will eventually fall as more weathering and erosion occurs.

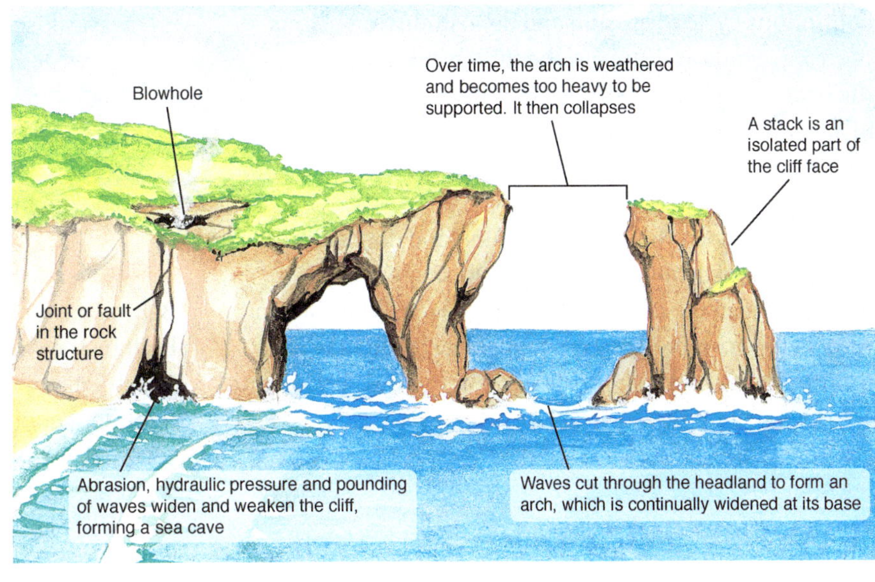

8.10 Formation of sea caves, arches and stacks

ACTIVITIES

8.2

Knowledge and understanding

1 Explain the power of waves and the effect of shallow water.

2 Explain why cliffs erode at different rates.

Applying and analysing

3 Study Figures 8.8 and 8.10. With knowledge gained from the text and your own research, write a report explaining the events and processes responsible for the formation of stacks and arches.

4 Construct a photo sketch of Figure 8.9. Annotate your sketch with the names of coastal landform features mentioned in this unit.

8.3 Depositional landforms

Beaches are familiar landforms seen on many coastlines. They can be made up of any deposited material, including shingle (small, rounded pebbles), but most beaches are made from sand. The sand is distributed along the beach by wave movements. It may be brought to the coast by rivers, or it may be deposited by waves from eroded rock material along the coast.

Beaches

Because beaches are made up of loose material, their location depends on the actions of the waves and rivers. Sand is continuously being deposited on beaches or taken away from beaches by waves. Beaches will form only in a place where there is a good supply of sand, and where they are protected enough to give the sand a chance to build up. On most beaches there is a continuous cycle of erosion and deposition. In seasons of strong winds and storms, the sand is taken from the beach and is often stored offshore as **sand bars** in shallow waters. In seasons of calmer seas and light winds, the sand is gradually deposited on the beach.

Figure 8.11 shows this cycle of erosion and deposition. The cycle affects not only the size and shape of the beach but also the dunes behind the beach. When storm waves are very large, they erode the beach and may also eat into the dunes.

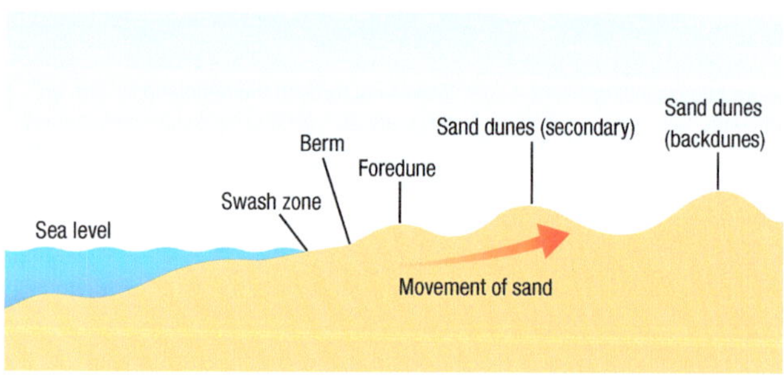

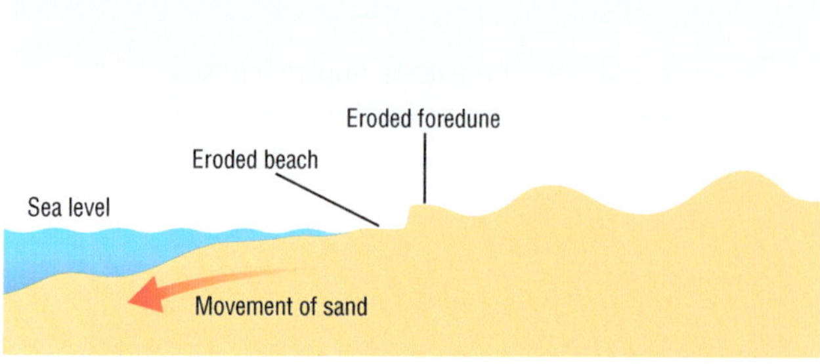

8.11 Sand deposition and erosion

Beach components

All beaches have three main components.

1 The swash zone—this is the most active part of the beach and is where the waves crash onto the beach and run up and back down. In this part of the beach the sand is always wet.
2 The **berm** and **foredune**—a ridge of sand called the berm develops at the top of the swash zone where waves lose their energy and begin the backwash. The waves deposit sand up to this point. The foredune begins after the berm.
3 Coastal sand dunes—dunes are landforms created by the wind. Dried sand is picked up by winds blowing from the sea and deposited over the land.

Dune formation

Behind a beach there are often **sand dunes**. These are built by the wind blowing loose, dry sand from the beach inland, where it accumulates to form a number of dunes called the foredune (or primary dune), secondary dune and backdune. The sand in the dunes is anchored, or stabilised, by plants, such as those shown in Figure 8.12.

Dunes are a very important part of the sand cycle. They prevent storm waves from extending into low-lying areas behind the beach. For a sand dune system to exist there has to be a large supply of sand, long periods of dry weather and frequent onshore winds.

As sand begins to accumulate in a dune, plants start to settle and grow. The first colonisers bind the sand together with their roots. This creates a more stable environment in which more and different plants can grow. Each following group of plants is taller and more complex, as the growing conditions are gradually improved by earlier plants.

8.12 The sand in these Tasmanian dunes is anchored, or stabilised, by plants.

Dunes are found along many beaches in Australia. They have been used for recreation, mined for sand and rare minerals, used as rubbish dumps, built over for housing and, in many areas, completely removed. People are now much more concerned about conserving dunes in their natural state. There is more public awareness and understanding of how easily dunes can be damaged. Planning rules now protect dune systems in many coastal areas.

Sand spits, bars and tombolos

Sand spits form in shallow water where waves dump sand, creating a narrow strip of land, as seen in Figure 8.13. The water behind the spit is usually very calm and it is common for mangroves or salt marshes to develop there. In some places, a sand spit forms across the mouth of a bay, joining two headlands. This is known as a bar. A tombolo is a rare form of spit that forms between the mainland and an island.

8.13 A sand spit, Fingal Bay Tombolo, Port Stephens, New South Wales

ACTIVITIES

Knowledge and understanding

1 Identify the conditions necessary for a dune build-up.

2 Outline how plants anchor the surface of a sand dune.

3 Explain what causes the changes in erosion and deposition on a beach.

4 Explain the difference between spits, bars and tombolos.

Applying and analysing

5 Describe the destructive effects on sand dunes of some human activities carried out on or near dunes, such as mining, building houses, walking, boating and planting trees.

8.4 Managing coastal landscapes

Coastal landforms need to be carefully managed. There are many natural events that can cause damage to coastlines. For example, large storms create powerful waves that can erode depositional landforms such as beaches and sand spits. Coasts have evolved to deal with these natural events and, over time, they repair themselves.

Human activities

Human activities have a great impact on coasts. Thirty-eight per cent of the world's population live within 100 kilometres of the coast and 44 per cent live within 150 kilometres. In addition, many of the world's largest cities are found on the coast. As a result, there is significant competition for the resources of coasts. As the world's population continues to grow, and more and more pressure is placed on coastal environments, the need for careful management becomes ever greater. Figure 8.14 outlines some of the main human activities that threaten coastal environments.

Managing beaches

After a day of heavy human activity at the beach, the tide gradually washes away most of the signs of usage and redistributes the sand. However, humans can permanently damage beaches in more indirect ways.

Longshore drift is a natural process that shifts sand along the beach. If interruptions to this movement are built along the beach, the sand piles up against them, while other parts are deprived of sand. If the supply of sand is reduced (for example by building on the dunes or by damming rivers, which stops the supply of sand from rivers), the beaches may not be replenished sufficiently by longshore drift. This means that sand is stripped away at a faster rate than it can be replaced.

8.14 Human activities that threaten coastal environments

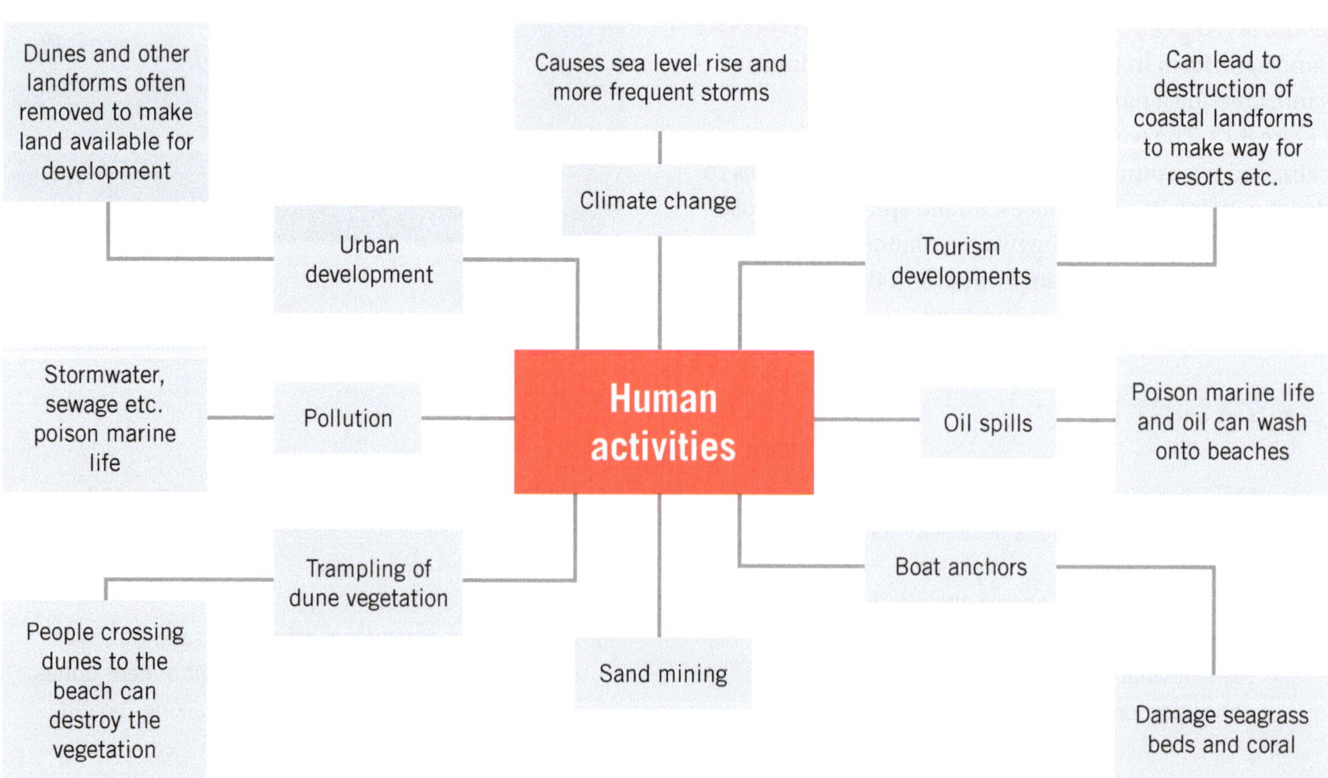

SPOTLIGHT

Climate change and coastal landforms

Climate change is one of the most significant threats to the coastal environment. Global temperatures have been steadily rising for many decades and the overwhelming majority of scientists now believe that this is the result of humans burning fossil fuels, such as oil and coal, and the resulting release of CO_2 gas into the atmosphere.

Climate change will have two main impacts on coastal environments.

1 ***Sea-level rise*** As sea temperatures rise due to global warming, the seawater expands. This is called thermal expansion. The melting ice caps in Antarctica and Greenland release huge quantities of water into the sea (Figure 8.15). High sea levels pose a real risk to low-lying coastal areas, making them susceptible to flooding.

2 ***Increased storm activity*** Warmer temperatures create ideal conditions for large storms, especially cyclones. Climate change is leading to more frequent and more intense storms, and this is creating more erosion and damage along coastlines, as shown in Figure 8.16.

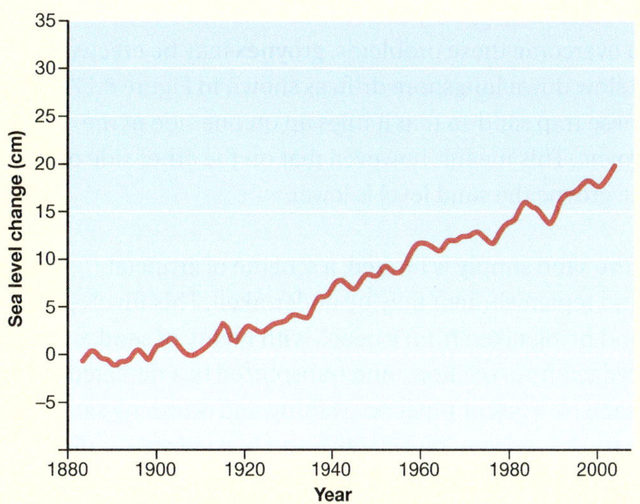

Source: United Nations

8.15 Rising sea levels are a major threat to coastal landforms.

8.16 Coastal erosion, Gold Coast, Queensland

Building groynes

To overcome these problems, **groynes** may be erected to slow down longshore drift, as shown in Figure 8.17. These trap sand so that it piles up on one side of the groyne. This means, however, that on the other side of the groyne the sand level is lower.

If the sand supply is limited, a scheme of artificial sand replenishment may be undertaken. This involves sand being taken from a beach with plenty of sand, or dredged from offshore, and transported to a depleted beach by truck or pipeline. Shifting and dumping sand by truck is reasonably effective and inexpensive in the short term. Building pipelines to bring sand from other beaches or from offshore is a more expensive approach but is often much more effective. Figure 8.18 shows a range of coastal management strategies.

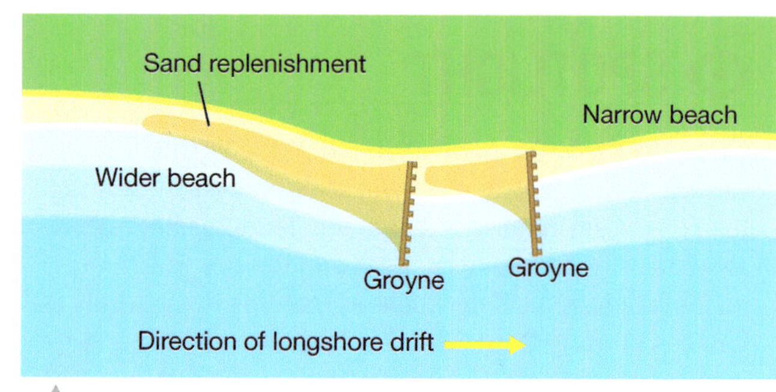

8.17 The natural movement of sand along a coast by longshore drift can be altered by constructing groynes or replenishing beaches with sand.

Managing dunes

Dunes are the most fragile part of the beach. Erosion of sand dunes is a major problem in many areas. It occurs whenever vegetation is removed, allowing the sand to blow away and the dune to shift. If a large section of a dune is blown away, a large, bare depression called a **blowout** forms. Strategies used to manage blowouts include the planting of new stabilising groundcover plants. To prevent the destruction of plant cover, walkways are often built through the dunes to keep people on set paths. The remainder of the dunes is then fenced off.

Managing coastal landuses

To ensure that the best use is made of coasts, careful planning and consistent management are required. In the past, most coastal areas were developed without any overall plan. Only when problems arose and it was recognised that the coastal land was under pressure was a plan developed. Therefore, any plans developed had to accept and take into account many of the existing landuses. Planners could only slowly change landuses due to the rights that owners had over their land.

Management now involves large-scale landuse plans, together with day-to-day administration of development permits to make sure that anything that is built or altered fits into the overall planning aims. Local councils usually administer these processes. Their aim is to make sure that new developments do not interfere with existing landuses and the rights of residents, and that whatever is built is for the future benefit of the area.

1

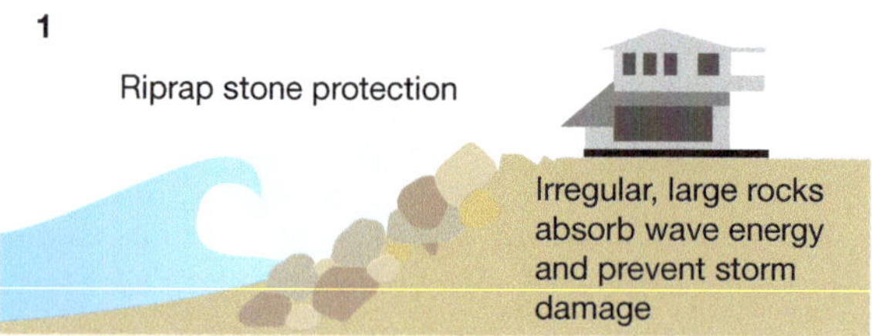

2

Shadecloth fencing traps sand

Revegetation

Build-up of sand

3

4

8.18 Some of the strategies used to manage coastal environments

State governments may have an overall management plan for the whole coast. The aim of this plan is to ensure that the physical and environmental character of the coast is not damaged.

Management options

There are four main options for managing coastal environments.

- ***Use hard engineering*** Building sea walls, groynes, **riprap** and other hard structures is very expensive and these structures are often unsightly. See Figure 8.19 for an example.
- ***Use soft engineering*** Building up beaches with sand-replenishment schemes can be expensive and must be maintained over the long term. It does not have as much negative visual impact as hard engineering.
- ***Discourage building and development*** This is a long-term measure. It needs the cooperation of local, state and national governments. It is often unpopular with local residents or developers.
- ***Do nothing*** This may be possible in sparsely populated areas but is increasingly impossible in highly populated areas.

8.19 Sea walls are examples of 'hard engineering' used to protect coastal development. While such walls absorb the power of the waves, they are expensive to build and maintain, and can cause more erosion because they reflect much of the energy of the waves; that is, as the wave bounces off the wall it carries away with it sand from the beach.

ACTIVITIES

8.3

Knowledge and understanding

1 Outline the damage that natural events can cause to coastal areas.

2 Describe the impact of climate change on coastal environments.

3 Explain why artificial sand replenishment is used on some beaches.

4 Explain why development plans and permits are necessary for coastal areas.

Applying and analysing

5 Study Figure 8.14.

a Describe one human impact on coastal environments and develop some strategies for dealing with the issue.

b List the controls that should be put on coastal development. Do you agree that governments and councils should be able to control all building and development along all coastal areas? Use a SWOT analysis to help you with your discussion.

6 Take on the role of an advertising agent. You have been commissioned by the state government to design a poster campaign to help reduce the damage done to sand dunes by people.

7 Write a report outlining the importance of dune systems and describing the strategies that can be used to manage dune systems.

8.5 In the field: Investigating coasts

The aim of this fieldwork is to investigate a coast. Coastlines are wonderful places to undertake fieldwork and there are many different activities you can do to investigate the geographical processes taking place along coastlines.

Preparation

The following equipment will be required to investigate a coast:

- ruler and measuring tape
- ranging pole (a pole on which measurements are marked)
- stopwatch
- camera
- compass
- three tennis balls
- anemometer, a hand-held weather device or a copy of a Beaufort scale
- clinometer
- hand lens

Select a beach that is easy to access at a variety of locations, as you will need a shoreline long enough to carry out activities.

Investigating waves

Waves can be either destructive or constructive. Destructive waves are powerful and cause erosion of the coastline. They are usually associated with storms. They commonly occur in winter in southern Australia but are more frequent during the summer cyclone season in northern Australia.

Constructive waves are much less powerful than destructive waves. Instead of eroding the coastline they deposit sediment on the beaches and in this way 'construct' the coast. They are usually associated with calmer weather, common during summer in southern Australia and winter in the north.

Constructive waves tend to be surging or spilling waves. Usually, the water is cloudier, as it contains sediment. You can often actually see the sand in the wave as it breaks.

Destructive waves are usually plunging and collapsing waves. The water is usually clear, although as these waves are very powerful there is often a lot of 'white water' associated with them.

Step 1

Choose a location high up, such as a headland, and look carefully at the waves.

Step 2

Take photos of the waves you observe and use Figure 8.20 to help you identify them.

Wind direction and speed

To measure wind direction and speed, complete the following steps. This enables you to identify the relationship between wind direction and speed.

Step 1

With the aid of a compass, record the direction from which both the waves are coming and the wind is blowing. Are they the same? What is the relationship between wave direction and wind direction?

Step 2

Using an anemometer, a hand-held weather device or a copy of a Beaufort scale, estimate the wind speed.

Step 3

Using a stopwatch or an appropriate app on your smart phone, count the number of waves breaking on the shore in one minute. Compare your count with the information in Table 8.21. Describe the potential impact of the waves observed.

A collapsing wave occurs when the whole wave becomes unstable and collapses on top of itself.

A plunging wave forms when the crest of the wave curls over the front face and falls into the base of the wave. This produces large amounts of foam and a high splash. It is often called a dumper.

A surging wave occurs when the crest of the wave remains unbroken while the base of the wave moves up the beach.

A spilling wave occurs when the wave crest becomes unstable and topples down the face of the wave. This wave is often referred to as a roller.

8.20 Types of waves

Investigating longshore drift

To investigate longshore drift and the direction of water and sediment movement along a beach, complete the following steps.

Step 1

Throw three tennis balls into the water in this order—one in the middle of the surf zone, one on the water's edge and one as far out as you can throw it. Before you do so, guess which ball will travel the furthest.

Step 2

Mark the point on the beach where you cast the balls into the water. It is important that the balls be in a straight line perpendicular to the beach.

Step 3

Follow the progress of the balls. Use a tape measure to measure the distance travelled along the beach by the balls. Take the measurement every five minutes. Record your data.

8.21 Potential impact of waves on beaches

Number of waves per minute	Potential impact
6–9	Beach is being built up
0–5	Beach is stable
11–15	Beach is being eroded

Step 4

Record the direction the balls travelled over the course of the activity. The results of this activity will show you the direction in which water (and therefore sediment) is moving along the beach.

Step 5

Using a camera, identify any landform features along the section of coastline that are a result of the process of longshore drift.

Identify any biophysical or constructed features of the landscape that would disrupt the movement of sediment associated with longshore drift.

Constructing a transect

A transect is a straight line that joins two points. Geographers use transects to show how things change. For example, you can construct a transect to see how the shape of the land (topography) changes. This is called a cross-section. You can also construct transects that show vegetation, soils and human activities.

By constructing a transect through a coastal dune you can see how they change as you move further inland away from the beach. You can investigate topography, vegetation and soil, as well as the impact of humans.

Remember, as you collect your data, stay on the paths through the dunes so you don't damage the fragile vegetation.

Step 1

Working in small groups, begin to record the data for your transect. To construct your cross-section you will need a tape measure and a **clinometer** (a instrument that measures the angles of slopes). Figures 8.22 and 8.23 show you how to make and use a simple clinometer.

Step 2

Use the tape measure to calculate the horizontal distance between the dune features, for example from the beach berm to the foredune. Then use the clinometer to measure the steepness of the incline (or decline). Record the distance and the angle for each feature.

Step 3

At each feature also investigate the soil and the vegetation. Using a hand lens and a ruler, measure the size of the grains of sand and note their colour. Usually, the grains become smaller the further away they are from the beach, and the soil becomes darker, as it contains more organic material. Record this data on your table. For each feature write a description of the vegetation and take a photo. Lastly, record any evidence of human impacts.

Step 4

Figure 8.23 shows how to use the clinometer. For your transect, measure the angles of the slope each time there is a change. For example, start at the berm and work backwards to measure the change in angle to the foredune, and so on. You will need to use a common marker, for example a metre ruler or even the top of your partner's head.

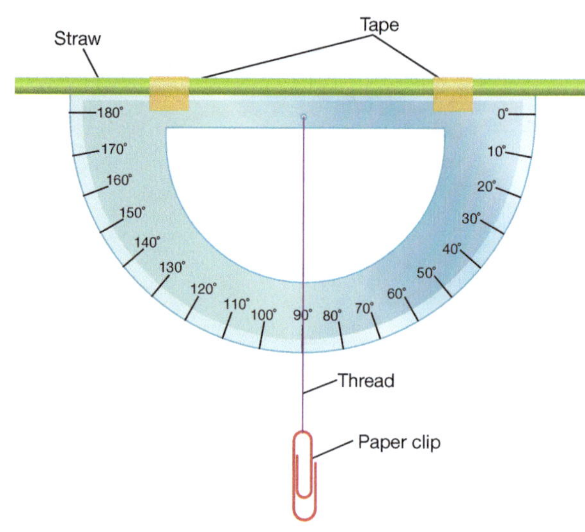

8.22 Making a simple clinometer. Clinometers are special instruments that show the steepness of a slope by measuring its angle. If you don't have access to a clinometer you can make one using a simple protractor, some string and a weight.

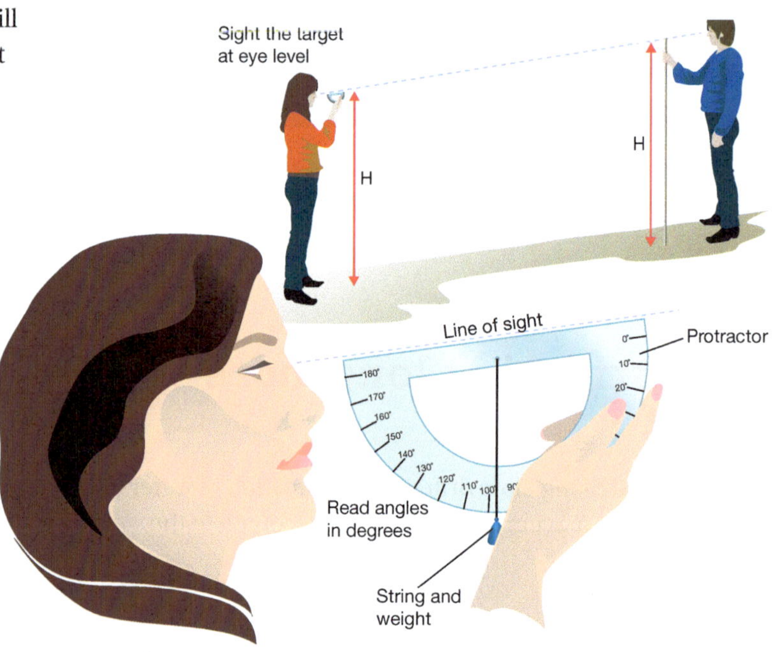

8.23 Using a simple clinometer

Step 5

Back at school you will need to draw up a frame in which to construct your transects. Along the base of the frame make a scale and then include scales for the horizontal distance as well as the changes in angle measured in degrees.

Step 6

Transfer the data from your field trip and join the dots with a line. Add boxes to record soil and vegetation. You will need to develop a key for this. Figure 8.24 shows an example of a series of simple transects.

8.24 Example of transects

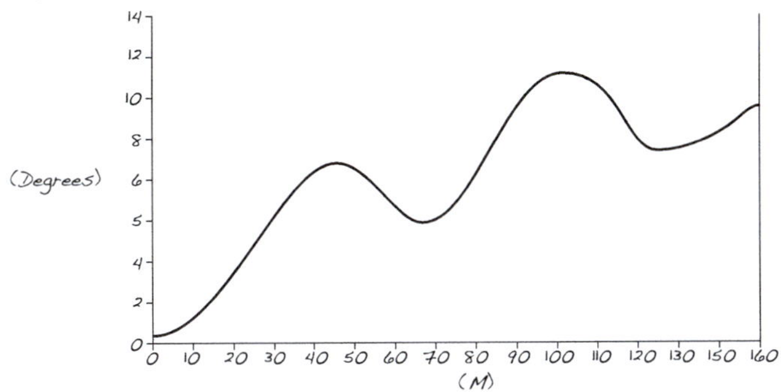

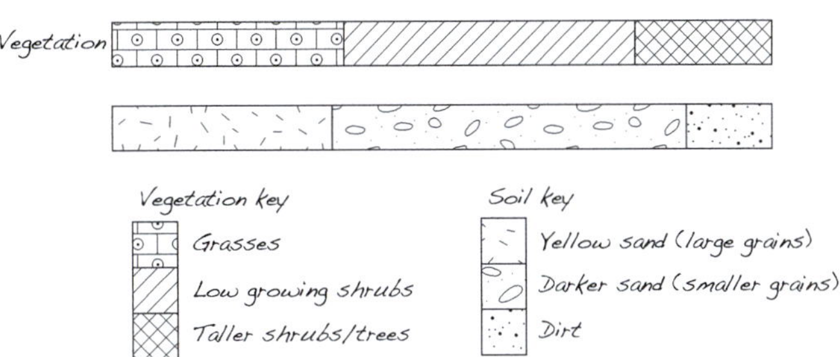

ACTIVITIES

Aim

To investigate a coastline.

Instructions

1 Select a suitable beach to investigate and obtain a map of the area investigated.

2 Investigate waves.
   a Observe and collect images.
   b Paste and annotate your images.

3 Record the wind speed on the beach.

4 Measure wave direction and speed.

5 Investigate longshore drift and record your data in a table like the one below.

Longshore drift record sheet					
Time (minutes):	10	20	30	40	60
Ball 1: Furthest out					
Ball 2: Surf zone					
Ball 3: Beach zone					

6 Construct a transect of the beach area being investigated. Use the table below as a guide.

Data record sheet						
Dune feature	Distance from the berm (m)	Change in angle	Soils		Vegetation description	Evidence of human impact?
			Grain size (mm)	Soil colour		

7 Provide a field sketch of the beach area being studied.

Evaluation

8 Comment on your findings, including the following information in your response.
   a Identify the relationship between wind speed and direction.
   b Describe the impact of the wave speed and direction on the beach. For example, is the beach being built up, is it stable or is it being eroded? Explain.
   c Describe the impact of the direction of water movement on the distribution of sediment (i.e. longshore drift) in this section of the coastline.
   d Identify any landform features along the section of coastline that are a result of the process of longshore drift.
   e List any natural or constructed features of the landscape that would disrupt the movement of sediment associated with longshore drift.

Conclusion

9 Describe what you have learnt about coasts.

8.6 Case study: Coorong

The Coorong is found along South Australia's southern coast, east of Adelaide. It is a complex coastal system that includes a very large shallow lake system that stretches along the coast for more than 140 kilometres.

8.25 The Coorong and the mouth of the Murray River

Coorong features

The Coorong's vast lagoon is saline, meaning that it contains high levels of salt water. It is separated from the Southern Ocean by a narrow barrier system called the Younghusband Peninsula. This barrier is made up of coastal sand dunes, which act like a wall, keeping the sea from flooding into the lagoon (see Figure 8.25).

A key element of the Coorong is the waters of the Murray River, Australia's largest and most important river system. The Murray–Darling Basin collects water that eventually flows into the sea in South Australia. This basin, which covers approximately one-third of the Australian mainland, stretches from Central Queensland down into New South Wales, Victoria and South Australia. As can be seen in Figure 8.26, the Murray flows into Lake Alexandrina just above the Coorong. There the fresh waters of the river mix with the salt water of the Southern Ocean to create the unique environment that is the Coorong.

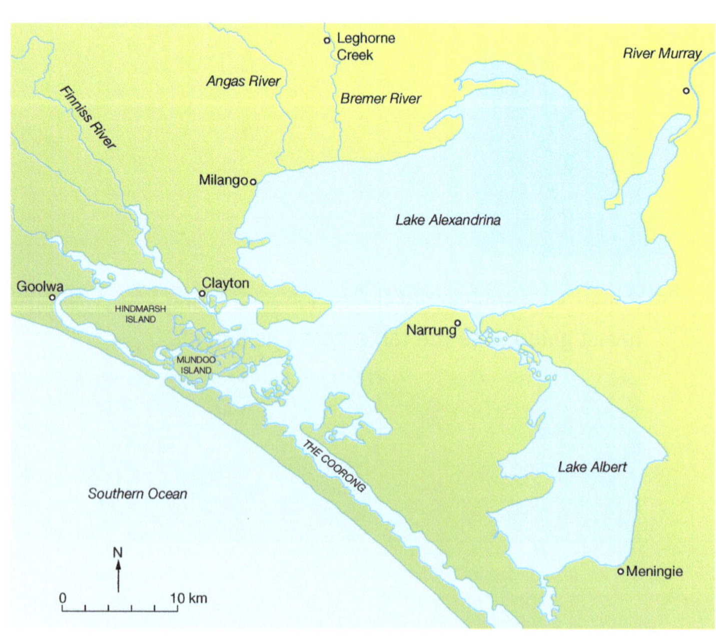

8.26 The Coorong is a long, thin lagoon separated from the Southern Ocean by a barrier system.

Younghusband Peninsula

Younghusband Peninsula is the sandy barrier system that separates the Coorong from the sea. The barrier was formed when rising sea levels, following the end of the last Ice Age, transported sediment landwards from the previously exposed Lacepede Shelf (see Figure 8.27). As the sea levels rose, they isolated a narrow back-barrier lagoon (the Coorong) from the direct impact of the Southern Ocean. The peninsula extends 180 kilometres from the mouth of the Murray towards the south-eastern town of Kingston. Many of the dunes that dominate the barrier system are not heavily vegetated. This makes for highly active dunes, as there is not much vegetation to secure the sand.

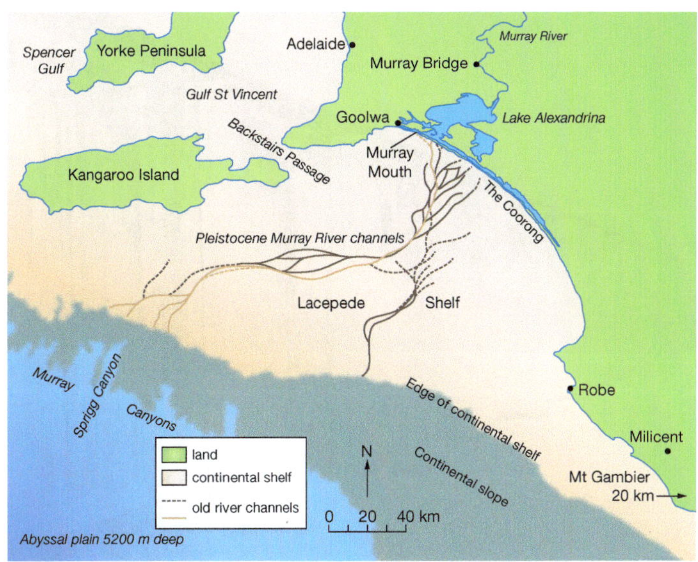

8.27 The sandy barrier system that separates the Coorong from the Southern Ocean

Environments of the Coorong

There are twenty-three different types of wetlands found along the length of Coorong. The wetlands, lagoon and lakes cover an area of more than 140 500 hectares and provide habitat for many threatened species.

These wetlands are so important that they are now protected by the Ramsar Convention. The Ramsar Convention is an international agreement to protect the habitats of migratory wading birds. These are birds that travel vast distances between breeding and feeding grounds. Many migratory birds travel from as far away as northern Siberia, in Russia, to Australia. Because they travel through so many different countries on their migration, an international agreement is needed to help protect their habitats.

One of the important migratory bird species that is found in the Coorong is the red-necked stint shown in Figure 8.28. These tiny birds weigh less than 50 grams and have a wing span of just 30 centimetres. Despite their small size they travel all the way to the Coorong in August and September from their summer breeding grounds in Siberia and Alaska, a journey of about 12 500 kilometres. These tiny birds can fly more than 3000 kilometres non-stop.

Human impacts

The impact of human activity is very evident in the Coorong. Some scientists now believe that the Coorong is being changed so much by human activity that the ecosystem is being transformed. The Coorong has long been a paradise for water birds. It is home not only to migratory birds but also to local birds, such as stilts, terns, pelicans and egrets, which seek refuge there during droughts.

SPOTLIGHT

The Ngarrindjeri people

The Coorong and the lower Murray are the traditional lands of the Ngarrindjeri people. The Yarluwar-Ruwe is the Sea Country of the Ngarrindjeri people and it is of great spiritual significance to them. The fresh water that flows into the Coorong from the Murray is seen as the life blood of their home, and the health of the Ngarrindjeri people is linked to this water.

For the Ngarrindjeri people, the impact of human activity on the Coorong is very distressing. They are working closely with local authorities to help return the Coorong to health.

8.28 The red-necked stint

8.29 Barrages stop the flow of salt water into the Coorong, gradually changing it.

Building barrages, dams & weirs

In its natural state the Coorong would receive fresh water from the Murray and salt water from the ocean. However, in the 1940s, a series of barrages were built across the Murray to stop the flow of sea water into the lake system (see Figure 8.29). This was so the water in the lake could be used for agriculture and other purposes. In addition to the barrages, a series of weirs and dams along the Murray and the other rivers that flow into the Murray have significantly reduced the flow of water down the river. These weirs and dams have also affected fish species, as they block their migratory paths. The result has been far less water in the lakes of the Coorong and also fewer fish. This has had a devastating effect on bird life in the Coorong.

Reduced river flow

In the last 25 years the situation has become especially bad for the Coorong. The Murray flows through some of the most intensely farmed areas in Australia. The farms use vast quantities of water. As water is diverted from the Murray for irrigation, the amount left to flow down the river and into the lakes of Coorong shrinks. In the droughts of the first decade of the twenty-first century, almost no water flowed into the lakes; water levels dropped and the lakes began to dry.

Increased salinity

Emergency dredging of the mouth of the river was necessary to allow salt water to flow in and keep water in the Coorong lakes. However, the lakes have evolved to have brackish water—a combination of salt and fresh water. The changing balance between salt and fresh water is affecting the ecosystem of the Coorong. In some parts, the water is actually saltier than the sea water. This has meant that whole species of plants and animals have died out, and been replaced by new species that live in saltier water.

8.30 Revegetation trials in the lower lakes

Changing ecosystem

The entire ecosystem is beginning to change. For example, brine shrimp have reached plague proportions in some parts of the Coorong. These tiny salt-loving animals were once eaten by the Murray hardyhead fish, but the hardyhead have less tolerance to salt and have all disappeared. The Murray hardyhead fish were one of the main sources of food for the pelicans and their disappearance is disrupting pelican breeding patterns. Other species that don't eat brine shrimp have also suffered.

Along with fresh water, the Murray brings large amounts of sediments, which are used to build depositional landforms within the Coorong. With water flows reduced, sediment levels have fallen. Some shorelines along the Coorong lakes are being eroded at a rate of about 1 metre per year. This is changing the shape and nature of the lakes, making them increasingly shallow and muddy.

Future of the Coorong

Despite the many problems facing the Coorong, strategies to restore and protect this important coastal environment have been implemented. These include:

- increasing the flow of fresh water
- buying water from irrigators to return an environmental flow to the river system
- using the extra water to flush out of the Coorong lakes, to reduce overall salinity levels
- establishing breeding programs for endangered species such as the Murray hardyhead
- planting vegetation to stabilise the soils in the Coorong, as shown in Figure 8.30.

SPOTLIGHT

The collapse of the Murray hardyhead

The decline of the Murray hardyhead (Figure 8.31) is an example of the problems facing the Coorong. This tiny fish, once abundant in the Coorong, is now critically endangered. It grows to just 5 centimetres in length but is a very important part of the Coorong ecosystem. The hardyhead is a special type of freshwater fish because it has evolved to live in waters with moderate levels of salt, such as those in Coorong.

At the height of the great drought in 2006, the number of hardyhead had fallen to such low levels that a captive breeding program was begun. This was needed because the salinity (salt) levels in the Coorong had become too high for the hardyhead. In addition, introduced species, such as the aggressive gambusia fish, have taken over Murray hardyhead territory.

8.31 The Murray hardyhead is now listed as critically endangered.

ACTIVITIES

Knowledge and understanding

1. Describe the importance of the River Murray to the Coorong.
2. Explain why the Coorong is protected by the Ramsar Convention.
3. Draw a flow chart showing the impacts of drought and water extraction.
4. Describe what is being done to help the Coorong overcome the problems caused by human activity.

Applying and analysing

5. Write a short report explaining the location and the nature of the Coorong environment.
6. Prepare a mind map to summarise the impacts of human activity on the Coorong.
7. You are an environmental scientist working for the South Australian Government. Prepare a presentation for the United Nations Educational, Scientific and Cultural Organization, to seek World Heritage status for the Coorong. Create a multimedia presentation explaining the importance and uniqueness of the Coorong.

8.7 Case study: Port Campbell

Victoria's Port Campbell National Park is home to one of the world's most spectacular coastlines. It features an amazing collection of soaring cliffs, sea stacks, arches, gorges and blowholes.

Origins

The coastline of Port Campbell National Park (Figure 8.32) originated 10 to 20 million years ago, when millions of tiny marine skeletons beneath the sea gradually broke down. The skeleton pieces became the foundation for the sedimentary rock limestone. When the sea retreated it left the soft limestone exposed above sea level. Over time, the raging seas of the Southern Ocean and the strong winds that batter the coast carved the remarkable landforms for which the coast is internationally famous, shown in Figure 8.33.

8.32 Victoria's Port Campbell National Park

8.33 The spectacular erosional coastline of the Port Campbell National Park

Twelve Apostles

The Twelve Apostles (actually a collection of nine sea stacks) were formed by the processes of weathering and erosion (Figure 8.34). Over millions of years the water and salt-laden air of the Southern Ocean gradually weathered and eroded the soft limestone of the area to carve caves in the cliff face. These, in turn, became arches. Eventually, the arches collapsed, leaving sea stacks, some of which are up to 50 metres high. This process is ongoing. In 1990, the arch known as London Bridge collapsed, isolating a previously connected section of headland and forming another sea stack. Eventually, many of the headlands seen in Figure 8.34 will become new limestone stacks.

SPOTLIGHT

London Bridge collapse, 1990

London Bridge was once among the most popular tourist attractions in Port Campbell National Park. The landform was, until 15 January 1990, a complete double-span arch formed by the process of erosion. The arch closest to the mainland collapsed without warning when the supporting rock strata could no longer support the weight of the arch. Two stranded tourists had to be rescued by helicopter.

8.35 London Bridge before and after its collapse in 1990

8.34 Port Campbell's world-famous Twelve Apostles

Limestone-based landforms

Limestone is typically formed from the remains (especially the shells) of the ancient sea life that lived in shallow marine environments. Limestone dissolves easily in water and weak acid solutions, making it vulnerable to weathering and erosion. This process, however, may take thousands to millions of years. Significantly, the top layers of the Port Campbell limestone are harder than the bottom layers. The erosive forces of wave, wind and rain eroded the softer rock first. This process created overhangs, arches and eventually new stacks. The surviving rock stacks, such as the Twelve Apostles, are generally composed of harder rock than the areas surrounding them.

Landforms that develop in areas where rock (usually limestone) is easily dissolved by water are known as karst landforms.

ACTIVITIES

Knowledge and understanding

1 Outline the origin and formation of the coastal landforms of Port Campbell National Park.
2 Describe how limestone is formed.

Geographical skills

3 Study Figure 8.32.
   a Which direction is Port Campbell from Melbourne?
   b What is the distance between Port Campbell and Melbourne?
   c Which town is the closest to Port Campbell?
4 Construct an annotated photo sketch of Figure 8.34.
5 Study Figure 8.35. Draw a sequence of annotated block diagrams to illustrate the formation of a sea stack.

CHAPTER 9

RIVERINE LANDFORMS

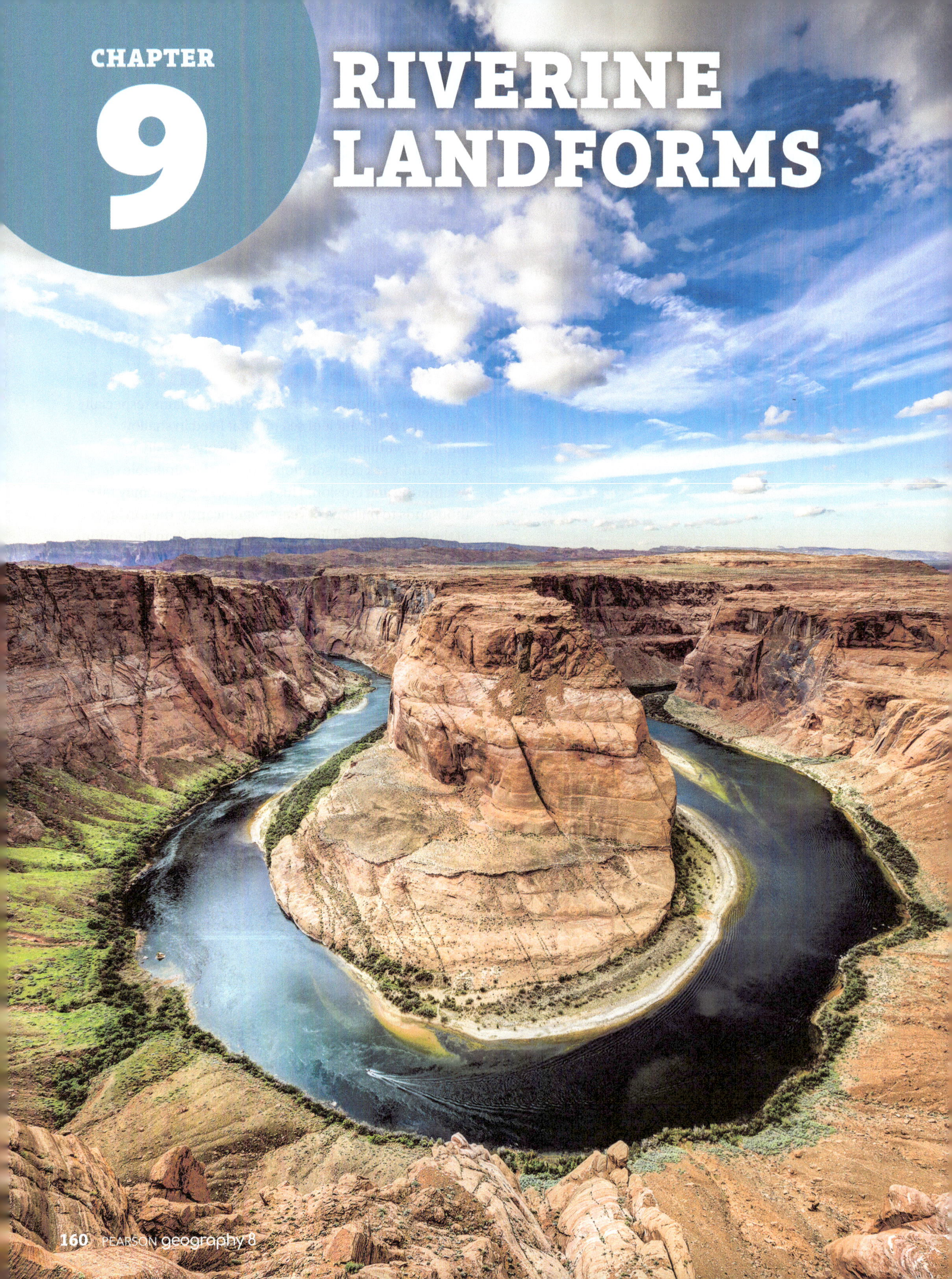

Water has travelled across landscapes for millions of years. As it has done so, it has sculpted the surface of the land. It has carved deep valleys and deposited sediments to form the rich alluvial plains that support our most productive agricultural systems.

The landforms produced by rivers are among the most spectacular on earth. The Colorado River, over a period of at least 17 million years, carved a deep path through layer after layer of sedimentary rock to produce the Grand Canyon, one of the natural wonders of the world. The Mekong River in South-East Asia flows 4350 kilometres from its headwaters on the Tibetan Plateau, through China, Burma, Laos, Thailand, Cambodia and Vietnam, where it enters the sea. The delta of the Mekong is so vast that it makes up a large portion of south-western Vietnam. The great rivers of Europe, the Danube and Rhine, are now popular holiday destinations.

In this chapter we examine the processes shaping riverine landscapes and landforms. We also look at how the activities of people impact on riverine landscapes. In particular, we look at the catchment of the Bow River in Canada to examine the landforms shaped by rivers and the ways in which people affect such landscapes and landforms.

KEY IDEAS

- To understand the processes responsible for the formation of riverine landforms
- To describe how riverine landforms change over the course of a river
- To investigate how the activities of people affect riverine landscapes and landforms

9.0 Horseshoe Bend, Colorado River, United States of America

GLOSSARY

abrasion	the wearing down or wearing away of rock by friction
attrition	the wearing away of material due to the friction caused by particles rubbing against each other
bed load	material transported along the bed of a river
braided channel	a river channel featuring a network of small channels separated by small and often temporary islands of sediment
catchment	the area drained by a river and its tributaries; an alternative term for 'river basin' or 'drainage basin'
delta	extensive deposit of alluvial material (sediment) at the mouth of a river
deposition	accumulation of sediment by the action of erosional agents, such as water and wind
discharge	the amount of water that flows from a river catchment and into another river system, the sea or a lake
environmental flow	the amount of water that is needed to maintain a healthy river ecosystem
flood plain	a nearly flat plain along the course of a river that is subject to flooding
gradient	a measure of the steepness of a slope
groundwater	water beneath the earth's surface that fills pores or tiny spaces in the earth or rock
infiltration	the movement of water from the land surface into the soil
meander	a bend or curve in the course of a river
natural levee	the build-up of sand along, and sloping away from, either side of a river
point bar	the accumulation of sediment on the inside of a river bend
rapid	a section of a river with a relatively steep gradient that causes the velocity and turbulence of water to increase
sediment	rock-based material that has been broken down by weathering and erosion and then transported by the action of wind, water or ice, and the force of gravity
suspended load	fine particles of silt and clay carried in river water
tributary	a river system flowing into a larger river
turbidity	muddiness; high turbidity occurs when there are high levels of suspended sediment in water
watershed	the boundary between catchments

9.1 Rivers

Rivers are responsible for shaping most of the landforms covering the earth's surface. They are powerful agents of erosion and they transport vast amounts of sediment that is eventually deposited to form a range of depositional landform features.

9.1 The Uvac canyon meander, Serbia

River erosion

Rivers erode the land over which they flow through the processes of:

- **abrasion**—sediment in the river grinds against the beds and banks of the river channel. This action dislodges material and carries it away
- **attrition**—dislodged materials collide with the river sides and bed, and each another. Over time, they become smaller and are eventually reduced to fine particles called silt
- *corrosion*—the solvent action of water dissolves soluble materials and carries them away in solution
- *hydraulic action*—the rocks are broken down, usually along lines of weakness in the rock lining the river channel.

Transporting sediment

The eroded sediment in a river is transported as either **bed load** (the larger fragments that move along the riverbed) or **suspended load** (the finer fragments carried in the water). This process is illustrated in Figure 9.2. Transported sediment is then deposited.

Transforming the land

As a river flows from its source towards where it empties into the sea or lake, it changes the land.

Figure 9.3 illustrates how the riverine landscapes changes from a narrow V-shaped valley in the upper reaches of the river to a broad, meander-dominated river valley in its lower reaches. In the upper reaches (A in Figure 9.3) the river is narrower and usually has a rapid, tumbling flow that cuts a narrow channel through rocky hills or mountains. Over time, a small flood plain develops and the first signs of river meandering appear (B). In the middle reaches of a river the flood plain develops further and a series of meanders form in the sediment deposits that have accumulated on the valley floor (C). In the lower reaches of a river the flood plain is broad and the meandering river channel is well developed (D). The presence of oxbow lakes suggests that the course of the river has changed over time.

9.2 The process of sediment transport in a river

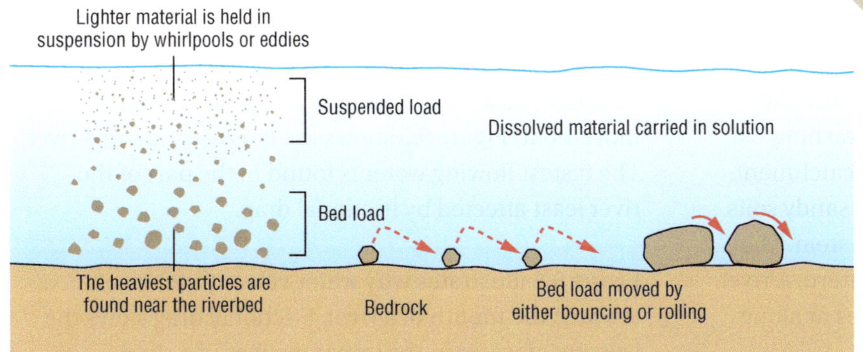

9.3 Riverine landscapes change over the course of the river.

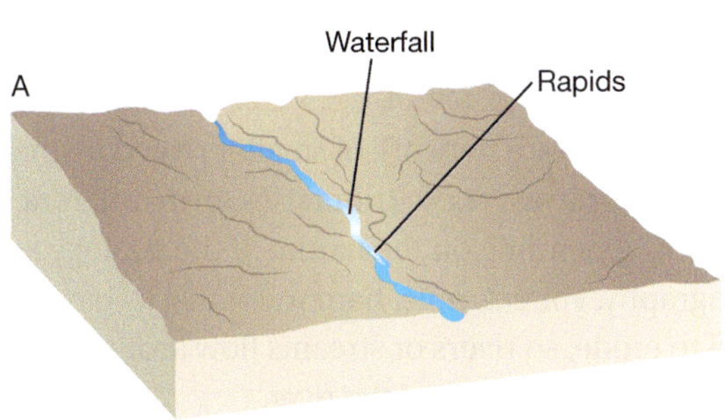

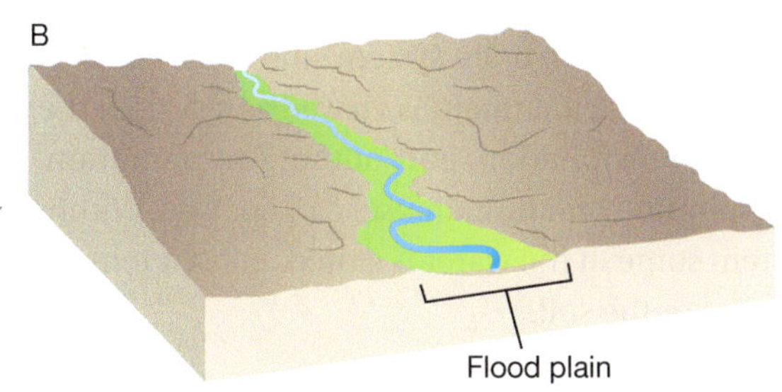

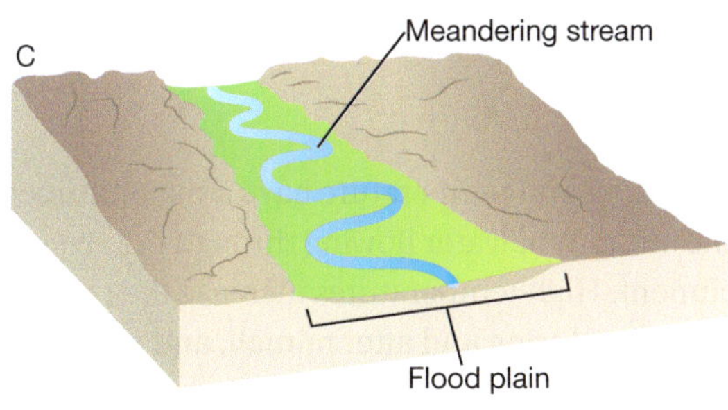

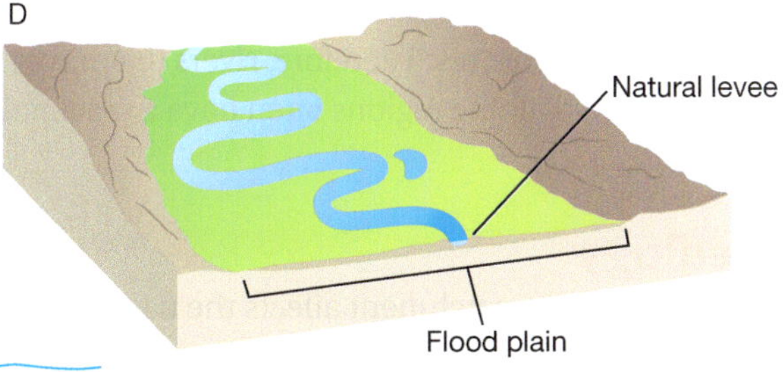

Catchments

When rain or snow falls, some water soaks into the ground, while the rest runs over the surface of the land. The high point in the landscape that determines which river system the water flows into is called a **watershed**, or divide. The watershed forms the boundary between river **catchments**, or drainage basins, as seen in Figure 9.4.

9.4 A watershed, or divide

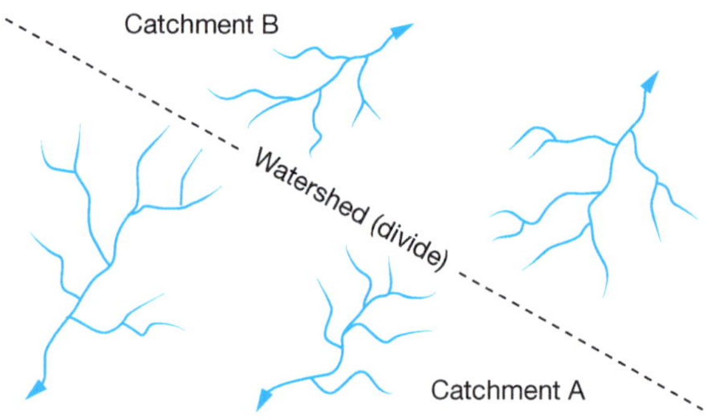

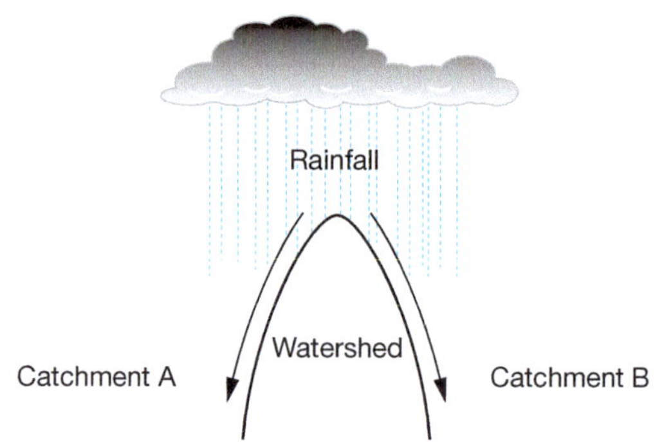

Drainage patterns

Rivers are fed by a network of smaller rivers or streams called **tributaries**. These tributaries form different patterns depending on the nature of the geology and topography. For example, hard rock such as basalt is hard to erode, so rivers or streams flow around this rock, influencing the course of the river.

Rivers as systems

A catchment is an open system, and is part of the water cycle. When a catchment is viewed as a system the terms 'inputs' (precipitation) and 'outputs' (evaporation, transpiration and human use) can be used. Within the system some of the water is stored, at least temporarily, in lakes or the soil.

Impacts on rivers

There are many factors that determine the nature of a river and its associated landforms.

Climate

Precipitation and temperature have a major influence on the amount of water flowing through the river catchment. High temperatures increase the rate of evaporation during and after rainfall, and so reduce water run-off. Under cold conditions, precipitation in the form of snow often accumulates at higher altitudes and does not provide much run-off until it melts in warmer temperatures. Therefore, rivers that have their source in cold alpine regions often have a seasonal flow of water.

Geology

The geology of a catchment affects the nature of the river in at least two important ways. The rate of **infiltration** (the movement of water soaking into the soil) and the type and amount of sediment washing into rivers are affected by the geology of the catchment. Sandstone landscapes produce very porous, sandy soils. Erosion of sandstone creates a sandy river system. The geology also influences the river channel pattern. A river will flow through an area of weaker rock types or along fault lines.

River basin topography

The size, shape and gradient (slope) of the catchment basin has a major impact on the **discharge** (the amount of water that flows from the river catchment) rate of the river. Large catchments tend to discharge more water, over a longer period than smaller ones. Also, a steeper gradient causes water to drain more quickly.

Soils and vegetation

A catchment's soils and vegetation affect the amount and rate of water run-off. If soils are sandy, much of the water soaks into the soils. Thick groundcover vegetation also increases water absorption. Clay soils are heavy and water is not easily absorbed. When clay soils with little vegetation cover have been saturated from previous rainfall, there is much greater water run-off and sometimes flooding.

River channels

As water passes over land, friction slows down its movement. Figure 9.5 shows the cross-section of a river. The fastest flowing water is found in the part of the river least affected by frictional drag.

Figure 9.6 illustrates why water velocity increases towards the mouth of a river. Frictional drag slows the velocity of water in the upper reaches of a river.

Velocity, capacity and discharge

Velocity

The velocity, or speed, of water within river channels is influenced by:

- the shape of the channel—the rate of flow is faster in channels that are as deep as they are wide thanit is in channels that are wide but very shallow, or very deep and narrow
- the roughness of the channel bed and banks—water flowing through a channel full of large rocks is more turbulent, but slower, than water flowing through a channel lined with fine silt
- the gradient or slope of the river—as a river approaches the sea its gradient decreases, but because more water is added to the channel, the velocity of the river actually increases.

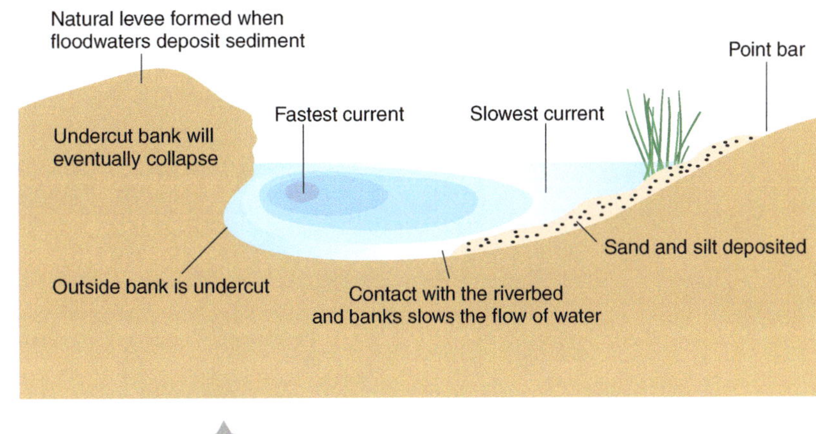

9.5 Water velocity in the cross-section of a river

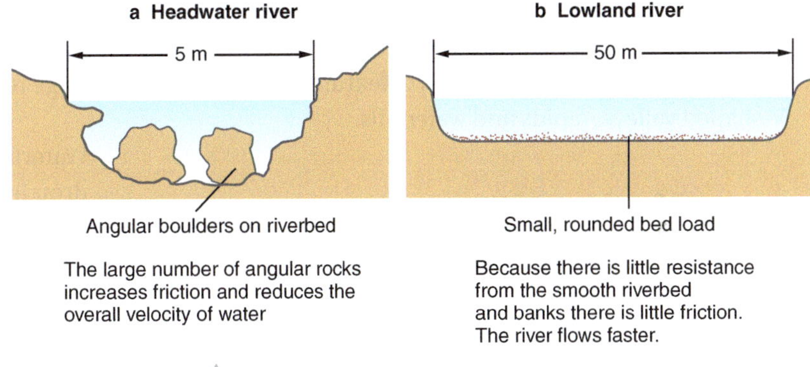

9.6 Why the water velocity of a river increases towards its mouth

Capacity

Stream capacity is the ability of a river to transport its load of sediment. It is expressed as the amount of sediment that can be moved past a particular point over a specified period of time. Stream capacity increases as sediment size within the channel decreases, and as stream discharge increases.

Discharge

The amount of water that flows from the catchment into another river system, the sea or a lake is called discharge. Discharge is expressed in cubic metres per second.

ACTIVITIES

Knowledge and understanding

1 Distinguish between abrasion and attrition.
2 Explain what corrosion and hydraulic action are.
3 Distinguish between bed load and suspended load.
4 State what a catchment is and outline the role of the watershed.
5 Outline the ways in which a river can be thought of as a 'system'.
6 Outline the factors that influence the velocity of water in a river channel.
7 State what is meant by the term 'stream capacity'.
8 Define the term 'discharge'.

Applying and analysing

9 Study Figure 9.3. Outline how and why the nature of riverine landscapes and landforms changes along the course of the river, beginning at its headwaters or source.
10 Construct a mind map illustrating the factors that determine the nature of rivers.
11 Study Figure 9.5. Explain why water velocity is fastest at the point highlighted in the illustration.
12 Study Figure 9.6. Explain why average river velocity is greater in the lowland stretches of a river.

9.2 Riverine landforms

There is a wide variety of riverine, sometimes referred to as 'fluvial', landform features. This diversity results from the ways the river interacts with the geology and topography of the land. The nature of the vegetation and the activities of people also play a role.

Upstream landforms

In the upper reaches near the river's source, the channel is often narrow and deep, and the **gradient** steeper than further downstream. This results in the development of a number of distinctive landform features, including V-shaped valleys, rapids and waterfalls.

V-shaped valleys

The turbulent fast-flowing waters of the upper reaches of a river generate a great deal of corrasion as rocks crash against each other and the riverbed. In this part of the river, erosion cuts downwards more than it does sideways. The development of a V-shaped cross-sectional profile is the result (see Figure 9.7).

9.7 V-shaped valleys develop in sections of the river where erosion cuts downwards more than it does sideways.

Rapids and waterfalls

A **rapid** is a section of a river with a relatively steep gradient that causes the velocity and turbulence of water to increase. In a rapid, the river becomes shallower and large rocks are exposed above the surface.

Waterfalls occur where water plunges over a vertical drop in the course of a river. They are most commonly found where the river channel is narrow and deep (that is, V-shaped). Figure 9.8 illustrates the formation of a waterfall. When a river comes into contact with a harder layer of rock, the rate of erosion slows. As the water plunges over the edge of the harder rock it creates turbulence, which erodes any underlying softer rock.

Tectonic influences

Tectonic forces can also impact on riverine landform features. Figure 9.9 shows what can happen when a section of land is uplifted slowly and rapidly. If the rate of uplift is slower than the downward rate of erosion,

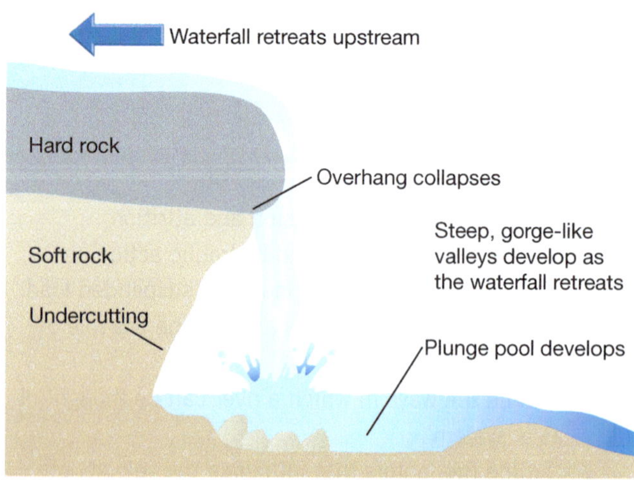

9.8 Waterfalls develop when a harder layer of rock slows the downward rate of erosion. Gradually, the waterfall retreats upstream.

a canyon will form. If the rate of uplift is faster than the downward rate of erosion, the path of a river can be blocked or a lake formed.

Downstream landforms

In the lower reaches of a river that is characterised by depositional landforms, the water speed slows.

Flood plains

A **flood plain** is the flat or nearly flat land on either side of a river. It stretches from the banks of the river channel over the area inundated (flooded) when the river is at its peak. The landform features of the flood plain change over time. During floods, the waters of a river are capable of eroding, transporting and depositing vast amounts of **sediment**. The landform features of the flood plain are shown in Figure 9.10.

Natural levees

As sediment-full floodwaters spread out across the flood plain they quickly lose speed. Much of the suspended silt and sand settles on the ground. The greatest amount of **deposition** occurs close to the channel. This causes a **natural levee** of elevated ground to develop on either side of the main channel (see Figure 9.10).

9.9 The impact of geological processes on river systems

a

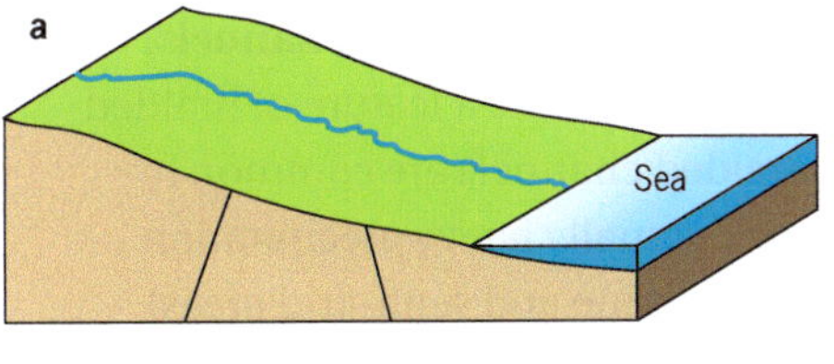

b

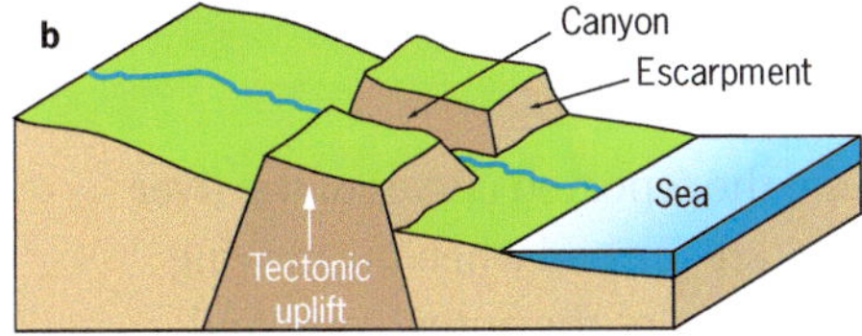

Slow tectonic uplifts may allow the river to maintain its existing path but result in the formation of canyons or valleys.

c

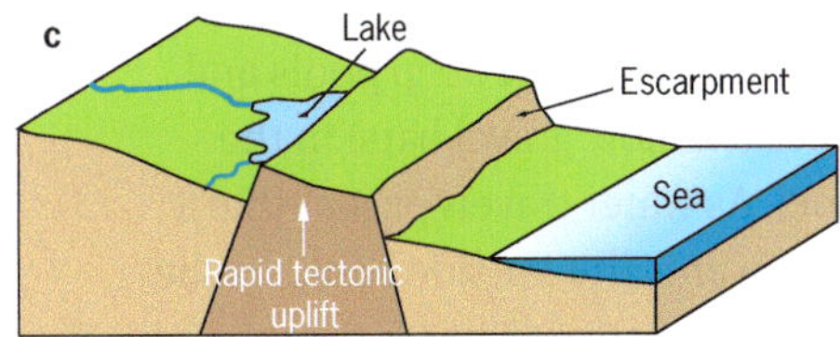

Rapid tectonic uplifts can result in changes to drainage patterns and the formation of lakes.

9.10 Typical riverine landform features along a river flood plain

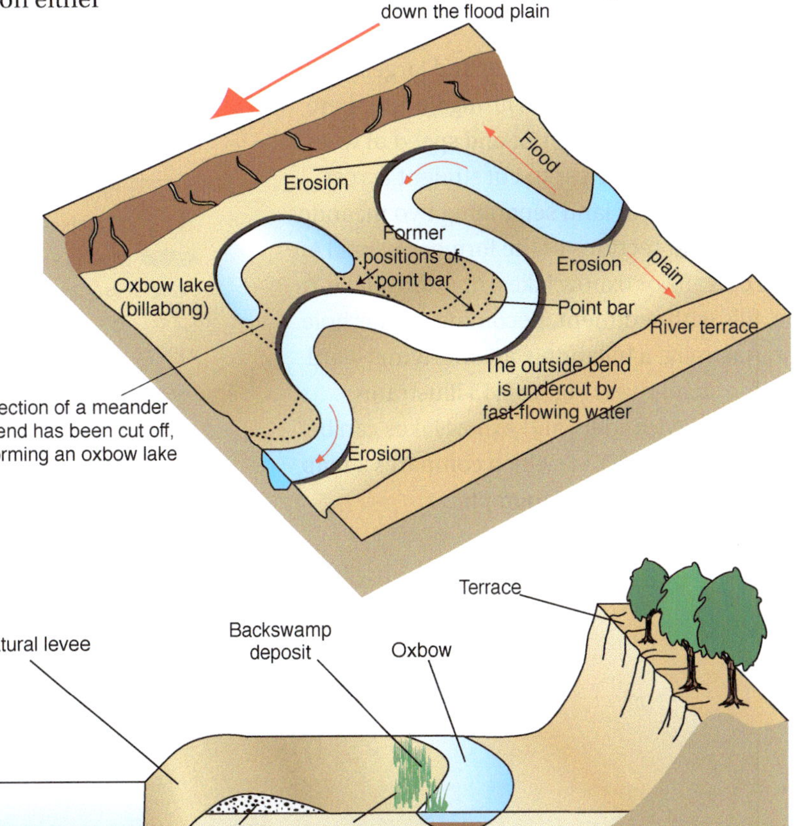

Braiding

When a river becomes choked with sediments it divides into a series of **braided channels**. The sediment 'islands' are a feature of braided channels. Braided channels are common in regions with seasonally heavy precipitation (for example in alpine and semi-arid areas) and heavy loads of sediment (see Figure 9.11).

9.11 Braided river channel, Murray River, Australia

Meanders

Meanders are the repeated curves of the river channel. They are formed when the moving water in a river erodes the outer banks and widens its valley. Meanders are more common in areas where rivers flow through flat land. Meanders are formed from deep pools and shallow riffles (shallow gravel bars) in the river channel. As water travels past a riffle it is deflected towards the outside bank. As the river undercuts the outside bank, the meander migrates outwards. **Point bar** deposits build up on the inside of the loop. This process is illustrated in Figure 9.12.

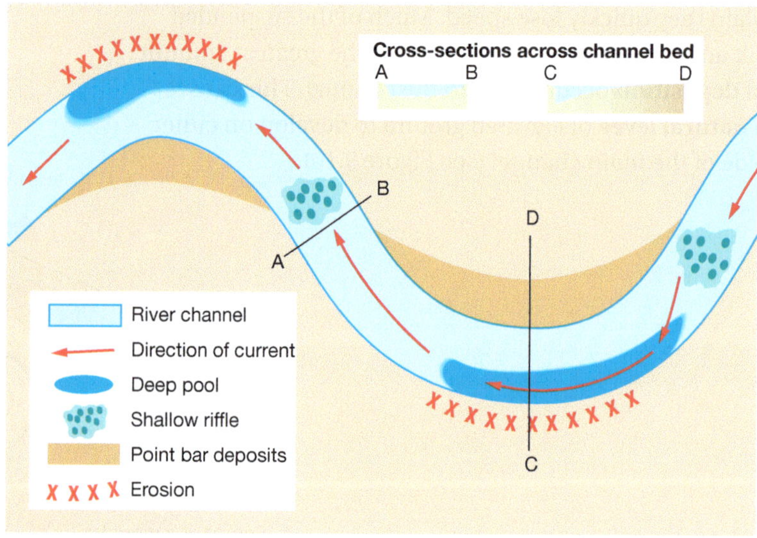

9.12 Features of a meandering river

Oxbow lakes

The meander loops grow larger and larger until the channels almost meet. During floods, there is an increase in the amount and speed of water. This may cause the river to cut straight across the narrow neck of land separating two meander loops. The old sections of the former channel are called cut-offs or oxbow lakes. An oxbow lake, in time, fills with sediment, becoming a waterhole or, in Australia, a billabong, an Aboriginal word meaning 'dead river'. Figure 9.13 illustrates this process. Meanders and oxbow lakes can be observed in Figure 9.14, which compares a map extract and an aerial photograph.

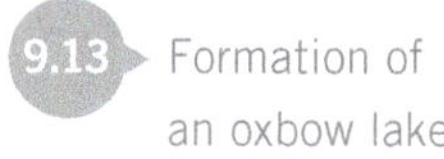

9.13 Formation of an oxbow lake

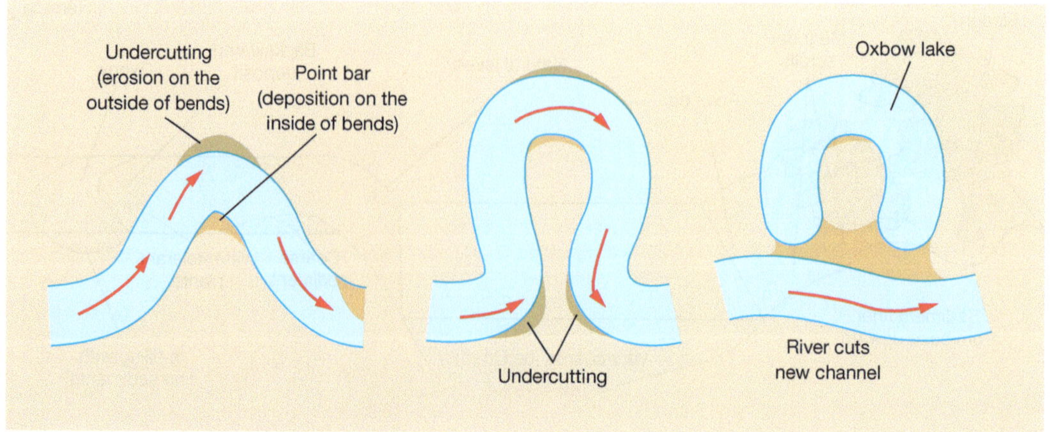

Deltas

Deltas are the landform features that develop at the mouth of a river. As the waters of a river reach the sea (or lake), there is a drop in its speed and the river's ability to support its suspended load declines. As a result, much the sediment load is deposited near or at the mouth of the river. Over time, a delta develops. Depending on factors such as wave action, currents and tides, different types of deltas develop.

Other rivers, particularly those located on coasts with a significant tidal range, do not form deltas. Rather, they enter the sea through an estuary—a partly enclosed coastal body of water.

9.14 Topographic map extract (left) and aerial photograph (right) showing the landform features associated with meandering.

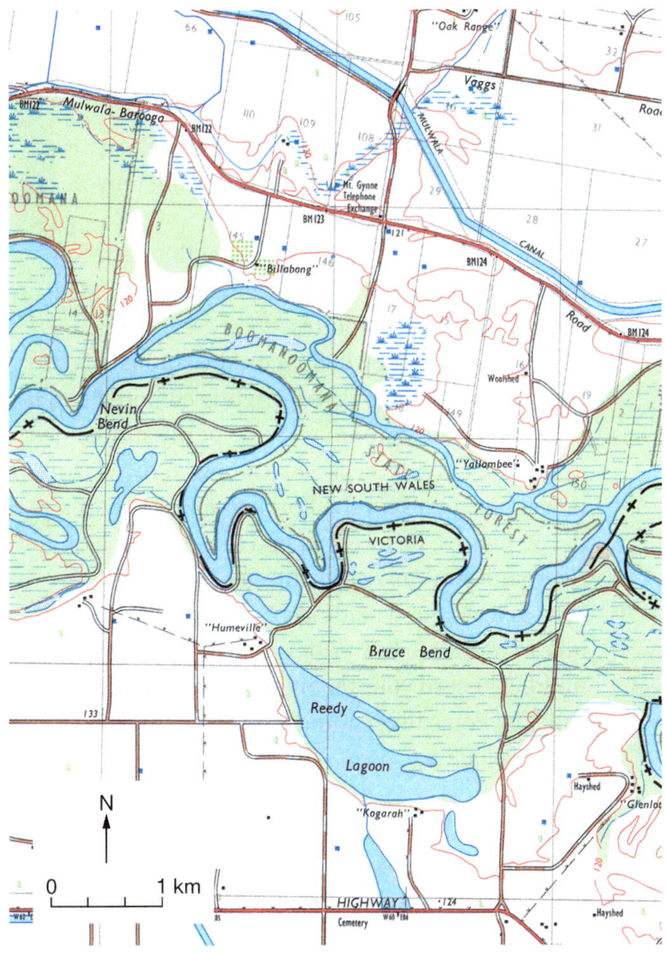

ACTIVITIES

Knowledge and understanding

1 a List the landform features associated with the upper and lower reaches of a river.

  b Identify and describe the main process responsible for their formation (that is, whether they are erosional or depositional landform features).

Geographical skills

2 Study Figure 9.14. Construct an annotated sketch map to identify the various meander-related landform features of the topographic map extract and the aerial photograph.

In the field: Investigating rivers

The aim of this fieldwork activity is to investigate a river. A fieldwork investigation of a length of river provides an opportunity to learn more about these important features of the physical environment. It also allows you to practise a range of geographical skills.

How to investigate a river

A variety of instruments will be required to investigate a river (see Figure 9.15):

- ruler
- tape measure
- ranging pole
- stopwatch
- clinometer
- flow meter.

You will need to select a river or steam location that has the following features:

- a bridge to cross
- easy access to riverbanks at a variety of locations

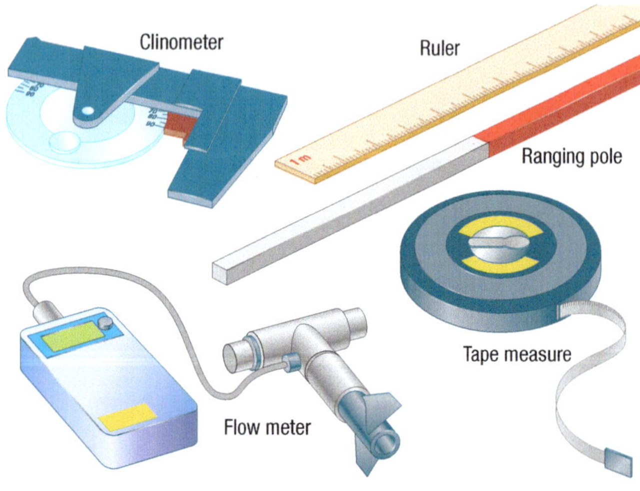

9.15 Tools used to collect data in the field

Calculating water velocity

The most accurate way to measure water velocity is to use a flow meter. If you don't have a flow meter you can use the following procedure.

1. Select a straight section of the river that is free of pools and/or shallow, fast-flowing sections. Measure out a distance of 50 metres.
2. Find an object that will float on the surface of the river. The object should be brightly coloured and it is best if it is heavy enough to be partly submerged in the water. With the aid of a stopwatch, measure how long it takes for the floating object to cover the 50 metres. Ensure that you select an area of riverbank you can easily and safely access.
3. Take at least three readings. For greater accuracy take readings on both sides of the river and in the middle. Record your readings on your data record sheet. Average the readings to determine the water velocity.

Calculating the cross-section

To calculate the cross-section of an area of river, carry out the following steps.

1. Use a long tape measure to determine the average width of the river.
2. Measure the depth of the water at regular intervals across the width of the river (for example every 100 centimetres). A 2-metre pole marked with 10-centimetre intervals will assist you in this task. If there is a low bridge over the river, stand on it to measure the depth of the river. Record your measurements on your data record sheet.
3. Calculate the average depth by adding all the depth readings and dividing by the number of readings. Using the data shown in Figure 9.16, the average would be (1.2 m + 1.4 m + 1.5 m + 1.3 m + 1.0 m) ÷ 5 = 1.28 m.

4 Multiply the average depth by the average width of the river to give the area. Using the data shown in Figure 9.16, the area would be 1.28 m × 16 = 20.48 m^2.

9.16 Data sample for a cross-section

Width of channel (bank to bank): 18 metres						
Average width of river: 16 metres						
Height of bank above the river:		• Left side: 50 centimetres • Right side: 75 centimetres				
Depth of river:		**Left bank**			**Right bank**	
	Reading	1	2	3	4	5
	Depth	1.2 m	1.4 m	1.5 m	1.3 m	1.0 m

Calculating river discharge

Discharge is the amount of water that flows from a river catchment and into another river system, the sea or a lake. The discharge can be calculated by using the following formula:

Discharge = velocity × cross-sectional area

Measuring turbidity

Turbidity refers to the cloudiness of water, which is caused by suspended sediment. To measure the turbidity of a river, carry out the following steps.

1 Collect a 1-litre sample of water at a number of sites along the course of a river.

2 Transfer the samples into separate glass containers. Allow the water to stand for at least 24 hours so the sediment will settle.

3 Using a ruler, measure the depth of sediment at the bottom of each container. Record your data in a spreadsheet file and present your data as a graph.

Measuring suspended load

The suspended load of a river comprises fine sand particles, silt and clay. The amount of suspended material in water is closely linked to the level of discharge. To measure the amount of suspended load in a river, carry out the following steps.

1 Use four 1-litre plastic bottles to collect water samples at four sampling sites along the course of a river. When preparing each plastic bottles, block the opening with a cork and then drill two holes through the cork. Push two flexible plastic tubes through the holes, as in Figure 9.17.

2 Anchor the bottle to the riverbed with two or three stones. When doing so, make sure that you stand downstream, so that you do not stir up too much sand and sediment.

3 When the bottle is full, remove it from the river and remove the cork and plastic tubes. Seal the bottle with its original screw cap. Repeat the exercise at your other sampling sites. Allow the bottles to stand overnight.

4 Observe the layer of sediment that has collected at the base of the bottle. Make note of the sediment's colour, the water's clarity (turbidity) and, if possible, the amount of time it takes for the sediment to settle.

5 Shake the bottle so that the sediment is again redistributed through the water sample and then very slowly pour the contents of the bottle through a previously weighed piece of dry filter paper. You could use a filter suction pump to assist in this process.

6 Allow the sediment-encrusted filter paper to dry for at least 48 hours, or dry it in an oven for 1–2 hours at 100°C. Subtract the weight of the dry filter paper to find the weight of the suspended sediment. Express your answer in grams per litre of water.

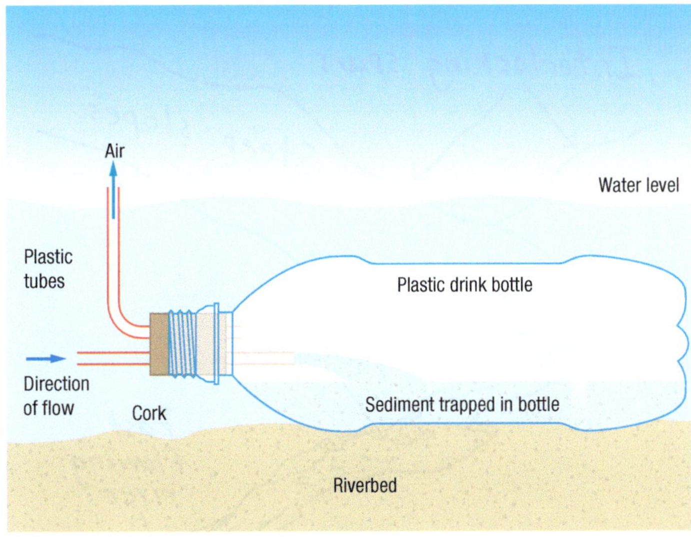

9.17 Sediment sampler

Drawing a cross-section

To draw a cross-section of a river, you need measurements of the:

- depth of the river
- width of the channel
- width of the river from bank to bank
- height of the bank above the river.

Using these measurements, construct your cross-section by carrying out the following steps.

1 Study the measurements you have collected and select a scale that will fit on your paper. Start your cross-section by drawing a line representing the width of the river. Make sure you leave enough space to draw in the river channel below it.

2 Look at the measurements for the height of the bank above the river level on both sides and mark the position of both banks. Measure the width of the channel from bank to bank. Now draw in the banks.

3 Mark in the riverbed by using your measurements of the depth of the river from the surface. Join the points together to show the shape of the riverbed. Add a scale and a heading, as shown in Figure 9.18.

Drawing a field sketch

To draw a field sketch of the river being studied, use Figures 9.19 and 9.20 as a guide.

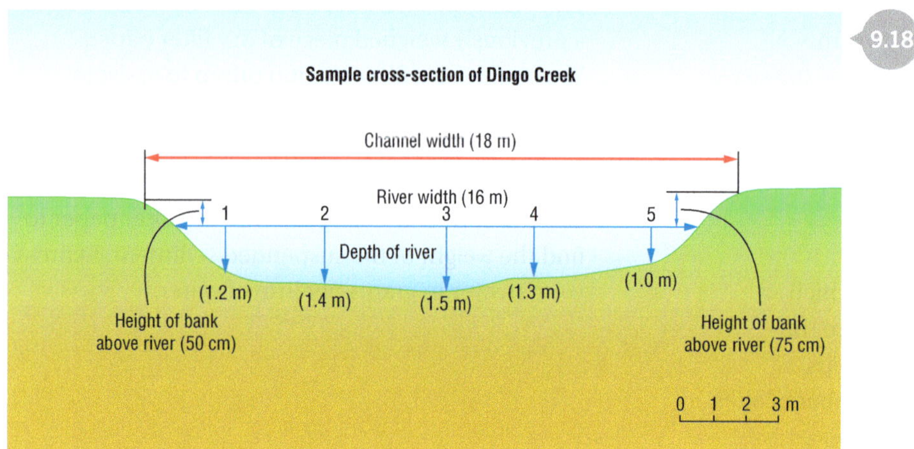

9.18 Sample cross-section of a river

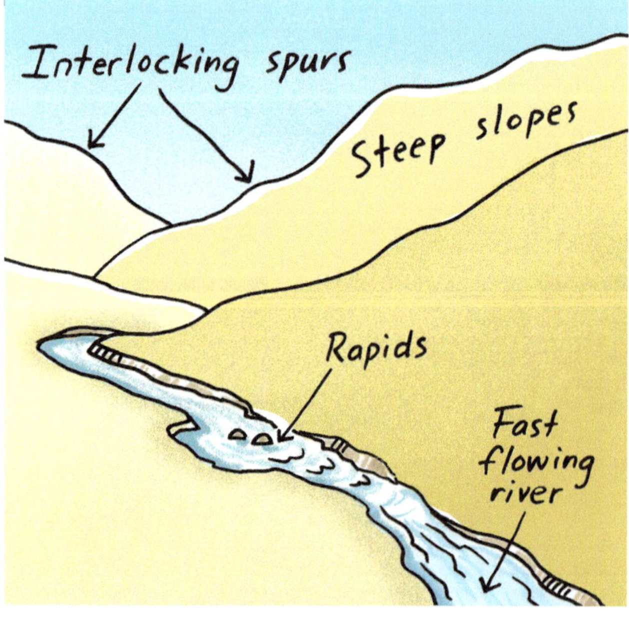

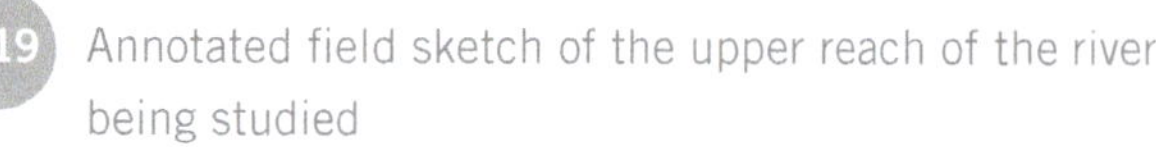

9.19 Annotated field sketch of the upper reach of the river being studied

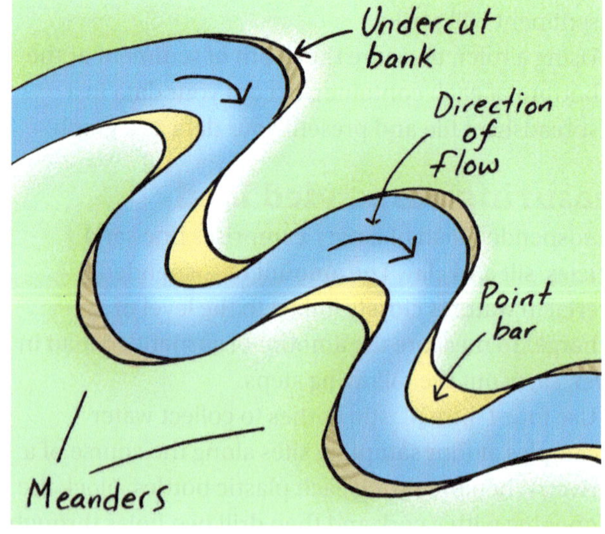

9.20 Sketch plan of the mid-reach of the river being studied

ACTIVITIES

Aim

To investigate the physical features of a river.

Instructions

1 Select a suitable river to investigate and include a map of the area investigated.

2 Calculate water velocity.

a Collect data and record it in a data record sheet similar to the sample below.

b Select data from two different locations, and follow the steps above.

River: Sandy Creek	**Location:** Site 1	**Date:** 23 May 2013
A. Water velocity		
Length of river section:	50 metres	
Measurement 1:	75 seconds	
Measurement 2:	90 seconds	
Measurement 3:	85 seconds	
Average:	83.3 seconds	
Average velocity:	50/83.3 = 0.6 metres per second (i)	

3 Calculate the cross-section of the river. Collect data and record it in a data record sheet similar to the sample below.

River: Sandy Creek	**Location:** Site 1	**Date:** 23 May 2013
B. Cross-sectional area		
Average width of the river	15 metres	
Channel depth at various points (in centimetres)		
Point 1:	20	
Point 2:	25	
Point 3:	35	
Point 4:	45	
Point 5:	30	
Point 6:	20	
Average depth:	175/6 = 29 centimetres	
Cross-sectional area: average river width × average depth 15 × 0.29 = 4.35 square metres (ii)		

4 Calculate river discharge.

River: Sandy Creek	**Location:** Site 1	**Date:** 23 May 2013
C. Discharge		
Velocity (i) × cross-sectional area (ii) 0.6 metres per second × 4.35 square metres = 2.61 cubic metres per second		

5 Measure turbidity. Collect data and record it in a data record sheet similar to the sample provided.

6 Measure suspended load.

7 Draw a cross-section of the river.

8 Draw a field sketch of the river area being studied.

Option 1

Create an annotated visual display of the area investigated.

a Take images of the river channel, upstream, downstream and land on either side of the river channel.

b Explain how each has or might influence the river channel.

c Annotate the photos and explanations around the map of the area of river investigated.

Option 2

Investigate the river from its source to the mouth of the river. Include the following information:

- topography, vegetation and landuse
- how the river water is being used (agriculture, irrigation, recreational, household use, etc.)
- problems associated with the river (pollution, reduced water flow, etc.).

Evaluation

9 Once you have collected enough data, display the results and comment on your findings. Include the following in your commentary:

a a description of the features of section of river that you investigated

b information about changes in speed flow along the river

c information about levels of turbidity and suspended load—was there a correlation between the two?

Conclusion

10 Describe what you have learnt about river profiles.

9.4 People's impact

There are many ways in which human activity can impact on river-based landforms and ecosystems. Any activity within a river's catchment has the potential to impact on the river.

The impacts

The impacts of people on riverine landforms and landscapes are outlined in Figure 9.21.

9.21 The impact of people on river systems

River modifications
Any change designed to control river flow will impact on the rate or erosion and deposition within the channel.

Reduced water flow
Water used for irrigation and household and industrial use reduces river discharge. This, in turn, affects the rate of erosion and deposition.

Disruption to migratory patterns
Dam walls can prevent fish from moving upstream to breed.

Temperature change
Storing water in large dams causes a drop in temperature (especially when the dam is very deep). Industries and thermal power stations can pollute rivers with very warm water.

Change in seasonal flows
Increases in the use of water in summer can reduce the amount of water available to support life in the river.

Pollution
Pesticides can kill aquatic plants and animals, and enter the food chain, causing damage throughout the food web. Fertilisers can encourage excessive plant growth, including toxic algal blooms.

Inappropriate landuse practices
This includes deforestation, overgrazing and excessive ploughing. These practices expose the soil to erosion, resulting in a increase in river sediment loads.

Overfishing
This results in a decrease in fish populations.

River improvements
The removal of logs and other material from rivers may destroy fish habitats.

Introduced species
Introduced fish compete for food and habitat with native species. Introduced plants can choke waterways and crowd out native species.

River modification works

River modification works are designed to change or control the natural course or flow of a river. This is usually done to improve drainage and reduce flooding.

River modification works usually involve rearranging the riverbed or bank material, as has been done along the River Lea in London, shown in Figure 9.22. It is commonly thought that straightening the channel, either by constructing a totally new river canal, or a flow path across a meander bend, increases river drainage capacity. In some cases, a river channel's capacity can be increased by enlarging and straightening the channel and then lining it with concrete. This usually happens in urban areas.

The construction of levee banks (see Figure 9.23), or retarding basins, is designed to control floodwaters.

Rivers can also be modified to reduce erosion. This may involve the removal or relocation of sediment bars and islands within the channel to prevent flows from being diverted to the banks and causing erosion.

Almost all river modification works involve the use of earth-moving machinery in or near the river channel. These machines damage the riverbed and/or banks, and can, at least in the short term, increase the amount of sediment flowing into the river.

London's River Lea is now a heavily modified urban waterway. The modifications date back hundreds of years. The most recent of these modifications occurred as part of the development of the Olympics site.

Large-scale levee construction in the catchment of the Missouri River, United States of America

Protection

The best way to protect riverine landforms and landscapes is to manage the whole catchment. Sources of soil erosion can be identified and addressed. Obstacles to the natural flow of water can be removed. Pollutants such as sewage, industrial pollution, agricultural fertilisers and pesticides, and salty **groundwater** can be identified and treated before they find their way into a catchment's waterways.

There is also a need to control and manage the amount of water taken from rivers for irrigation and urban and industrial uses. Experts now talk of maintaining an adequate **environmental flow**, ensuring that there is enough water in the river to help maintain a river's natural processes and to protect the health of the river and its wetland ecosystems.

Figure 9.24 illustrates how farmland next to a river can be managed to control the quality of water entering a river.

9.24 Careful management of riverbanks can help to protect riverine landforms and water quality.

Urban areas

People in cities can also protect riverine landforms. Many stormwater drains in cities and built-up areas flow into rivers and creeks before flowing into the sea. Figure 9.25 shows how people can reduce the amount of wastes entering stormwater drains.

Raingardens

A raingarden looks like a normal garden but is designed and built to capture stormwater. When water falls on hard surfaces such as roads, footpaths and roofs, the raingarden captures the water and filters it through layers of sandy soil. As the water soaks through the layers, pollutants such as animal droppings are removed. Figure 9.26 shows a raingarden in action after rainfall.

Flood detention basins

Flood detention or retention basins capture a large amount of stormwater and slow the amount of water entering a stream or river after heavy rainfall. A basin can remove pollutants and sediment flowing into a river or stream. Basins like the one shown in Figure 9.27 are being constructed throughout Australia.

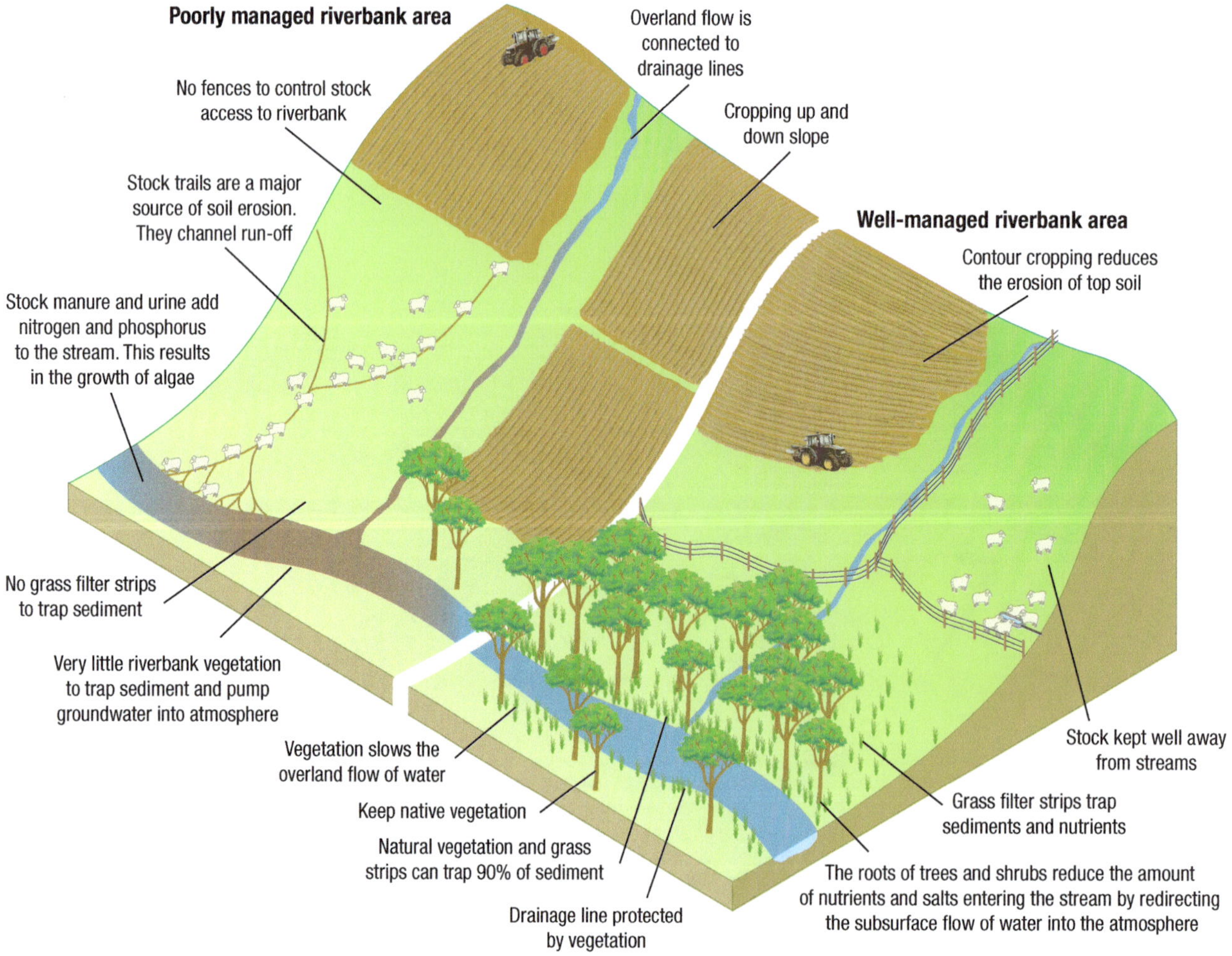

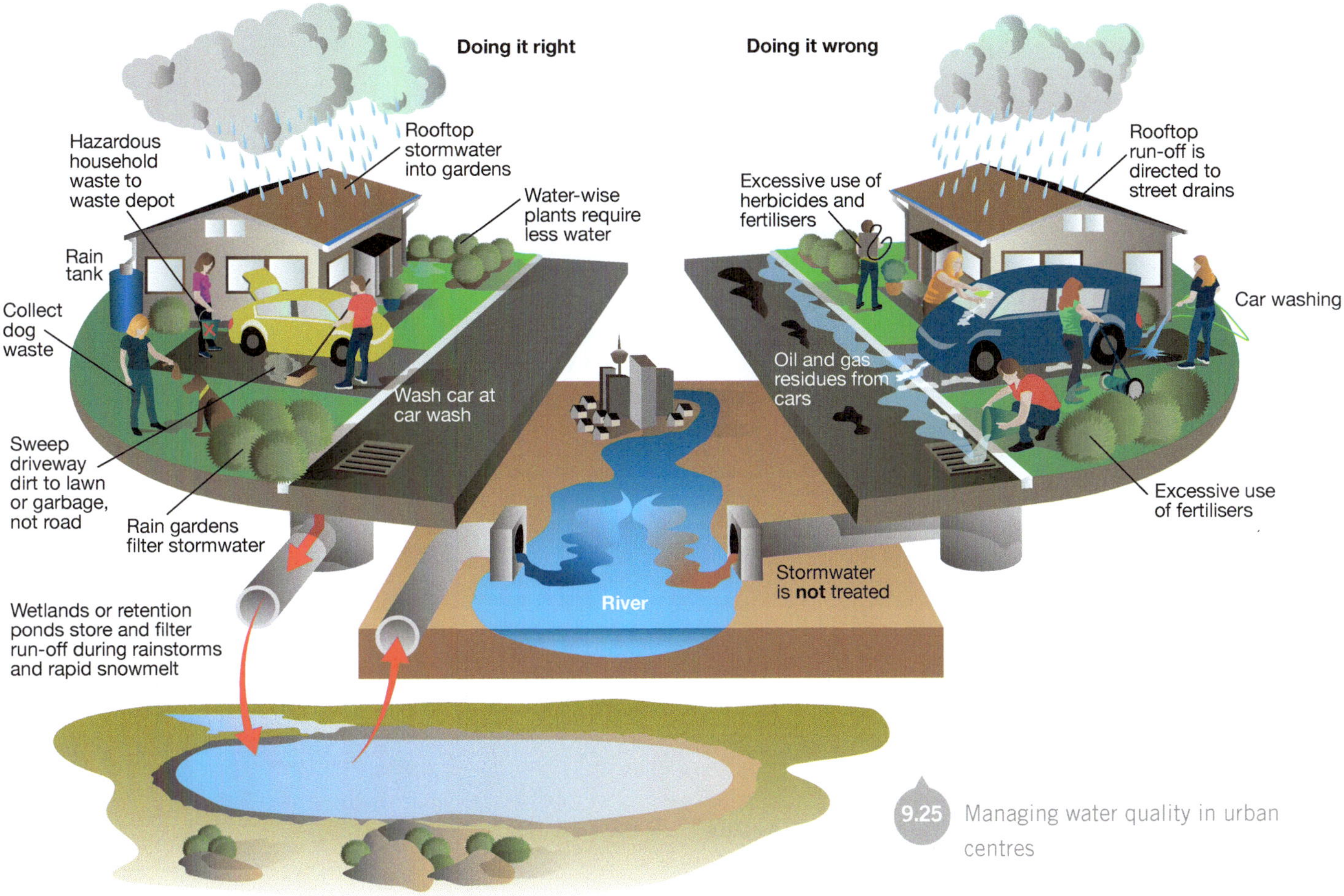

Doing it right

Hazardous household waste to waste depot
Rooftop stormwater into gardens
Water-wise plants require less water
Rain tank
Collect dog waste
Wash car at car wash
Sweep driveway dirt to lawn or garbage, not road
Rain gardens filter stormwater
Wetlands or retention ponds store and filter run-off during rainstorms and rapid snowmelt

Doing it wrong

Excessive use of herbicides and fertilisers
Rooftop run-off is directed to street drains
Car washing
Oil and gas residues from cars
Excessive use of fertilisers
Stormwater is **not** treated
River

9.25 Managing water quality in urban centres

9.26 A suburban raingarden

9.27 Karkarook Park—a flood detention basin aimed at removing pollutants

ACTIVITIES

Knowledge and understanding

1 Describe the ways in which river modification works can impact on a river.
2 Explain how riverine landforms and landscapes can be protected.
3 Define the term 'environmental flow'.

Applying and analysing

4 List the ways your household could manage its wastewater better.
5 In pairs or small groups, prepare a presentation for the School Council and Principal, outlining a plan for a new rain garden for your school. Your presentation should include the following information:
- map of the school showing the location of the rain garden
- benefits of the raingarden
- a sketch or diagram showing what the raingarden will look like.

9.5 Case study: Bow River catchment

The Bow River is located in the Canadian province of Alberta. Over the length of the river can be seen the full range of riverine landforms. The river is an important source of drinking water and water for irrigation and hydro-electric power. It also provides habitat for wildlife and opportunities for recreational activities such as fishing and boating.

9.28 Bow River catchment

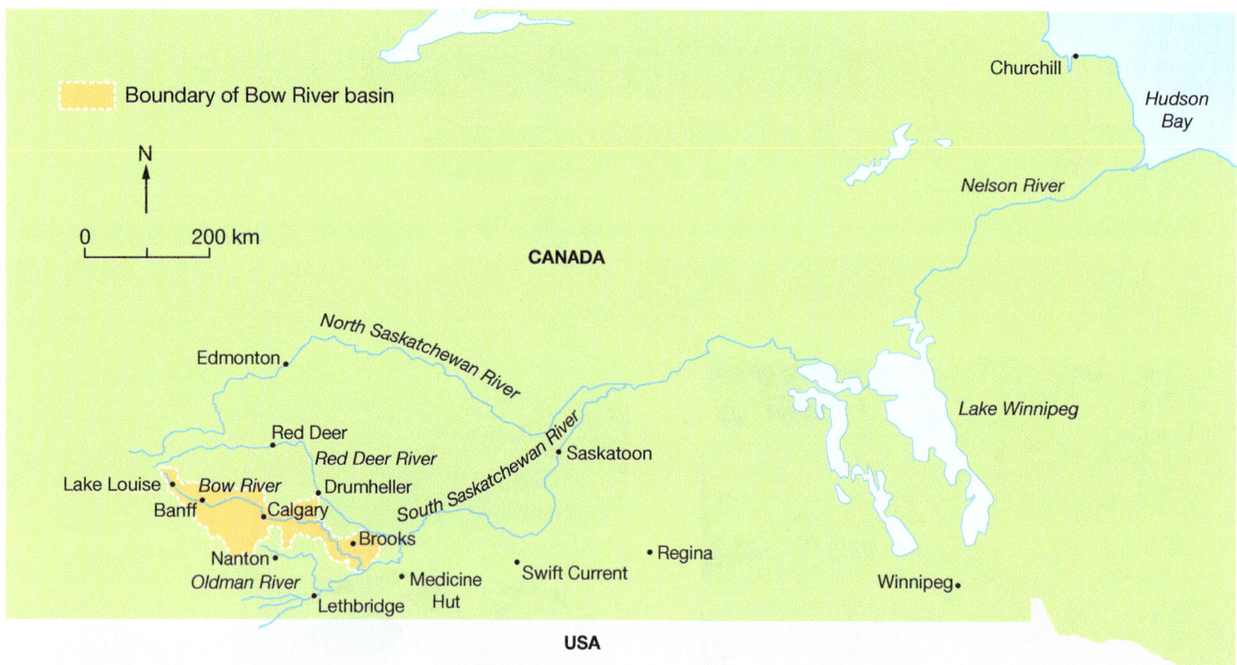

Riverine landforms

The Bow River's source is the meltwater of the Bow Glacier high in the Wapta Icefield of the Canadian Rockies. Figure 9.28 shows the location of the Bow River. The meltwater first finds it way into Bow Lake, shown in Figure 9.29, before it flows to the south through the rocky moraine debris on the floor of a former glacial (U-shaped) valley. It then flows through the village of Lake Louise and town of Banff to Ghost Lake reservoir. It then flows onto the prairies to the city of Calgary, which has a of population 1.1 million. Downstream from Calgary are a range of depositional river landform features including meanders, cut-offs and oxbow lakes. Further downstream the Bow River flows into the South Saskatchewan River, which continues to wind its way across the Canadian prairies before flowing into Lake Winnipeg and then into Hudson Bay via the Nelson River.

9.29 The Bow River has its source in the meltwater of the Bow Glacier high in the Canadian Rocky Mountains. Bow Lake is in the foreground.

9.30 The Bow River in its upper reaches

The water cycle

The Rocky Mountains force air to rise and cool, causing moisture to condense and fall as rain or snow. This precipitation, together with the meltwaters from glaciers, feeds the Bow River through its many mountain tributaries. Figure 9.30 shows the upper reaches of the river.

The grasslands, or prairies, of Canada's interior have a relatively dry climate. Due to their height, the Rocky Mountains strip moisture from eastward-moving air masses in what is referred to as a 'rain shadow' effect. As a result, little of the moisture from the eastward-moving air masses reaches the prairies of Alberta. The region therefore relies on the Bow River and groundwater for irrigation in agriculture and human use.

The water cycle of the Bow River catchment is illustrated in Figure 9.31.

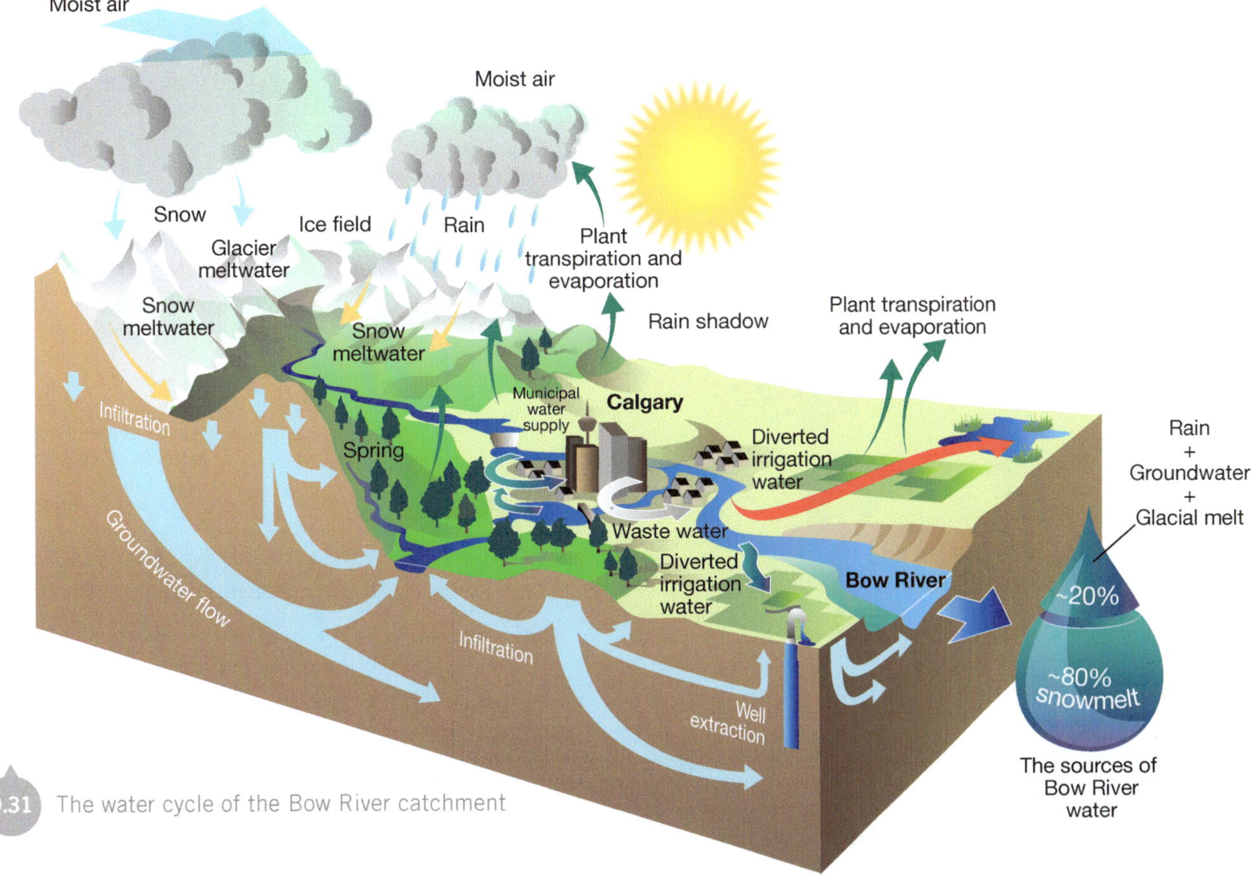

9.31 The water cycle of the Bow River catchment

Human modifications

In their efforts to use the water resources of the Bow River for irrigation, people have significantly changed, or modified, the river. Water is used for agriculture (irrigation), urban water supplies, hydro-electric power, industry and recreation.

Irrigation

Early in the last century, European settlers developed irrigation systems to move the waters of the Bow River out onto the prairie. Water was supplied to farmers through a system of canals and storage reservoirs. Over time, communities and industries developed across Alberta's prairies.

Figure 9.32 shows the extent of the modifications made to the river. Irrigation is used during the growing season, which begins in May and continues through to October. Improved irrigation techniques have greatly reduced the water required to grow crops. The canals and reservoirs provide important wetland habitat for waterfowl and fish.

Urban water use

Municipalities (towns and cities) in the area return over 90 per cent of the water they use as treated sewage back to the Bow River. Calgary, a rapidly growing city on a relatively small river, is by far the largest urban centre in the Bow River catchment. As a result, it has limited capacity to absorb wastewater without reducing water quality. Because of this, Calgary's wastewater treatment standards are among the highest in Canada. During summer, Calgary's residential water use rises by 50 per cent, largely due to garden and lawn watering. Calgary's water usage is shown in Table 9.33.

Protecting the Bow River

There are a variety of ways in which the riverine landscape of the Bow River can be protected.

Of particular importance is the health of the riparian zone—the area bordering streams and wetlands where moist soils and shallow water tables allow water-loving plant communities to establish. These zones are important ecosystems. They stabilise stream banks and protect water quality. They also provide habitat for wildlife. Cattle grazing in riparian areas must be managed carefully so that river-related landforms are not degraded.

9.32 The waters of the Bow River have been diverted, dammed and used for irrigation.

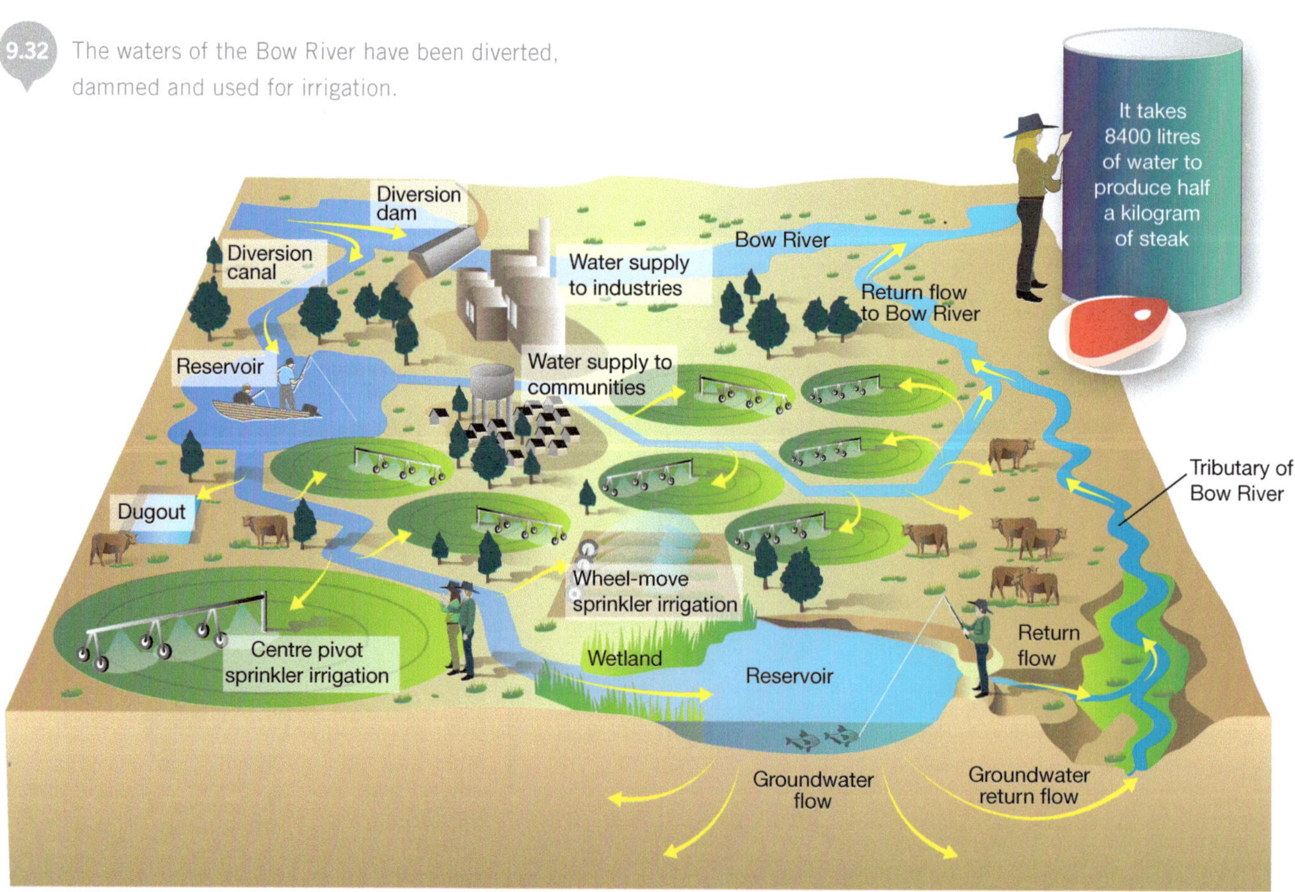

9.33 Municipal and residential water use

Calgary's municipal water use	
Residential	52%
Industrial commercial infrastructure	34%
Non-revenue (e.g. leakage)	12%
Other communities	2%
Calgary's residential water use	
Toilet	29%
Clothes washer	20%
Taps	16%
Shower	13%
Leaks	10%
Water softener	9%
Baths	2%
Dishwasher	1%

Impacts of climate change on the Bow River catchment

The potential impacts of climate change on the Bow River catchment include:

- reduced snowfall and shrinking glaciers, causing a decline in the amount of river flow, which in turn will:
  - affect water supply, water quality and both the extent and type of recreational activities that people can do
  - lead to decreased hydro-electric power generation
- warmer, drier summers and earlier springs, resulting in an increase in the frequency and intensity of forest fires
- an increase in the incidence of extreme weather events such as tornadoes, hailstorms, heatwaves, droughts, dust storms, floods and blizzards
- increased demand for irrigation water and a change in crop types due to a longer growing season
- warmer river temperatures, resulting in increased stress for cold-water fish species such as trout
- reduced groundwater recharge, resulting in lower water tables and the drying up of shallow wells.

Many people are concerned about the rate at which the glaciers high in the Rocky Mountains are retreating. While this is a serious problem and one of the major impacts of climate change, the contribution of meltwater to the annual flow of the Bow River is relatively small (less than 1 per cent). However, the portion of Bow River water that comes from glaciers rises during the summer as snowmelt declines. During a drought year, the relative contribution of glacier meltwater to the discharge of the Bow River is higher. Without glaciers in the Bow River basin, water supply during drought years would be less secure. However, as long as it snows and rains every year, we can expect the river to keep moving.

ACTIVITIES

Knowledge and understanding

1 With the aid of Figure 9.31, describe the geography of the Bow River catchment.

2 Explain why it is important to protect the catchment's riparian zone.

3 Outline the strategies used to manage landuse in the Bow River catchment.

4 Describe the potential impacts of climate change on the Bow River catchment.

Applying and analysing

5 Study Figure 9.32.
  - a Describe the extent to which people have modified the flow of water in the Bow River catchment.
  - b Reflect on how this might impact on the river itself and list the ways the riverine landscape can be managed better.

Geographical skills

6 Study Figure 9.28.
  - a List the rivers that flow into the Saskatchewan River and the cities on each river.
  - b Name the lake into which the Saskatchewan River flows.
  - c True or false: All rivers that flow into the Nelson River originate in Canada. Explain your answer.

7 Study Table 9.33. Construct two proportional pie graphs—one showing Calgary's municipal water use and the other showing Calgary's residential water use.

CHAPTER 10

DESERT LANDFORMS

Deserts are among the world's most spectacular landscapes and feature amazing landforms that have been shaped by wind and water.

Deserts are found where precipitation is low and evaporation is high. Contrary to popular belief, deserts are not necessarily hot. All deserts have a number of features in common, including very low and irregular rainfall, highly specialised plants and animals, and few, if any, people.

In this chapter you study the geographical processes responsible for shaping the desert landforms. You will study the landforms of the Sahara and investigate how human activities are causing the world's deserts to expand—a process called 'desertification'.

KEY IDEAS

- To understand why deserts are dry and how the distinctive landform features of deserts are formed
- To investigate desertification
- To describe how the activities of people contribute to desertification

10.0 Sand dunes in the desert, Morocco

GLOSSARY

alluvial fan	a low, cone-shaped deposit of sand and rock formed when a stream deposits its load on a low-lying plain as a result of an abrupt change of slope
aquifer	a layer of permeable rock that is capable of storing significant quantities of water
bolson	an inland desert basin surrounded by mountains
butte	a flat-topped landform feature that is taller than it is wide
deflation	the process whereby material is removed from a surface by wind
desert	a very dry environment in which evaporation is greater than precipitation or precipitation is less than 250 millimetres a year
desertification	the process by which productive land in arid areas is changed into desert
erg	a desert surface covered with sand dunes
escarpment	a long, steep slope that defines the edge of a plateau or separates areas of land at different heights
hamada	a desert landscape dominated by exposed bedrock
inselberg	a large erosion-resistant rock feature surrounded by an eroded plain
mesa	a flat-topped landform feature that is wider than it is high
oasis	a fertile or well-watered area within a desert
playa lake	a dried-up salt lake found in arid or semi-arid environments
reg	a stone-covered desert
sand dune	a hill of sand shaped by the wind
succulent	a plant capable of storing water in its tissues
transpiration	the loss of water vapour from the leaves of plants
wadi	a watercourse or gully that is dry except during periods of rainfall
xerophyte	a drought-resistant plant adapted to the dry desert environment

10.1 Deserts

Deserts occur where evaporation is greater than precipitation or where precipitation is less than 250 millimetres a year. These areas cover about 30 per cent of the earth's surface. Approximately 13 per cent of the world's population live in deserts.

Desert features

Deserts are the driest places on earth. They usually have few plants and hard, wind-blown surfaces covered with rocks and sand. The range of temperatures experienced each day is often very broad, with very high daytime temperatures and very cold night temperatures. There is no cloud cover, so heat is lost into the atmosphere very quickly. The lack of moisture in the air (combined with the lack of vegetation) also allows the surface of the desert to lose heat rapidly after the sun goes down.

Desert location

Deserts occur in regions where the climate is dominated by high-pressure systems. High-pressure systems are areas in which there is little rainfall, as outlined in Figure 10.1. Evaporation rates in deserts are often twenty times the annual precipitation.

Air near the Equator rises into the atmosphere because it is hot. This air then travels towards either the Tropic of Cancer or the Tropic of Capricorn. As it does so, the temperature of the air decreases and it sinks back towards the earth. This sinking air creates an area of high pressure. There is little chance of precipitation in these areas because air must be rising before condensation and therefore precipitation can develop.

Distribution of deserts

Figure 10.2 shows the distribution of deserts. Deserts are located either in the interior of continents or near the coast next to cold ocean currents, excluding the Sahara. The distribution of deserts is influenced by:

- cold ocean currents—as evaporation rates from cold water are low, the amount of water in the air and therefore the amount of rainfall are reduced, as shown in Figure 10.3
- wind direction—in the mid-latitudes, hot dry winds blow mainly across land towards the sea
- mountain rain shadow—mountains can block the inland movement of rainfall further inland. This is shown in Figure 10.4.

10.1 Deserts occur in areas where the climate is dominated by high-pressure systems.

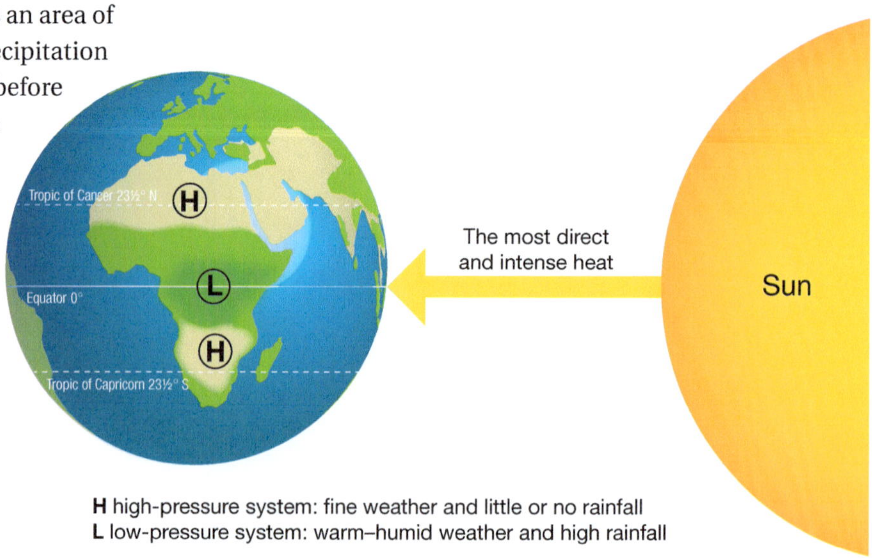

10.2 Distribution of the world's desert lands

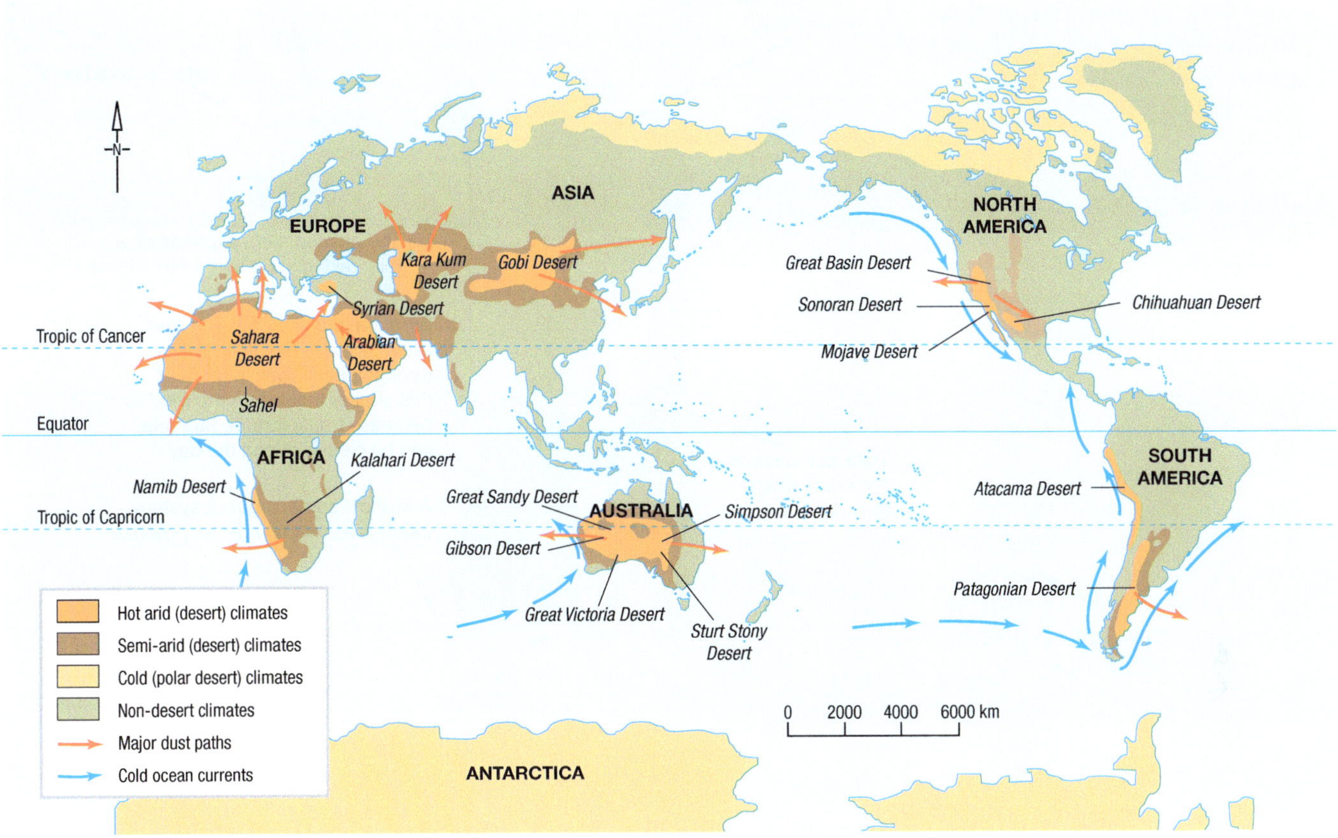

N
ASIA
EUROPE
NORTH AMERICA
Kara Kum Desert
Gobi Desert
Great Basin Desert
Syrian Desert
Sonoran Desert
Chihuahuan Desert
Tropic of Cancer
Sahara Desert
Arabian Desert
Mojave Desert
Sahel
Equator
AFRICA
Kalahari Desert
SOUTH AMERICA
Namib Desert
Great Sandy Desert
Atacama Desert
AUSTRALIA
Simpson Desert
Tropic of Capricorn
Gibson Desert
Patagonian Desert
Great Victoria Desert
Sturt Stony Desert
Hot arid (desert) climates
Semi-arid (desert) climates
Cold (polar desert) climates
Non-desert climates
Major dust paths
Cold ocean currents
0 2000 4000 6000 km
ANTARCTICA

Relatively dry air and, therefore, little rain
Prevailing wind direction
N
Sand dunes
Cold ocean currents

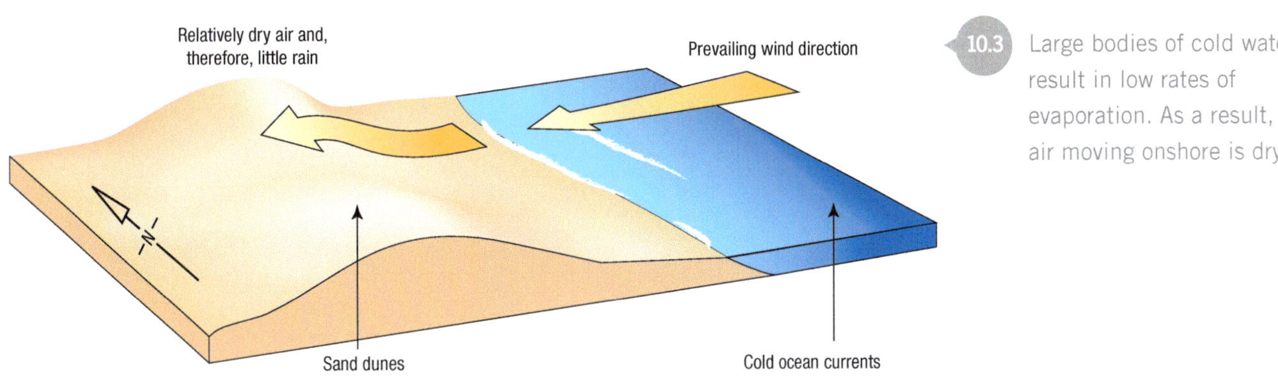

10.3 Large bodies of cold water result in low rates of evaporation. As a result, the air moving onshore is dry.

Moist winds from sea
Dry winds blow inland
Desert area in rain shadow
Heavy rainfall

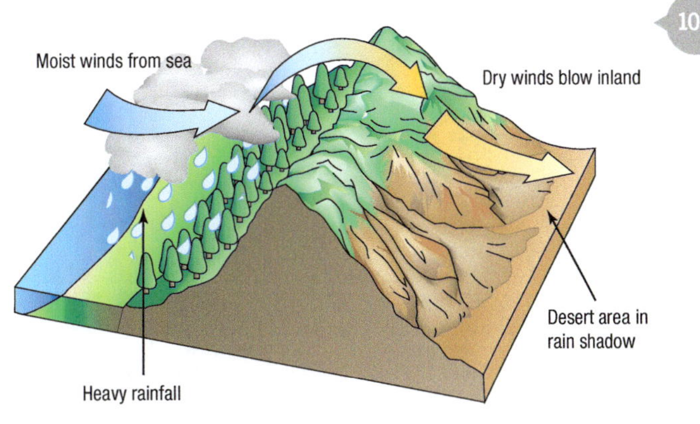

10.4 Mountain ranges can block the inland movement of moisture, producing desert environments.

SPOTLIGHT

Surviving aridity: The cactus

Drought-resistant plants are known as **xerophytes**. Many xerophytes are **succulents**. This means that they are able to store water in their tissues. Cacti (Figure 10.5), for example, absorb large amounts of water following rainfall. Their fleshy stems swell up, then slowly shrink as moisture is lost through **transpiration** (the loss of water vapour from the leaves of plants). Many succulents further reduce the amount of moisture lost by transpiring only at night.

Most desert plants, including cacti, have small leaves that are spiky or wax-coated. This reduces transpiration. Cacti also spread their shallow roots wide to take advantage of any rain or dew; other adaptations are shown in Figure 10.6.

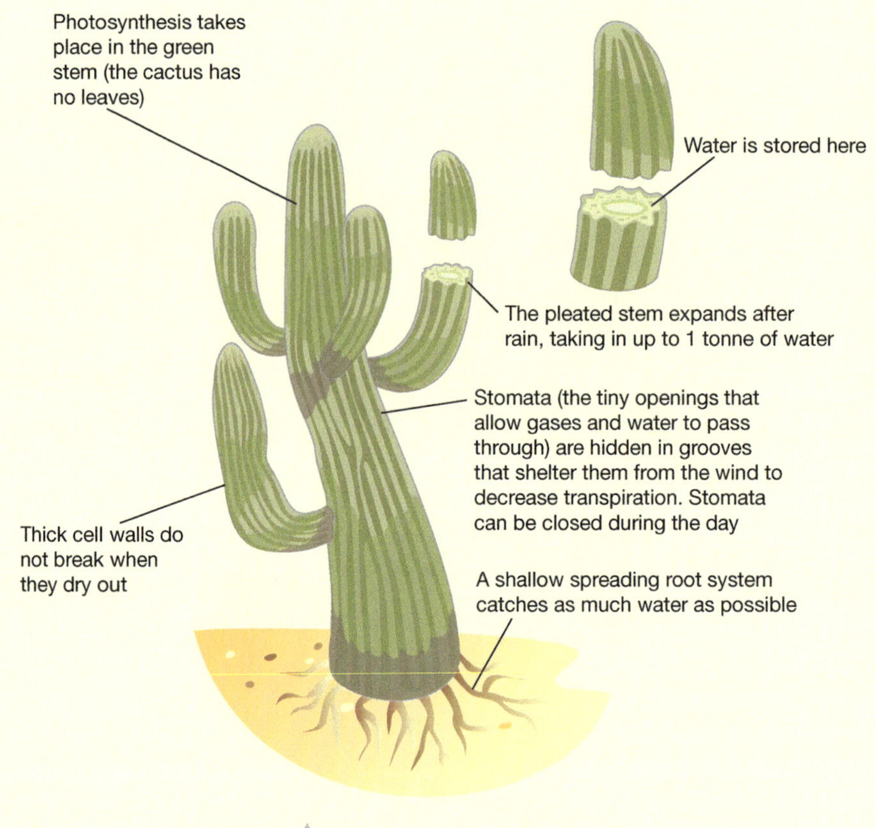

10.5 The cactus is a well-known symbol of North American deserts.

10.6 The saguaro cactus

Cold deserts

Antarctica and the Arctic regions are also classified as deserts. Although covered with a thick layer of ice, Antarctica is, in fact, a very dry place. The continent's average annual snowfall equals just 50 millimetres of rainfall.

Desert size

The Sahara is the world's largest desert, although the Antarctic Desert (Antarctica), at 13 829 430 square kilometres, is sometimes referred to as the largest. Table 10.7 shows the largest deserts in area.

10.7 The world's greatest deserts

Desert	Area (km²)
Sahara (Africa)	8600000
Arabian (Middle East)	2330000
Gobi (Asia)	1200000
Kalahari (Africa)	900000
Patagonian (South America)	673000
Great Victoria (Australia)	647000
Syrian (Middle East)	520000
Great Basin (North America)	492000
Chihuahuan (USA)	450000
Great Sandy (Australia)	407000

DID YOU KNOW?

If someone spent a day in the desert with no shade, food, water or clothes, their temperature would be about 46°C by sunset, and they would have lost 2 to 3.5 litres of water. By nightfall they would be dead.

Desert plants

Vegetation in deserts depends on the desert's location. The cactus is a well-known symbol of the desert, but is, in fact, found only in North American deserts. In southern African deserts, aloe plants are widespread and in northern African and Middle Eastern deserts, date palms are typical plants. In Australia, grasslands cover large parts of the arid zones.

Desert mirages

Mirages are often seen in the world's hot deserts. They occur when a shallow layer of warm air next to the ground is trapped by cooler air above. Light bends towards the horizontal line of vision, and eventually travels upwards: the mirage is an upside-down 'virtual' image (see Figure 10.8).

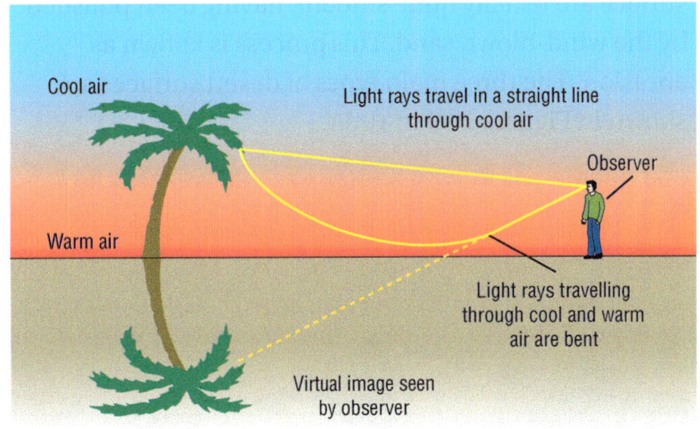

10.8 A mirage is an optical illusion.

ACTIVITIES

Knowledge and understanding

1 Define the term 'desert'.
2 State the proportion of the earth's surface that is covered by desert.
3 Explain why deserts are often very cold at night.
4 Identify the conditions under which deserts occur.
5 Identify the types of plants that are common in the deserts in the following locations:
  a southern Africa
  b northern Africa and the Middle East
  c North America.

Geography skills

6 Study Figure 10.2.
  a Describe the distribution of the world's hot deserts.
  b Describe the distribution of the world's cold deserts.
  c Explain why desert-like conditions are found in these areas.
  d Are there any deserts along the Equator? Explain.

10.2 Desert landforms

Desert landforms are some of the most spectacular landforms on the surface of the earth. The agents of erosion responsible for shaping these landforms are wind and running water.

Desert surfaces

Sand dunes cover only about 25 per cent of the world's deserts. The remaining 75 per cent are either exposed rock or stone-covered plains. The rocks of the desert surface are usually quite smooth, having been polished by the wind-blown sand. This process is known as abrasion. The three main types of desert surfaces are shown in Figures 10.9 to 10.11.

10.9 Sandy desert or erg surface, Namibia, Africa

10.10 Stony desert or reg surface (known as a gibber plain in Australia) of the Jbel Bani mountain range of western Sahara, Africa

10.11 Rocky desert or hamada surface, Double Arch, Utah, United States of America

Landforms shaped by wind

The erosion of fine surface sand by wind is known as **deflation**. Once airborne, sand can act as an abrasive tool, wearing down and shaping exposed rock. Pedestal rocks, for example, are formed when wind-blown sand cuts away the base of rock structures but leaves their tops intact.

When the wind-blown material is deposited it accumulates and forms **sand dunes**. The shape of these dunes varies according to the strength and direction of the wind, the amount of sand available and the type and extent of vegetation cover in the area. Dunes are named according to their shape and direction. Some of the more common types are shown in Figure 10.12.

Landforms shaped by water

Although it does not rain often in deserts, when it does, the rain is often very heavy and results in flash flooding. Because there is little or no vegetation, run-off is extremely rapid and can erode large amounts of material.

Surface run-off is channelled into dry riverbeds (**wadis**) that cut through plateaus, forming canyons. As plateaus are eroded, **mesas** and **buttes** are left isolated from the retreating **escarpment**, or steep cliff. Mesas are wider than they are high, while buttes are higher than they are wide.

The eroded material is carried by the water through the wadis. It is eventually deposited on the lowlands, forming **alluvial fans**. These spread out across the desert basin, or **bolson**, where the fine particles can be shaped into dunes by the wind.

Where water flows into a desert depression, **playa lakes** form. When the water eventually evaporates, salt pans or clay pans are formed.

Inselbergs are large masses of resistant rock that rise abruptly from the surrounding plain. They are exposed when the softer surrounding rock material is eroded. Uluru/Ayers Rock, in central Australia, is probably the world's best known inselberg. The landforms of the desert are shown in Figure 10.13.

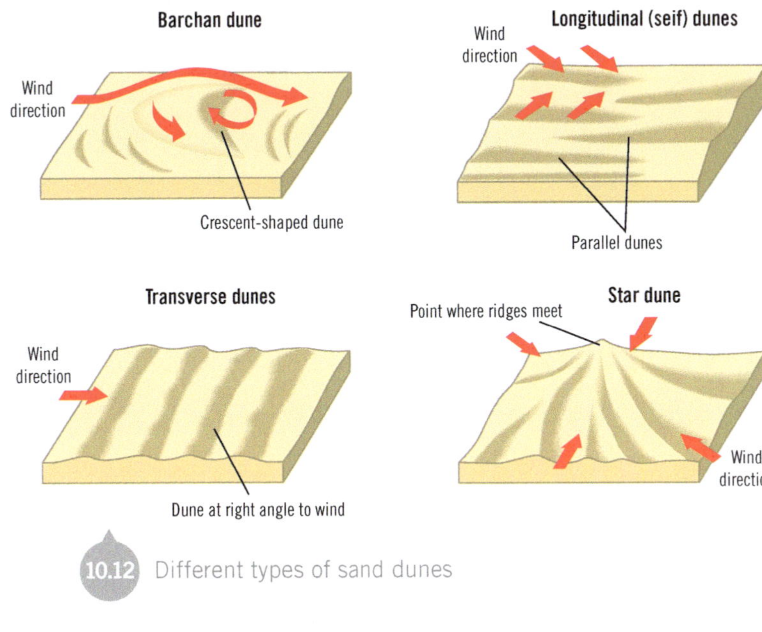

10.12 Different types of sand dunes

10.13 Landforms of the desert

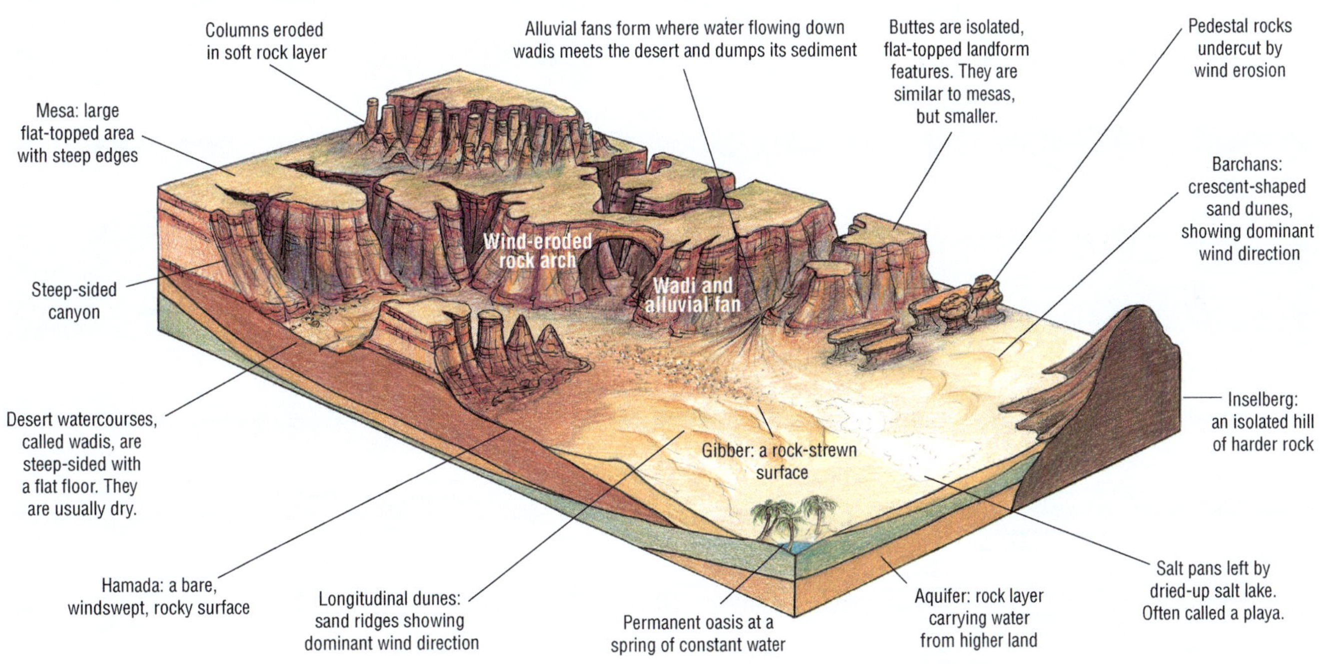

ACTIVITIES

Knowledge and understanding

1. Name and describe the three types of desert surfaces.
2. Explain what abrasion is. What impact does it have on rocks?
3. Define the term 'deflation'.
4. Explain why rainfall is such an effective agent of erosion in desert environments.

10.3 Case study: The Sahara

The Sahara is the world's largest hot desert and the second-largest desert after Antarctica. It is about as large as the area of mainland Australia.

Location

The Sahara covers almost one-third of Africa (see Figure 10.14), stretching from the Red Sea to the Atlantic Ocean. Its boundaries are the Atlas Mountains, the Mediterranean Sea, Egypt, Sudan and valley of the Niger River. It is a very hostile environment in which to live, with extremes of temperature, frequent sandstorms, little vegetation and few surface sources of water.

10.14 The Sahara occupies most of North Africa.

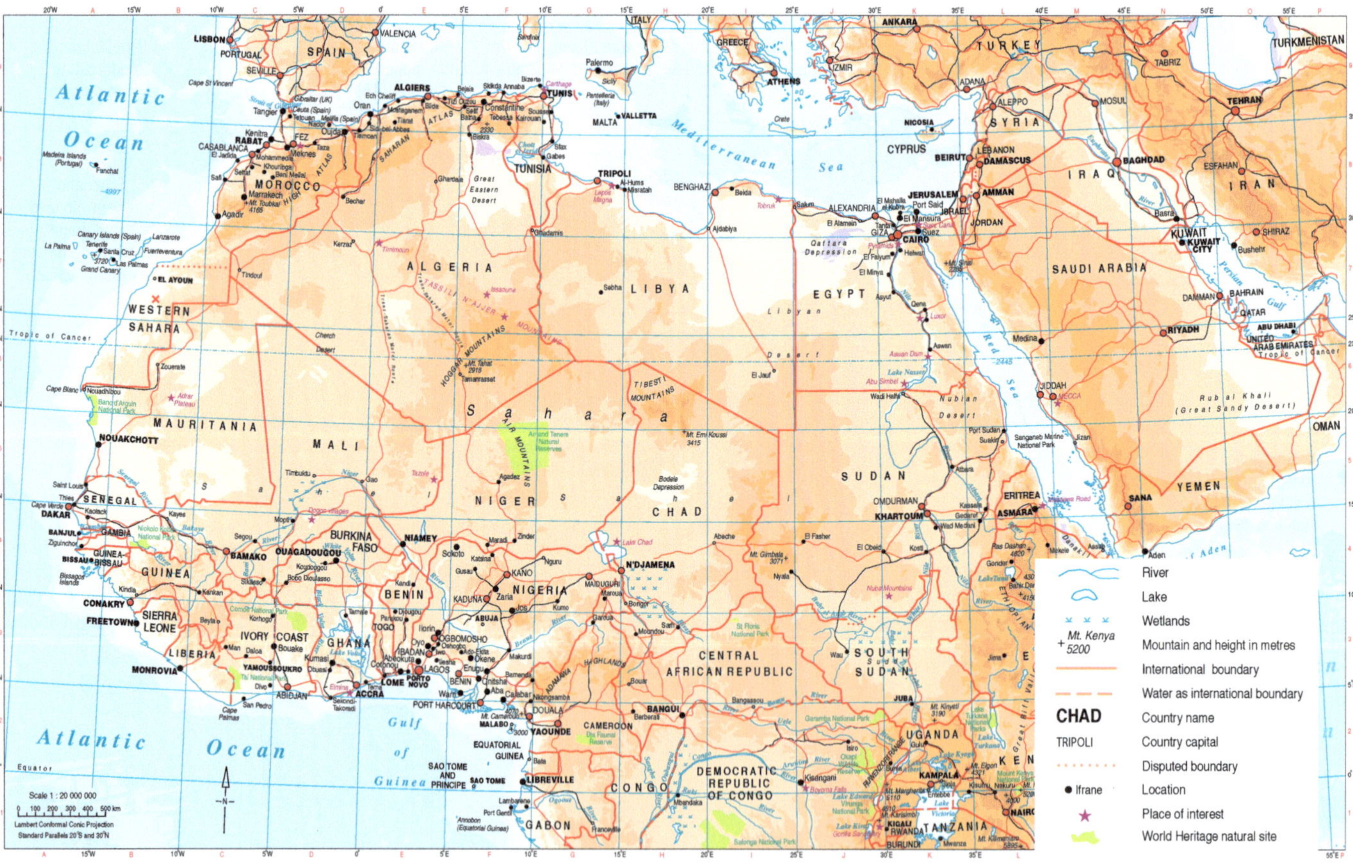

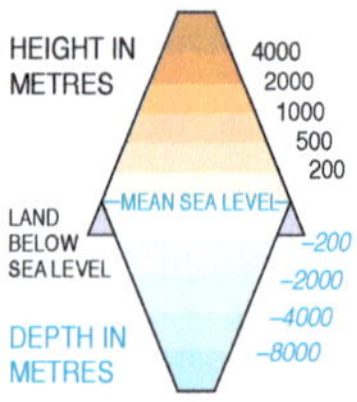

Source: *Heinemann Atlas*, 5th edition

Landforms of the Sahara

One-quarter of the Sahara consists of sand sheets (flat or gently undulating plots of sand) and dunes. These areas are called **ergs**. Some of the dunes are as high as 190 metres and they are constantly being reshaped by the wind. In other parts of the Sahara, there are large areas of stony plains, called **regs**, and high, rocky plateaus and mountains, called **hamadas**, shown in Figure 10.15.

At one point, in the Qattara Depression, the land dips to 133 metres below sea level. Other parts are quite mountainous. Emi Koussi, in the Tibesti Mountains, is the highest peak in the Sahara, with a height of 3415 metres. The Tibesti Mountains, one of the most prominent features of the central Sahara region, is range of inactive volcanoes located in northern Chad.

10.15 Massive sand dunes, some as high as 190 metres, cover a quarter of the Sahara's land surface. In the foreground is a reg surface. A hamada can be seen in the background.

The landforms of the Sahara are undergoing constant change. They are shaped by the direction of the wind and the occasional rainfalls.

The 'Eye of the Sahara'

The Richat Structure, more commonly known as the Eye of the Sahara, is a circular landform feature in Mauritania (Figure 10.16). The landform is a deeply eroded dome, nearly 50 kilometres wide. The Richat Structure formed when an anticline (Figure 10.17), a product of the folding of rock strata, eroded.

10.16 The Richat Structure, Mauritania

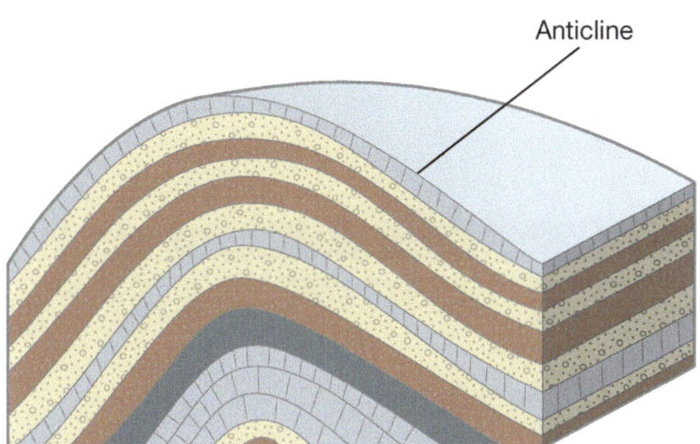

10.17 The formation of the Richat Structure

Water in the Sahara

In the Sahara there is very little water, as rainfall is very low (see Figure 10.18). Some plants and animals are able to make use of the rainwater that filters through the sand or collects briefly on the rocky surfaces. However, humans can find permanent water supplies only in rivers or underground water-bearing rocks. The major rivers in the Sahara are the Niger and the Nile. The Nile River dominates the north-eastern part of the Sahara.

The most widespread source of water in the Sahara is in underground water-bearing rocks called **aquifers** (see Figure 10.19). These lie in layers below the sand and surface rocks. An **oasis** (see Figure 10.20) is formed when these water-bearing rocks reach the surface of a depression in the desert. Due to increasing human use of underground water, the supply is declining, causing some springs to dry up.

10.18 Vast areas of North Africa receive less than 250 millimetres of rainfall a year.

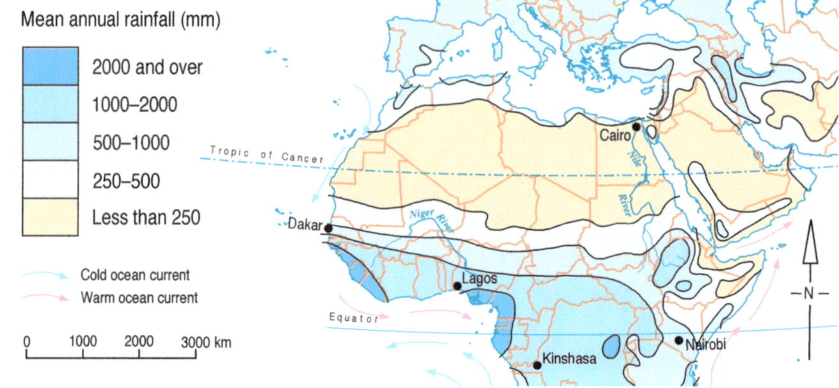

Source: *Heinemann Atlas*, 5th edition

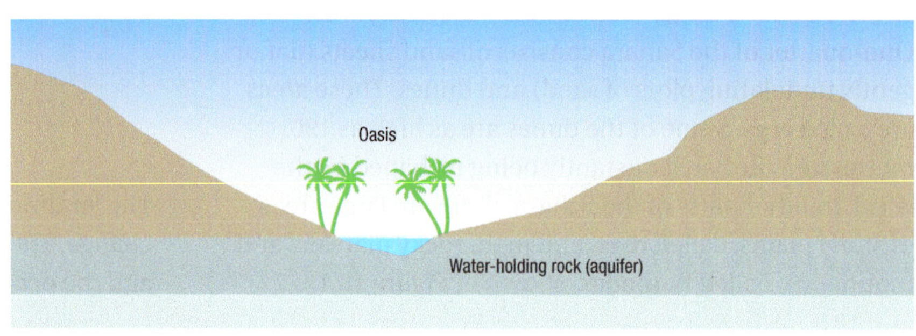

10.19 An oasis develops when an underlying aquifer (a saturated layer of rock) meets a depression in the desert surface.

10.20 A palm tree oasis, Sahara, Chad

SPOTLIGHT

Camels

There are millions of camels living in Africa, most of them in the countries that border the Sahara and the Horn of Africa. The African camels are of the single-humped (dromedary) type. They have a range of adaptations that allow them to cope with the harsh conditions of the desert.

These include:

- a woolly coat on the upper part of the body to provide good insulation from the sun
- a hump that stores fat reserves, which can keep the camel alive when there is no food available
- the production of dry dung, which reduces water loss
- the ability to carry large amounts of water (equal to about 25 per cent of its body weight) in its stomach
- a lack of wool on the underside of the body, which enables heat loss
- a web of tissue between the two toes, which stops the camel from sinking in the sand
- a feathery mouth that enables the camel to eat thorny desert plants
- nostrils that can be closed during sandstorms

Camels are used for transport, meat, milk, wool and leather. They have also made it possible to establish trade routes across the Sahara linking West Africa to North Africa (see Figure 10.21).

10.21 Bedouin camel train among the Sahara's sand dunes

Peoples of the Sahara

The peoples of the Sahara have adapted to the harsh environment and developed ways of obtaining the food and water required for survival.

The Tuareg

'Tuareg' is a term used to identify the diverse groups of nomadic people living in parts of the Sahara. The Tuareg share a common language and history.

For thousands of years, the livelihood of the Tuareg revolved around trans-Saharan trade. There were five main trade routes that crossed the Sahara from the Mediterranean coast of Africa to the great cities on the southern edge of the Sahara. Tuareg merchants transported goods from these cities to the north. From there they were distributed throughout the world.

Tuareg camel caravans continued to transport goods across the Sahara until the mid-twentieth century, when European trains and trucks took over. Today, many Tuareg live in the cities bordering the Sahara.

10.22 Tuareg man dressed in traditional blue robe

ACTIVITIES

Knowledge and understanding

1 Describe the location and extent of the Sahara.
2 Identify and describe the principal landform types found in the Sahara.
3 Explain why the Sahara is such a difficult environment for humans and animals.

Geographical skills

4 Study Figure 10.18. Write a paragraph describing the rainfall pattern of Africa.
5 Construct a photo sketch of 10.15. Annotate your sketch to highlight each of the types of desert landform depicted.

10.4 Case study: Monument Valley, USA

Monument Valley is one of the most famous landscapes in the United States of America and its unique sandstone landforms are instantly recognisable throughout the world. Monument Valley forms part of the Colorado Plateau and is located on the border between Arizona and Utah, the lands of the Navajo Nation.

Origins of Monument Valley

Monument Valley (Figure 10.23) was once a vast lowland basin. For hundreds of millions of years, eroded material was deposited in the basin. Over time, tectonic forces gently lifted the basin up, which elevated the horizontal sedimentary rock by almost 5 kilometres above sea level. The basin became a plateau. Over the next 50 million years, wind and water eroded the surface of the plateau, creating Monument Valley. Figure 10.24 illustrates the formation of Monument Valley's main landform features.

10.23 Monument Valley, United States of America

10.24 The formation of Monument Valley's landform features

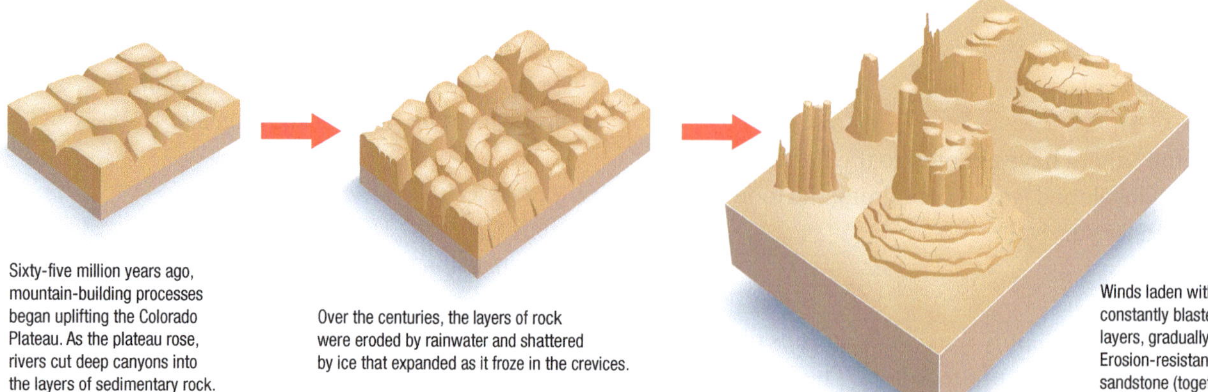

Skillsbuilder

Creating a 3-D model of a mesa and butte

Materials required

A4 paper; ruler; pencil; thick cardboard; scissors; yellow, orange, light-brown, mid-brown and dark-brown paint; glue; plywood suitable for the base; light-yellow paint or sand.

Procedure

1 Create a cross-section of a topographic map, using Figure 10.25.

2 Enlarge the map onto a sheet of A4 paper. Copy this enlargement five times: one for each contour height on the map.

3 Cut around each set of contour lines to create templates.

4 Using the templates, trace each set of contours on to a piece of thick cardboard. Carefully cut around the shape of the contour layer you have traced onto the cardboard. Paint each shape a different colour, starting with the darkest colour for the largest shapes through to the lightest for the smallest.

5 Using glue, assemble the mesa and butte and arrange them on the plywood base so that they are located in the same position as in the map. Paint the base a light-yellow colour or sprinkle sand onto a thin coating of glue. Refer to Figure 10.26 to see what the finished model should look like.

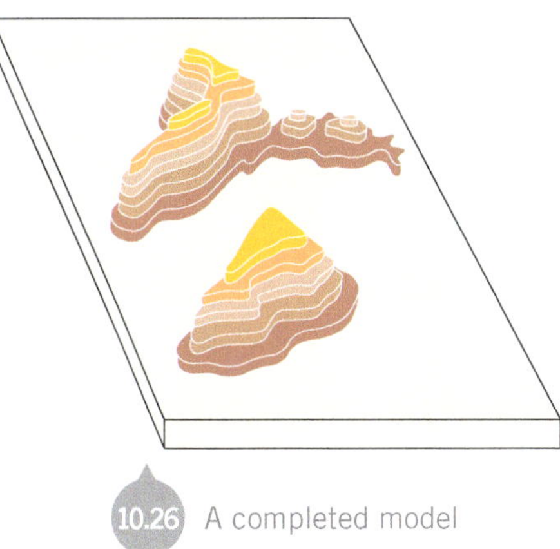

10.26 A completed model

10.25 Constructing cross-sections of mesas and buttes

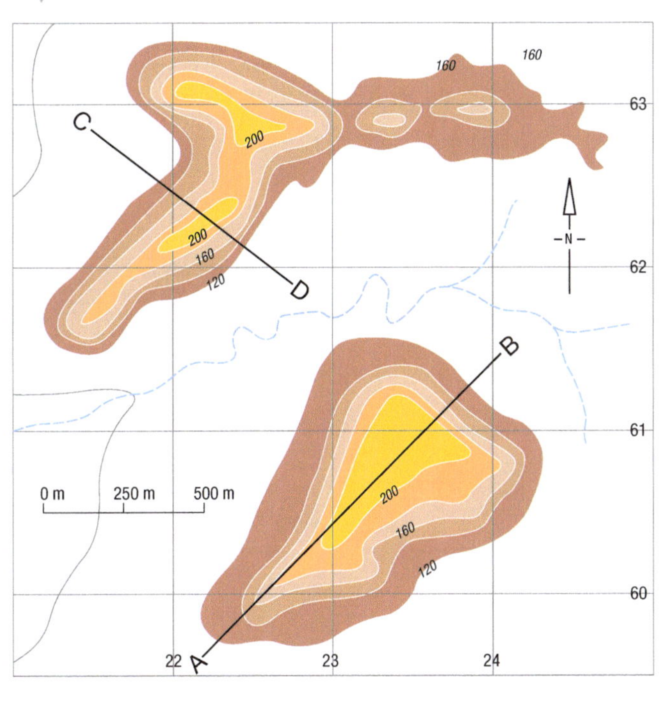

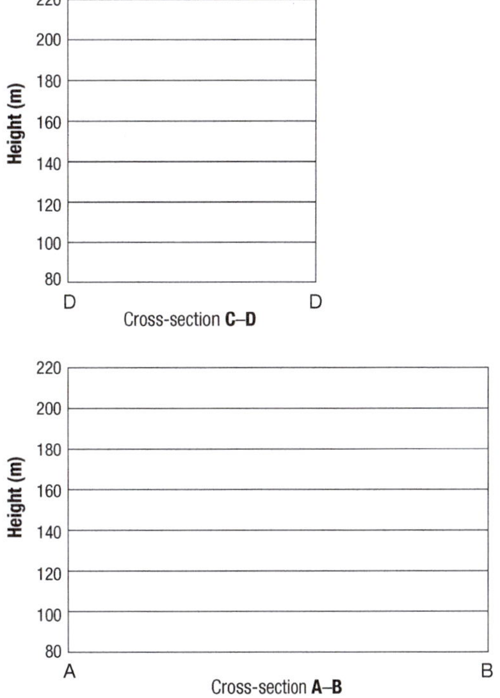

DID YOU KNOW?

Navajo Indians call Monument Valley *Tsé Bii' Ndzisgaii*, meaning 'valley of the rocks'.

ACTIVITIES

Knowledge and understanding

1 Describe the location of Monument Valley.

2 Outline the processes responsible for the formation of the landforms of Monument Valley.

10.5 Australian deserts

Australia's deserts cover approximately 81 per cent (or 1 371 000 square kilometres) of the Australian mainland, and desert is the largest landscape type in the country. Deserts are mainly located on the western plateau and interior lowlands of Australia.

10.27 Australia's deserts

Great Sandy Desert (WA) 267 250 square kilometres (3.5 per cent). The desert is located near the Pilbara and southern Kimberley regions. While the region has an average rainfall greater than 250 millimetres, it is very unreliable. Thunderstorms develop on 20 to 30 days a year.

Little Sandy Desert (WA) 111 500 square kilometres (1.5 per cent). The landforms of the region are similar to the Great Sandy Desert.

Gibson Desert (WA) 156 000 square kilometres (2.0 per cent). The Gibson is dominated by gravel-covered surface with tussock desert grasses. There are also large areas of red sand plains and dunefields. Several saltwater lakes are to be found in the centre of the region. Large areas of the desert remain in a near-natural state.

Great Victoria Desert (WA/SA) 348 750 square kilometres (4.5 per cent of the Australian mainland). The Great Victoria is dominated by many small sand hills, grassland plains, gibber plains and salt lakes. Average annual rainfall is low, ranging from 200 to 250 millimetres per year. Thunderstorms are relatively common.

Tanami Desert (WA/NT), 184 500 square kilometres (2.4 per cent). The Tanami Desert is dominated by a rocky terrain with small hills.

Tirari Desert (SA), 15 250 square kilometres (0.2 per cent). The Tirari is dominated by windswept white sand dunes and numerous salt lakes. The climate is characterised by high temperatures and very low rainfall. Average annual rainfall is below 125 millimetres.

Source: *Heinemann Atlas*, 5th edition

Simpson Desert (NT/Qld/SA) 176 500 square kilometres (2.3 per cent). The Simpson is an erg (a sand-dominated desert). It contains the world's longest parallel sand dunes. These north–south oriented dunes vary in height from 3 metres in the west, to around 20 metres in the east. The tallest dune, Nappanerica, is 40 metres high. Under the desert lies the Great Artesian Basin—a vast underground reservoir of groundwater.

The Simpson Desert contains the world's longest parallel sand dunes.

Strzelecki Desert (SA/Qld/NSW) 80 250 square kilometres (1.0 per cent). The Strzelecki is located to the north-east of the Lake Eyre Basin and to the north of South Australia's Flinders Ranges. The Tirari and Simpson deserts are also in the Lake Eyre Basin.

Pedirka Desert (SA) 1250 square kilometres (0.016 per cent). Pedirka is Australia's smallest desert. The landscape is dominated by a deep red longitudinal (north–south) dunefield. Dense scrub and grasses grow in between the dunes.

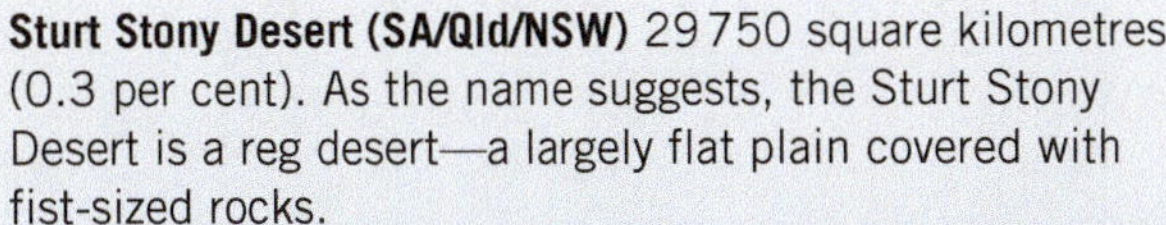

Sturt Stony Desert (SA/Qld/NSW) 29 750 square kilometres (0.3 per cent). As the name suggests, the Sturt Stony Desert is a reg desert—a largely flat plain covered with fist-sized rocks.

The Sturt Stony Desert

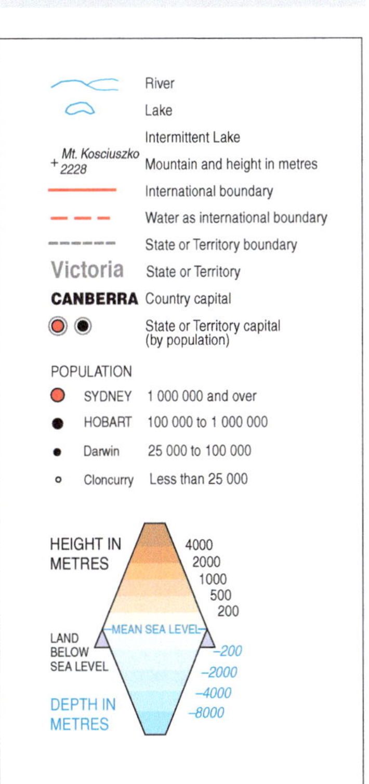

Famous Australian desert landforms

10.30 Wave Rock, a 14-metre granite cliff face near Hyden, Western Australia, weathered into a perfect wave formation.

10.31 Uluṟu/Ayers Rock, the world's second largest monolith (only Mt Augustus, WA, is larger), central Australia. Uluṟu is an inselberg, an isolated residual landform that rises abruptly from and is surrounded by extensive and relatively flat erosion lowlands.

10.32 Kata Tjuṯa (the Olgas), a large group of rock domes in central Australia

10.33 The Pinnacles, Western Australia—a vast desert of upright sandstone formations, weathered over time into weird shapes

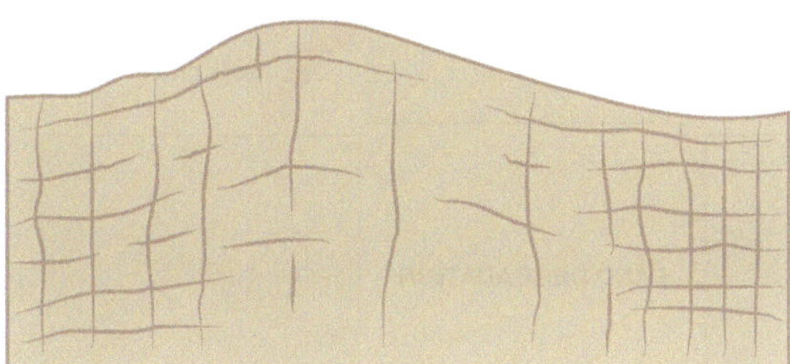

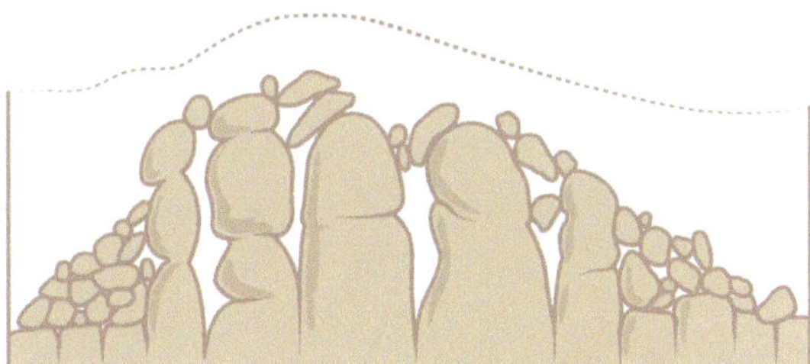

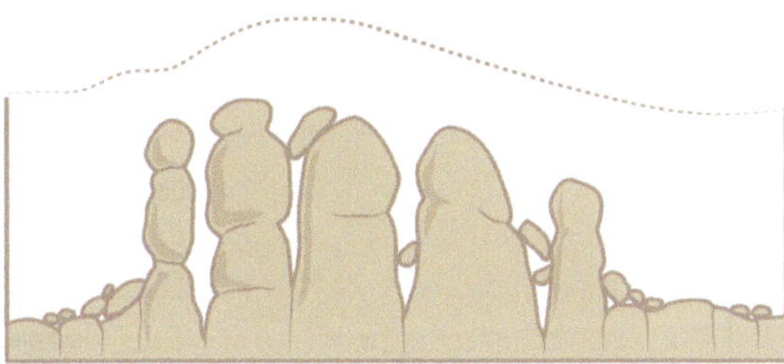

10.35 How inselbergs such as Kata Tjuta were formed

10.34 Kati Thanda–Lake Eyre, 15 metres below sea level in the South Australian outback

ACTIVITIES

Knowledge and understanding

1 Study Figure 10.27. List the Australian states and/or territories that have no desert regions.
2 Study Figures 10.30 to 10.34. Write a description of the different types of deserts. Comment on the sand, rocks and vegetation.
3 Study Figure 10.35. Write a paragraph outlining how inselbergs are formed.

10.6 Desertification

Since the early 1970s, deserts have been increasing in size. Semi-arid areas on the edges of deserts are getting drier and becoming deserts. This process is called desertification.

Causes of desertification

There are two main theories about the causes of desertification: it is due to natural processes or it is a result of human behaviour. Figure 10.36 shows the process of human and natural causes of land degradation.

Natural causes

Climate change has reduced the amount of rainfall and shortened the wet season. As a result, vegetation dies, soil is blown away and the land becomes degraded.

Human causes

Deserts and semi-desert lands have become degraded because of poorly managed landuse on the edges of deserts. In these areas, too many cattle graze the land, crops are grown on land that is not really suited to agriculture and people cut down trees to use as fuel for cooking and providing warmth.

Vegetation binds the soil and adds nutrients.
Vegetation lost due to overgrazing, fuel collection, poor landuse practices and climate change

No vegetation to intercept rain. Soil eroded. Soil moisture is lost. Soils crack

LAND DEGRADATION

Land is degraded. Soil loses fertility and structure

Soil is left exposed. Wind blows away soil

10.36 The process of land degradation

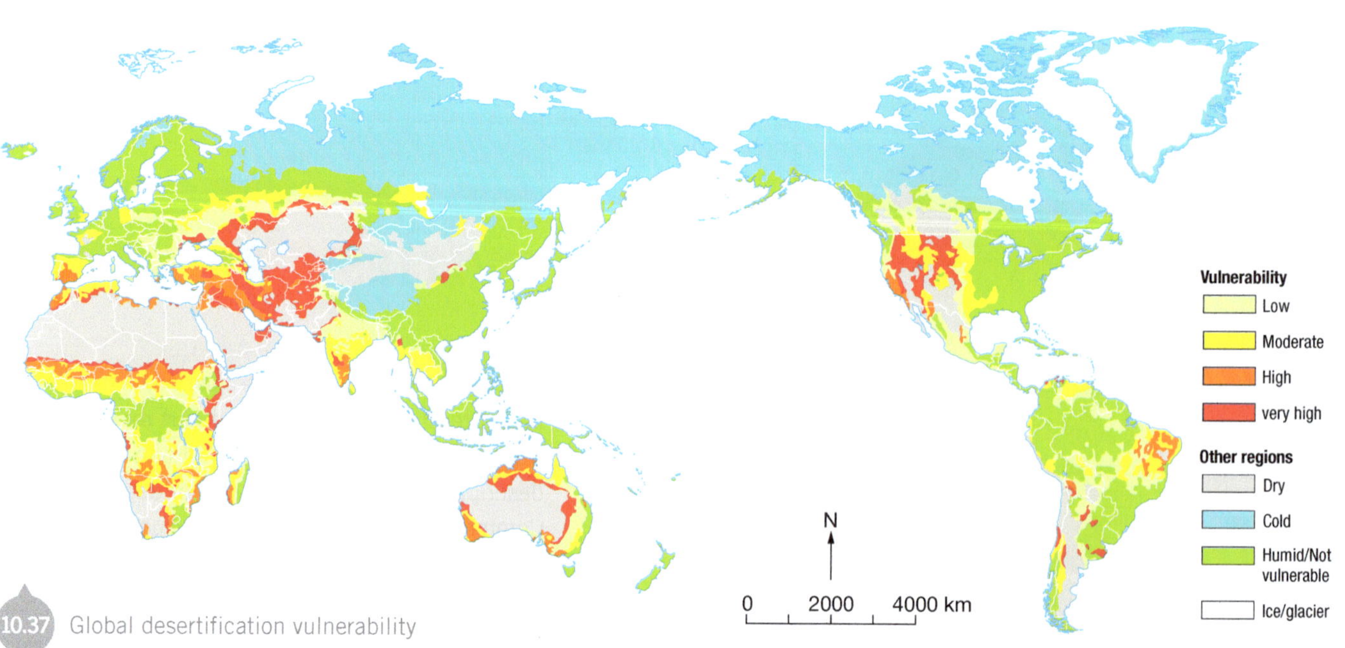

10.37 Global desertification vulnerability

Some scientists argue that desertification is caused by a combination of both natural and human factors and may also be linked to the process of global warming.

Impacts of desertification

When land becomes degraded, fewer crops are able to be grown and fewer animals are able to be grazed. An estimated 8.1 million square kilometres worldwide—an area the size of Brazil in South America—have been affected by desertification in the last 50 years (Figure 10.37). If this continues, the wellbeing of at least 1.2 billion people could be threatened.

Responding to desertification

The most effective way to slow desertification is to drastically reduce overgrazing, deforestation and excessive cultivation. Reforestation (tree-planting) programs will help to bind the soil and hold water while slowing the process of desertification.

SPOTLIGHT

The Green Wall of China

The Green Wall of China is a massive environmental project that involves creating a series of human-planted forests to prevent the expansion of the Gobi (Figure 10.38). By 2050, the project will feature a 4500-kilometre forested strip that will increase forest cover in northern China from 5 to 15 per cent.

The Chinese have used aerial seeding to promote tree growth over this vast area. They have also offered cash to farmers to plant trees.

The Green Wall of China involves the planting of sand-tolerant vegetation in a checkerboard pattern. The aim is to stabilise the developing sand dunes and reduce sandstorms by planting windbreaks (Figure 10.39).

10.38 Desertification in China. The country now has more than 2.62 million square kilometres of land under desertification, twice the amount of the total available farmland in China.

10.39 Massive tree-planting programs aim to halt desertification on the edges of the Gobi. Here, workers use hay to create grid patterns that stabilise sand dunes.

ACTIVITIES

Knowledge and understanding

1. Define desertification and outline the two theories about why desertification is occurring.
2. State how much land has been affected by desertification worldwide in the last 50 years.
3. Explain how the process of desertification can be slowed.
4. Explain the Green Wall of China.

Geographical skills

5. Study Figure 10.37.
   a. With the aid of an atlas, outline the location of places that are under threat of desertification.
   b. Describe the location of the areas that are not under threat of desertification.
   c. Describe the location of the areas in Australia that are under threat of desertification.

Review and reflect 2

Activity 1

Topographic mapping 1

Study the Laurieton topographic map extract in Figure R&R 2.1 and then answer the following questions.

a State the scale of the Laurieton topographic map extract.

b State the contour interval of the Laurieton topographic map extract.

c Name the feature of the natural environment located at the following grid references:
  - i GR 801974
  - ii GR 813994
  - iii GR 831979
  - iv GR 845986

d Name the feature of the constructed environment located at the following grid references:
  - i GR 848994
  - ii GR 812987
  - ii GR 842001
  - iv GR 810976

e State the direction of Camden Head (AR 8498) from the summit Oval (AR 8198).

f State the direction in which the Camden River is flowing in AR 8399.

g State the direction in which the creeks are flowing in AR 8097.

h State the bearing of Hamey Lookout (AR 8498) from the survey landmark at Dicks Hill (GR 831979).

i State the aspect of the slope in AR 8096.

j State the straight-line distance between Hamey Lookout (AR 8498) and Dicks Hill (AR 8397).

k State the length of the bridge in AR 8197.

l State the length of the Camden Haven River's northern breakwater.

m Estimate the area of Gogleys Lagoon.

n State the difference in elevation between Laurieton (GR848987) and Dicks Hill (AR 8397).

o Identify the type of vegetation found at the following grid and area references:
  - i AR 8096
  - ii GR 830990
  - iii GR 820984

p Identify the main economic activity in Godleys Lagoon.

q Construct the cross-section between points A and B using a vertical scale of 1 cm = 100 m.

r Calculate the vertical exaggeration of your cross-section.

s State the local relief in AR 8498.

Activity 2

Developing a natural hazard information brochure

Using internet-based resources and relevant software, develop a multiple-page website promoting community awareness about a selected geomorphological hazard. Include links to relevant government authorities, for example Geoscience Australia.

Your website should include information on:

- the nature of the geomorphological hazard
- the geographical processes involved
- the economic, environmental and social impacts of the geomorphological hazard
- the responsibilities of the various levels of government in respect to the geomorphological hazard
- the community-based groups involved in responding to the geomorphological hazard
- the strategies individuals can use to protect themselves and their property.

Good sources of information are:

- Emergency Management Australia
- Geoscience Australia
- US Geological Survey
- British Geological Survey.

Laurieton topographic map extract

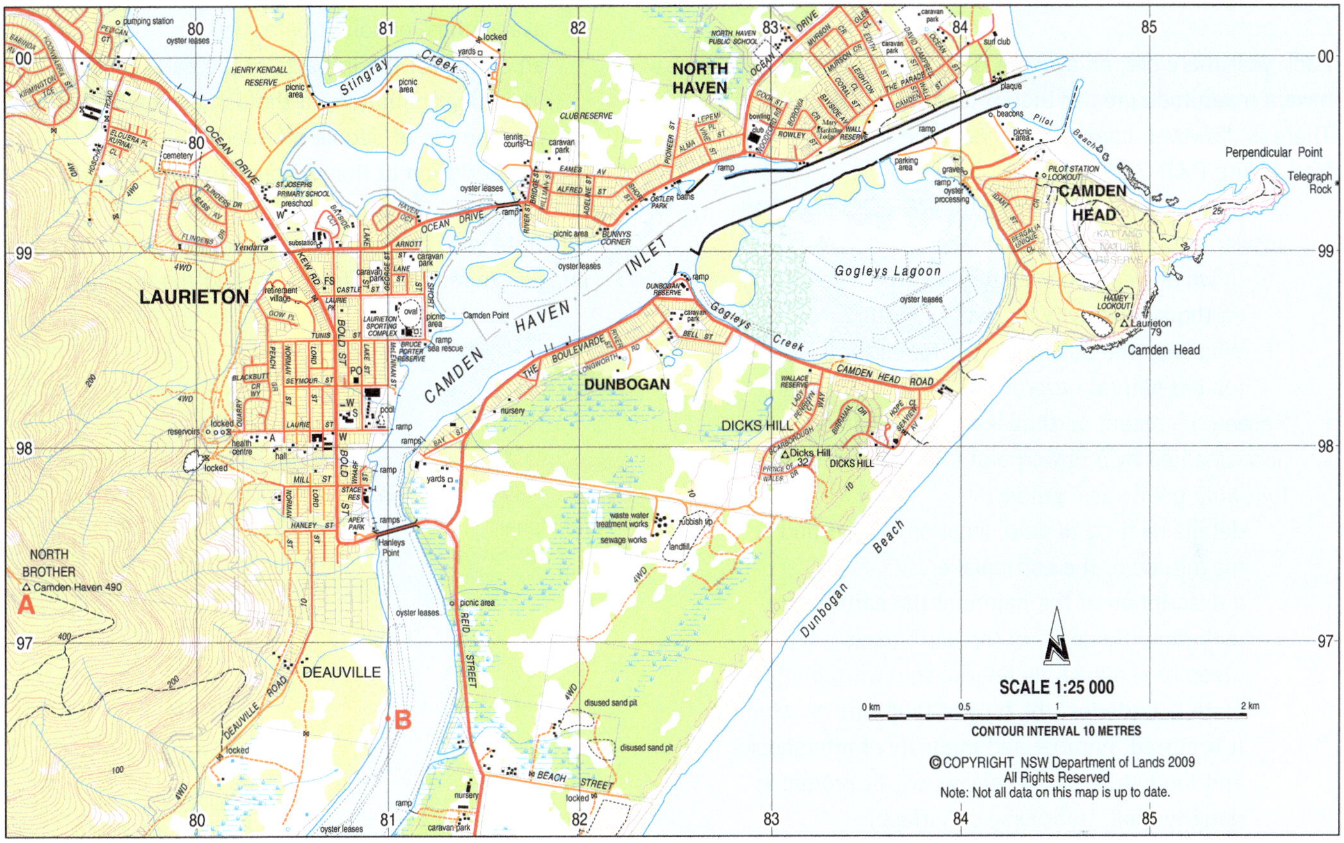

LAURIETON
NORTH HAVEN
DUNBOGAN
CAMDEN HEAD
DICKS HILL
DEAUVILLE
NORTH BROTHER
Camden Haven 490
CAMDEN HAVEN INLET
Gogleys Lagoon
Gogleys Creek
Stingray Creek
Dunbogan Beach
Perpendicular Point
Telegraph Rock
Laurieton 79
Camden Head
A
B

SCALE 1:25 000

0 km 0.5 1 2 km

CONTOUR INTERVAL 10 METRES

Note: Not all data on this map is up to date.

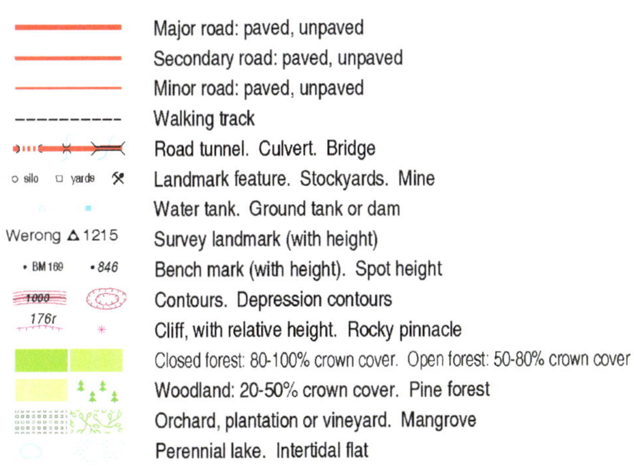

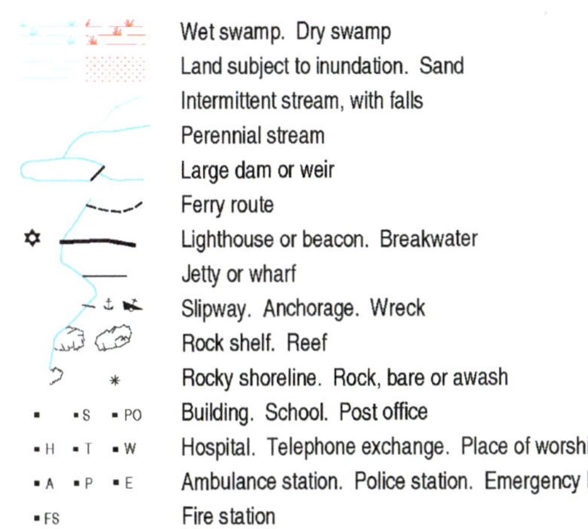

Review and reflect 2

Activity 3

Earthquake watch

Each year there are, on average, 150 earthquakes that have a magnitude greater than 6 (see Table R&R 2.2). There is, however, considerable variation from year to year (see Figure R&R 2.3)

a Study Table R&R 2.2 and Figure R&R 2.3 and then answer the following questions.
   i In which years this century was the level of earthquake activity highest?
   ii What was the deadliest year for earthquake-related natural hazards?

b Research a recent earthquake. Develop an oral report accompanied by a PowerPoint presentation using the following points as a guide:
   - details relating to date, location, extent and magnitude of the earthquake
   - a description of the nature of the earthquake, including the depth of the earthquake's focus (deep or shallow), its relationship with crustal plate boundaries, the type of fault line on which it occurred, number and intensity of aftershocks, and the nature of the seismic waves produced (longitudinal, transverse or surface)
   - a description of any associated geomorphological hazards, for example tsunamis, landslides and avalanches, or soil liquefaction
   - an outline of the impacts of the earthquake, including the nature of the damage caused and the number of people affected
   - a description of the response to the disaster.

Illustrate your presentation with appropriate maps and images sourced from the media and the internet.

Average annual occurrence of earthquakes by magnitude

Magnitude	Number of earthquakes (per year)
8 and higher	1
7–7.9	15
6–6.9	134
5–5.9	1319
4–4.9	13000
3–3.9	130000
2–2.9	1300000

Worldwide earthquakes, 2000–12

Activity 4

Mining, people and landscapes: You be the judge!

The rush to meet the growing global demand for coal and energy sources such as Coal Seam Gas (CSG) has caused conflict, especially in regional Australia. There is also a concern that mining and gas extraction will transform Australian landscapes and landforms.

There has been considerable hostility in some regions as landholders have realised that while they may own their land, they do not own the mineral rights to that land. Farmers in the United States of America can sell their mineral rights, but in Australia these rights belong to the state. So the state can issue a licence to a mining or energy company wanting to extract resources.

Arguments for coal mining and CSG extraction

Many landholders and communities in areas affected by coal mining and CSG extraction are supportive of such projects as long as environmental concerns can be satisfactorily resolved. Coal mines generate many jobs and earn substantial export earnings. Miners are capable of rehabilitating mined land, and gas extraction, if kept some distance from homesteads, does not significantly impact on people's wellbeing. The income from access fees over the life of the gas wells can supplement the landholder's income. Coal and gas are important in maintaining our material wellbeing. Without CSG extraction household energy costs would rise significantly.

State governments rely on the royalties from coal and gas exploitation. The money raised is used to fund a wide range of government activities, including in the areas of healthcare, education, law and order and infrastructure.

Arguments against coal mining and CSG extraction

People who are against coal mining and CSG extraction are concerned about the environment. They see mining and extraction as a threat to food production and the unique Australian rural landscape.

Issues raised in objection to coal mining and CSG extraction include the following.

- *Landscapes:* Coal mining and CSG extraction can transform landscapes. Many people feel the loss of the distinctive Australian rural landscape. This has an impact on the aesthetics of the landscape and, in some cases, its spiritual and recreational value.
- *High-value agricultural land:* Landholders of fertile cropping country, or specialised farms such as wineries and olive groves, oppose open-cut coal mining and CSG extraction because the necessary infrastructure, noise and traffic would have a significant impact on their operations.
- *Lifestyle:* Many rural communities, especially those where several generations of families have worked the land, believe that their identities and ways of life are at risk.
- *Global climate:* The burning of fossil fuels produces carbon dioxide gas.
- *Aquifers:* Coal mining can disrupt the movement of water in aquifers, as can CSG extraction. There is a fear that they will alter groundwater levels and the bores that farmers rely on may fail.
- *Water:* In some coals seams it is necessary to create fractures that provide pathways through which the gas can flow. The technique used to do this is known as hydraulic fracturing or fracking. This is done by injecting fluid made up of water, sand and chemicals under high pressure into the well. The fluid is then pumped back to the surface, where it has to be disposed of. There are concerns that some gas could escape and that the chemicals used could contaminate both ground and surface water.

Hypothetical

A large transnational corporation has been granted a licence to search for CSG in the coal body underlying one of Australia's most productive agricultural regions. A large Australian-owned coal company already operates a large open-cut mine in the area and has lobbied the government to expand its operations.

a As a class, make a list of statements that are in favour of the proposals to develop a CSG field in the area and expand the existing open-cut coal mine. Make a separate list of statements against the development.

b In groups of four or five, discuss the statements about the development proposal.

c Determine which point of view you agree with. Write an exposition outlining the arguments you would use to justify your position.

d Brainstorm the strategies you could use to influence public opinion and the government's decision-making processes.

CHAPTER 11

URBANISATION

The world has never before experienced urbanisation at the scale and speed that we see now in the countries of the developing world. Megacities are emerging from Jakarta to Istanbul, São Paulo to Cairo. Poor rural families are flooding into the world's urban centres, bringing challenges that have never before been seen—or met. Even the large cities of the developed world are facing challenges.

In this chapter we look at the process of urbanisation and its causes and consequences. We also look at the economic, environmental and social advantages and disadvantages of living in large cities.

KEY IDEAS

- To understand the causes and consequences of urbanisation
- To describe the economic, environmental and social advantages and disadvantages of living in large cities

11.0 Istanbul, Turkey

GLOSSARY

commute	to travel regularly between home and work, school or university
developing world	poor and middle-income countries of the world
formal economy	economic activities that are regulated and taxed by government
global economy	integrated world economy in which there are few restrictions on the free movement of goods, services and labour across borders
hukou system	a Chinese household registration system that entitles the holder to a range of services such as healthcare and education, used to regulate the movement of people in China
informal economy	economic activities that are not taxed or regulated by any form of government
infrastructure	physical structures such as buildings, roads, water pipelines, sewers, electricity distribution systems, railways and airports
megacity	a city with more than 10 million people
push and pull factors	the factors that cause people to leave the places where they live and the factors that draw people to places
quality of life	the happiness, wellbeing and satisfaction that a person experiences
regional centre	a rural city that supports smaller surrounding towns with services
sea change	a relocation from the city to the coast
squatter settlement	an informal, often illegal, settlement, built by the poor using material from the streets
standard of living	a measure of the economic wellbeing of people
tree change	a relocation from the city to a rural or regional area
urban sprawl	the outward spread of a city and its suburbs as they grow
urbanisation	the process by which an increasing proportion of a population lives in towns and cities
world city	a city considered to be an important centre of global economic activity, such as New York, Tokyo, London and Paris

11.1 World cities

In 2007, for the first time in human history, more than half of the world's population were living in towns and cities. This was a big increase from 1800, when just 3 per cent of people lived in cities.

Growth

Today's urban population of 3.2 billion will rise to nearly 5 billion by 2030, when three out of five people will live in cities. Current urban population statistics are shown in Table 11.1. In the future, it is estimated that 93 per cent of urban growth will occur in developing nations, with 80 per cent occurring in Asia and Africa. In 1950, there were 83 cities with populations exceeding one million; by 2008, this number had risen to 468. Figure 11.2 shows the increases in urban populations for different world regions, both historical and predicted.

Urbanisation

Urbanisation is the movement of people from rural areas to large cities, and occurs because of a number of **push and pull factors** (see Figure 11.3).

The most rapid rates of urbanisation are found in developing countries. Due to the enormous economic costs of providing employment, housing, transport, clean water, electricity and sewerage systems, developing countries are experiencing difficulties in meeting the infrastructure needs of their rapidly growing populations. The world pattern of urbanisation is shown in Figure 11.4.

11.1 The proportion of people living in urban areas

Area	Percentage of people living in urban areas in 2011
Africa	39
North America	80
South America	80
Asia	44
Europe	71
Oceania	66
World	51
Developed countries	75
Less developed countries	46
Least developed countries	28

Source: Population Reference Bureau World Population Data Sheet, 2011

11.2 Urban and rural populations, 1950–2050

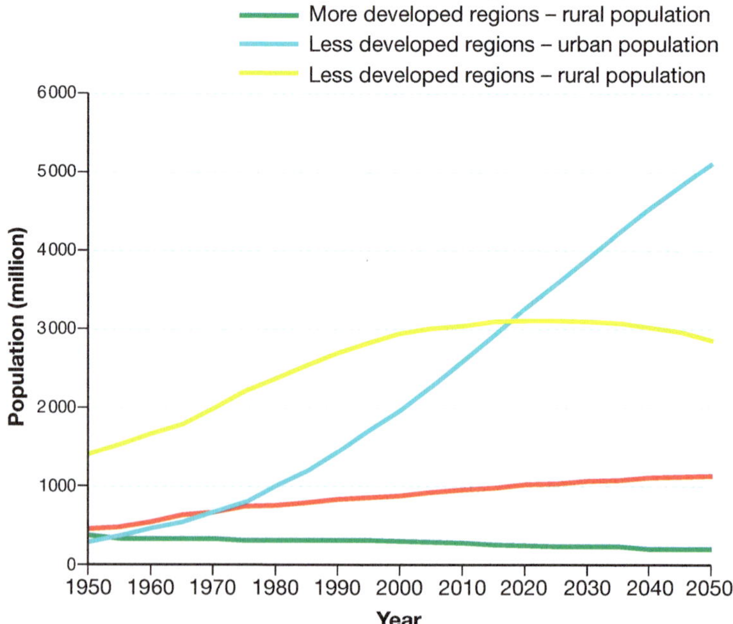

Source: World Urbanization Prospects, The 2011 Revision

11.3 The decision to move to large cities often involves both push and pull factors.

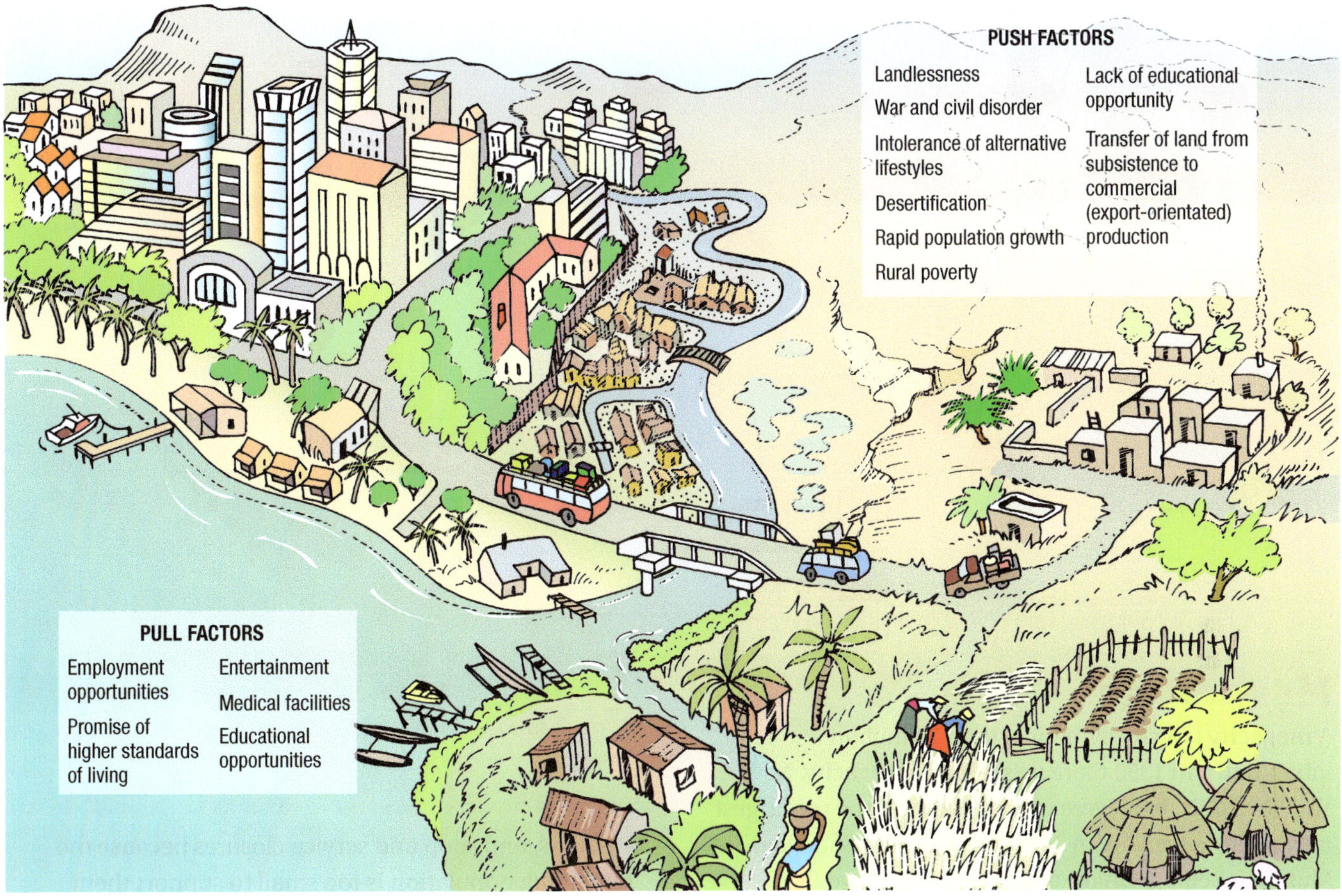

PUSH FACTORS
Landlessness
War and civil disorder
Intolerance of alternative lifestyles
Desertification
Rapid population growth
Rural poverty
Lack of educational opportunity
Transfer of land from subsistence to commercial (export-orientated) production
PULL FACTORS
Employment opportunities
Promise of higher standards of living
Entertainment
Medical facilities
Educational opportunities

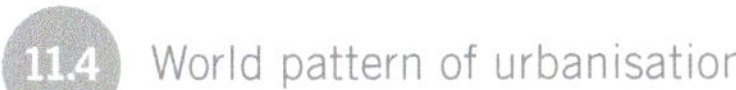

11.4 World pattern of urbanisation

Urban population
65–100%
35–65%
0–35%
N/A
N
0 1000 2000 3000 4000 km

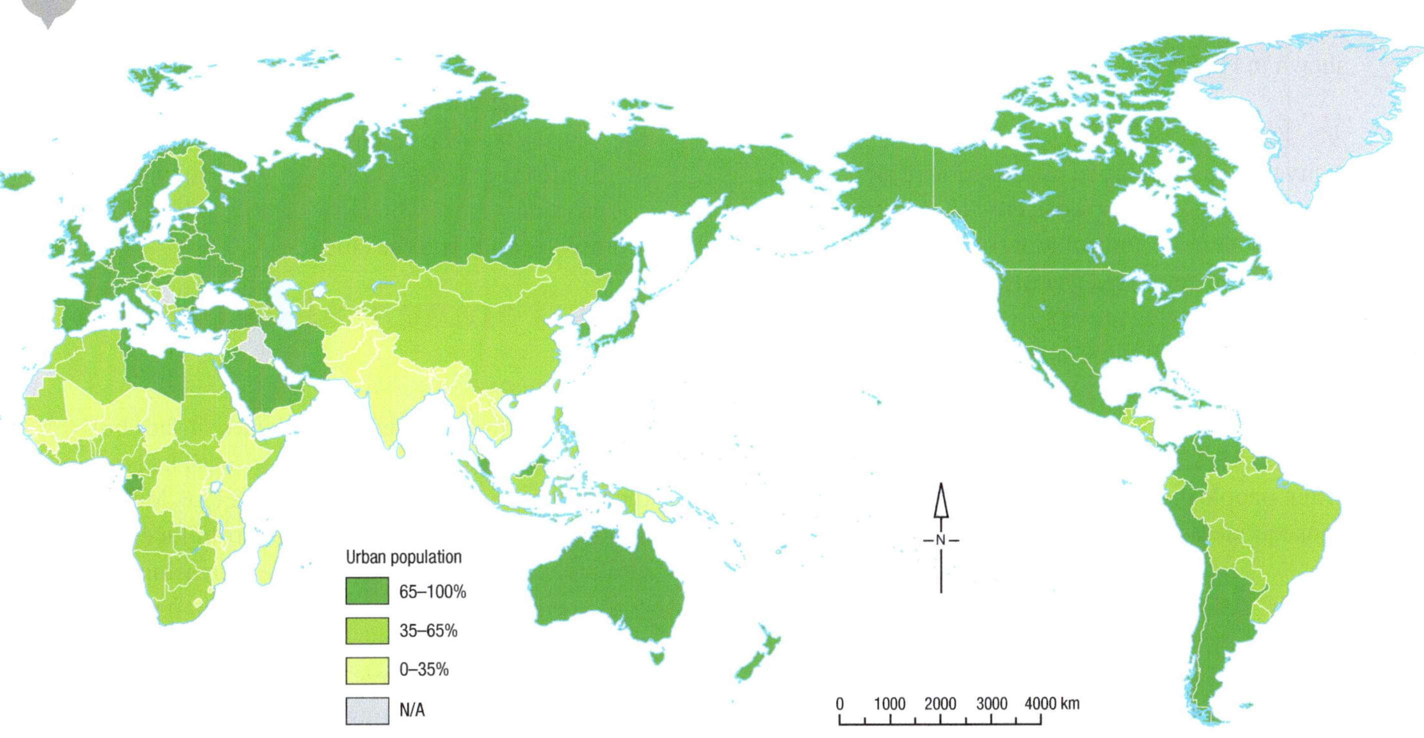

Source: World Urbanization Prospects, The 2011 Revision

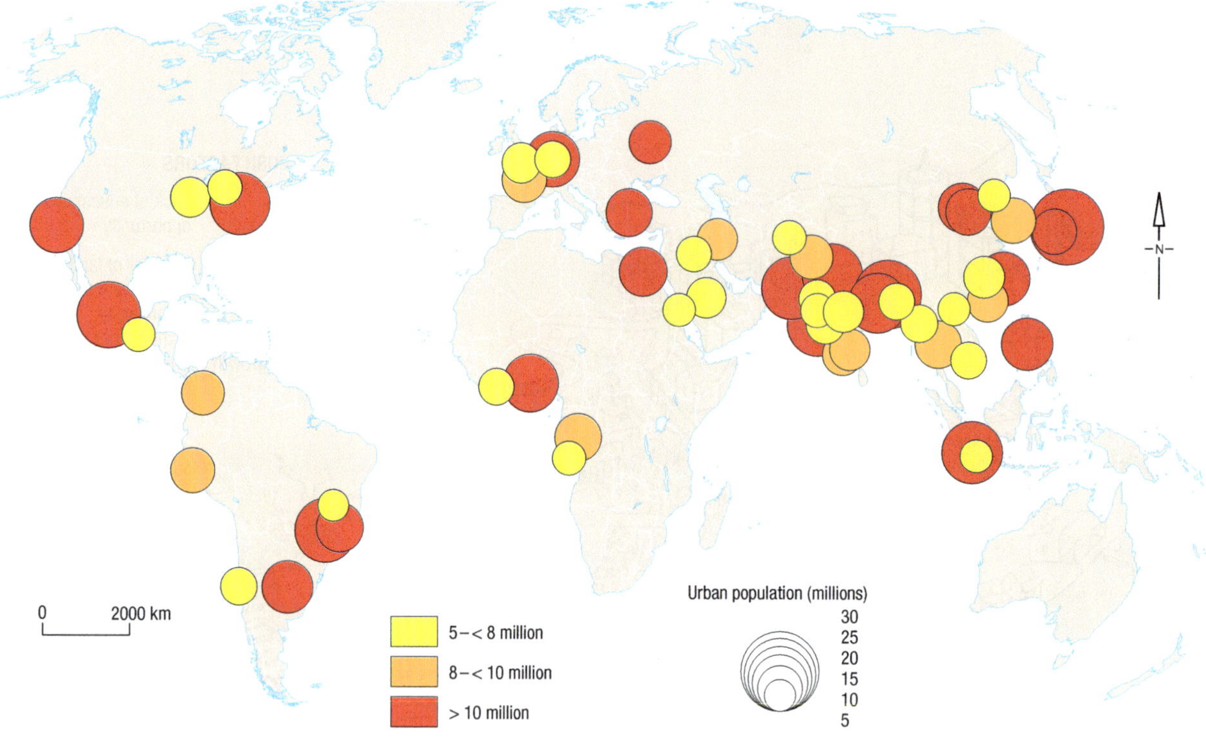

11.5 The distribution of the world's largest cities. The biggest concentration of world cities is found in southern and eastern Asia.

Megacities

A **megacity** is a city with more than 10 million inhabitants. In 1950 there was just one megacity, New York; by 2011, there were twenty-three. Of the ten largest megacities, seven are in Asia (Tokyo, Delhi, Shanghai, Mumbai, Beijing, Dhaka and Kolkata) and two (São Paulo and Mexico City) are in South America. New York City is the only megacity in the developed world in the top ten. The distribution of the world's largest cities is shown in Figure 11.5.

World cities

A city that has developed into a centre of global significance is called a **world city**. These cities have a level of economic and cultural control that extends well beyond their countries. Decisions made in these cities affect the wellbeing of people throughout the world. The most important of these cities are New York, London, Paris and Tokyo. Sydney is also classified as a world city.

Impacts on rural areas

When people move from rural areas to the city there are consequences in rural areas.

More developed

In more developed countries, such as Australia, it is usually young people and people with young families who migrate to the cities. There can be severe effects on rural areas when families move to the cities. These effects include primary school closures due to lack of students, and shop and service closures because the remaining population is too small to support them.

Less developed

In less developed countries it is usually men between the ages of 15 and 45 who migrate to the cities. This means that families are split up. If the family goes too, the loss of a family can have an even more severe effect on the area they leave. Often the elderly are left in the countryside—those who are least able to look after themselves and tend the land.

Environmental impacts

Carbon emissions

Cities cover three per cent of the land's surface but are responsible for 70–75 per cent of fossil fuel carbon dioxide emissions, as shown in Figure 11.6. Megacities are the biggest contributors to human-induced carbon emissions. Populations in cities are increasing and carbon emissions are increasing faster than population growth. The World Bank predicts that in the developing world megacities will grow by 4 per cent while their carbon emissions will grow by 10 per cent in the next 20 years. People living in cities on average use 5–10 times more energy than those living in rural areas.

11.6 Cities responsible for 70 per cent of fossil-fuel emissions

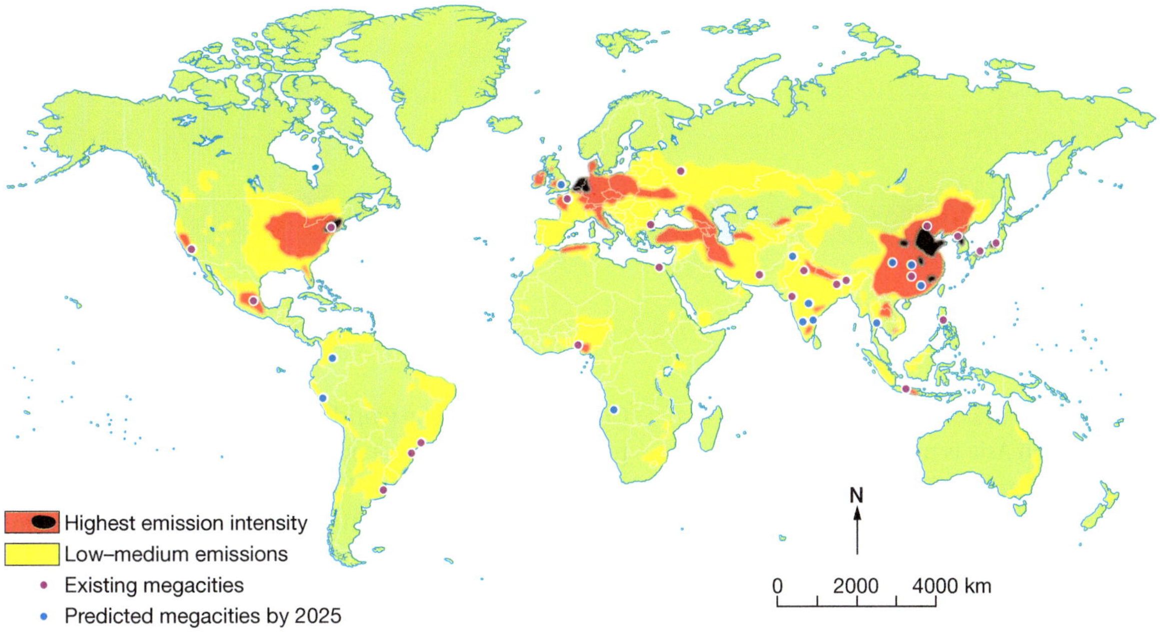

Highest emission intensity
Low–medium emissions
• Existing megacities
• Predicted megacities by 2025

N

0 2000 4000 km

Source: NASA

Urban heat island effect

The heat island effect occurs when cities are warmer than the surrounding areas. The heat comes from the built surfaces such as roads and buildings, which absorb heat during the day and slowly release it at night.

Water consumption

Cities use large amounts of water and produce large quantities of wastewater. As cities increase in size by population and area there is increased pressure on:

- access to safe water
- proper sanitation
- stormwater and wastewater disposal.

ACTIVITIES

Knowledge and understanding

1. Explain the process of urbanisation in your own words.
2. State where the rates of urbanisation are greatest. What problems does this create?
3. Outline the impact that urbanisation has on rural areas in 'more developed' and 'less developed' countries.
4. Explain the difference between a megacity and a world city. Give examples of each.

Geographical skills

5. Study Table 11.1 and Figure 11.2 and do the following tasks.
   a. Identify those parts of the world with the highest and lowest levels of urbanisation.
   b. Describe the trends in rural and urban populations in the developed and developing worlds using data from Figure 11.2.
6. Study Figure 11.4. With the aid of an atlas, do the following tasks:
   a. List the countries with more than 65 per cent of their population living in urban areas.
   b. Name ten countries in which 0 to 35 per cent of the population live in urban areas. Where are the majority of such countries located?
7. Study Figures 11.5 and 11.6 and explain the following statement. *There is a strong link between the location of megacities and increased carbon dioxide emissions.*

11.2 Urbanisation in Asia

Urban populations in Asia are huge. Twelve of the world's twenty-three megacities are found in Asia. The two largest cities in the world are in Asia, and each has a population greater than 20 million. The rapid growth of urban populations has presented governments and people living in such cities with a variety of challenges.

Growth in urban areas

Urban population growth in Asia is predicted to grow until 2025, as shown in Figure 11.7. The demographic characteristics of a range of Asian countries and aspects of their quality of life are outlined in Table 11.8.

China now leads the world in urbanisation. By 2050, 75 per cent of the Chinese population—1.1 billion people—will live in cities. In the Philippines and Indonesia, two-thirds of the population are expected to be living in urban areas by 2025.

Scale of urbanisation

The level and rate of urbanisation vary across Asia. For example, Singapore, shown in Figure 11.9, is 100 per cent urban. In the Lao People's Democratic Republic (Laos), just 33 per cent of the population live in urban centres (see Figure 11.10). The annual rate of increase in the urban population of Laos is relatively high at 4.9 per cent. Compare this to Japan, which has a level of urbanisation of 67 per cent and an annual rate of increase of just 0.2 per cent.

It is important to remember that even though cities can have low rates of growth, the actual numbers involved can be very large. Seoul, South Korea, must make room for over 350 000 new residents each year, even though the actual growth rate is low.

11.7 The growth of urban areas in Asia, 1950–2025

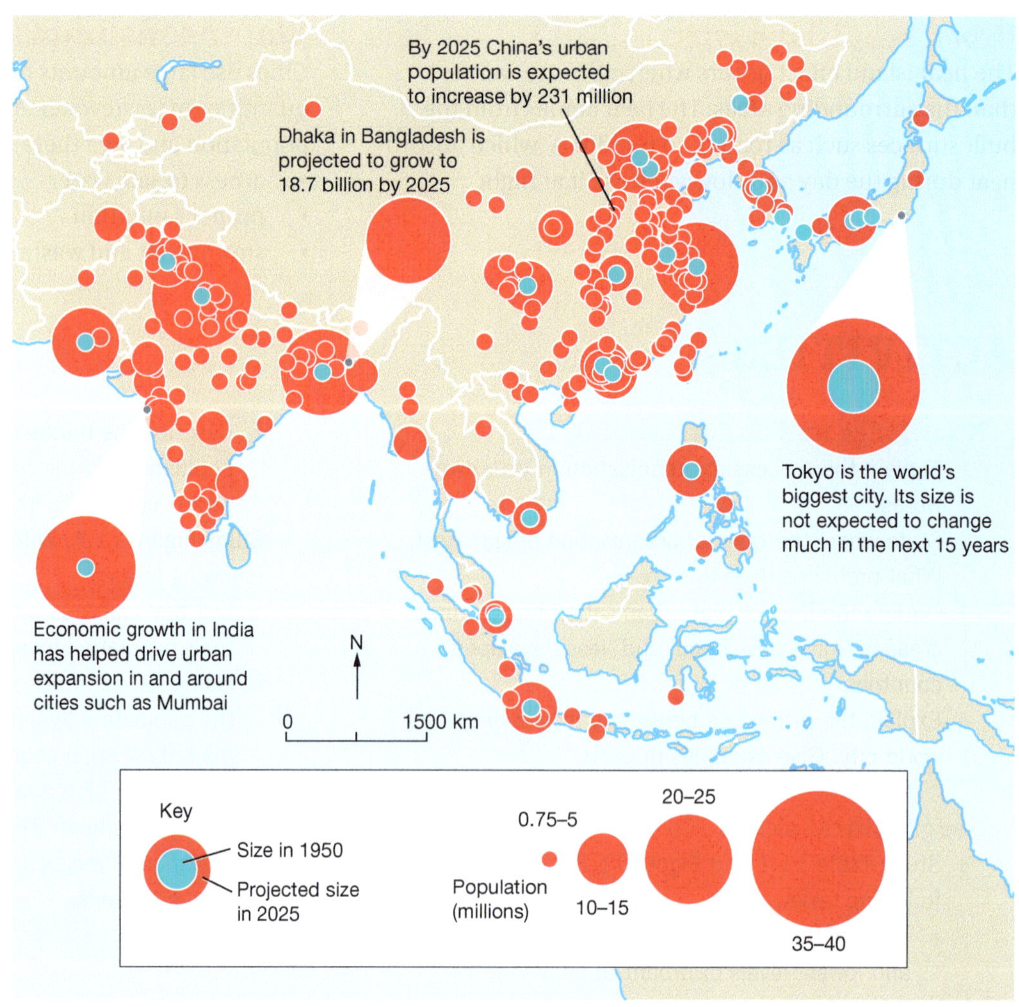

11.8 Demographic characteristics of Asian countries

Country	Total population (millions)	Population growth rate 2010–15 (%)*	Urban population (%)	Life expectancy at birth 2010–15* male (years)	Life expectancy at birth 2010–15* female (years)	Population using improved sanitation (%)	Population living below US$1.25 per day (%)
China	1347.6	0.4	47	72	76	55	16
India	1241.5	1.3	30	64	68	31	42
Indonesia	242.3	1.0	44	68	72	52	29
Japan	126.5	–0.1	67	80	87	100	0
Kazakhstan	16.2	1.0	59	62	73	97	2
Lao PDR	6.3	1.3	33	66	69	53	44
Malaysia	28.9	1.6	72	73	77	96	2
Philippines	94.9	1.7	49	66	73	76	23
Qatar	1.9	2.9	96	79	78	100	0
Singapore	5.2	1.1	100	79	84	100	0
South Korea	48.4	0.4	83	77	84	100	0
Thailand	69.5	0.5	34	71	78	96	2
United Arab Emirates	7.9	2.2	84	76	78	97	0
Vietnam	88.8	1.0	30	73	77	75	22

* estimated

Source: Adapted from data at UNFPA State of the World Population 2011

11.9 Massive public housing developments have been built to fix housing shortages in Singapore.

11.10 Vientiane, the capital and largest city in the Lao PDR.

SPOTLIGHT

Urbanisation in Indonesia

Indonesia has experienced rapid urbanisation. Rural towns have been transformed into cities and expanding cities have absorbed neighbouring rural towns. These rural communities, while not originally urban, had very high population densities. Recently, the Indonesian government has tried to restrict migration to large cities by redirecting people to smaller communities or rural areas. Jakarta, with a population of 18.9 million, has a closed city policy. Residency permits are given only to those who can provide evidence of housing and employment. In addition, a new law was introduced in September 2009 that prohibits people from giving money to beggars and roadside workers. The new law also bans squatter settlements along riverbanks and highways. Jakarta is experiencing many negative effects of rapid urbanisation. Traffic jams in Jakarta, as shown in Figure 11.11, are a common event. Jakarta is built on a flood plain. Due to removal of groundwater for human use, Jakarta is experiencing land subsidence, or sinking. In response to the traffic congestion, a monorail is being planned for the city.

11.11 Average traffic speed in Jakarta is just 8.4 kilometres per hour. In 2011, a total of 474 new cars and 2946 motorbikes per day were added to the estimated 14.4 million vehicles in the city.

11.12 Population growth in rural and urban Asia, 1950–2050

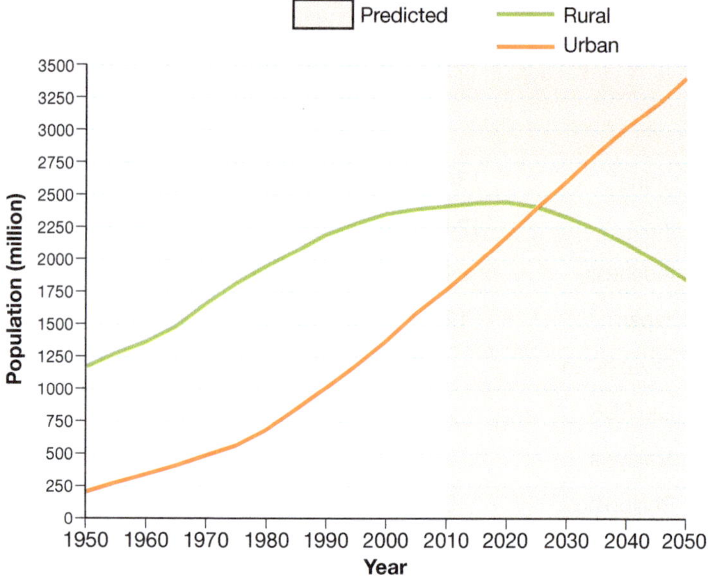

Rates of urbanisation

The highest rates of urbanisation are often found in Asia's smaller cities. While the growth rates of cities with populations between 5 and 10 million have stabilised, and the growth rates of cities of over 10 million people slowed to 1.7 per cent a year between 2000 and 2010, cities with populations between 500 000 and 5 million have experienced the greatest growth, especially cities in China, India and the Philippines. Figure 11.12 shows the increase in population in rural areas declining, while urban populations continue to grow.

Reasons for rapid urbanisation

One of the main reasons for the rapid urbanisation occurring in Asia is the emergence of the **global economy**, which has resulted in the move of labour-intensive manufacturing to countries in South and East Asia and Central and South America. The process has created a huge demand for labour. The opportunity for a relatively well-paid job in a factory is a powerful pull factor for people in rural areas. The growing urban centres also offer the possibility of better healthcare and education. Cities are also seen as more exciting and dynamic places, especially for the young.

For those unable to access the **formal economy** (economic activities subject to regulation and taxation), in cities there is the opportunity to supply goods and services to those who do. This has led to the growth of the urban-based **informal economy** (economic activities not regulated by government)—for example street vendors.

Declining death rates and, at least in some parts of Asia, high fertility rates have also fuelled urban population growth. Where villages have been absorbed into expanding urban areas there is often a lack of distinction between the urban environment and the previously rural community—people do not necessarily move to the city; they simply find themselves suddenly part of one.

Positive outcomes

Economic outcomes

Urbanisation has contributed to economic growth and increased living standards by raising incomes and reducing poverty. For example, factory workers are paid more than rural workers. Growing incomes contribute to local demand for goods and services, which further fuels economic growth

Social outcomes

Urbanisation in Asia has influenced people's aspirations, lifestyles and social relationships. The more diverse mix of cultures, castes and religions found in cities is said to create greater harmony and break down social and cultural barriers.

Negative outcomes

Economic outcomes

The benefits of urbanisation are not evenly distributed. Up to 40 per cent of Asia's urban population live in slums and **squatter settlements**. The lack of access to clean water and sanitation means waterborne diseases spread rapidly. The **infrastructure** needs of cities (roads, schools, hospitals for urban residents) are often met at the expense of rural communities.

Social outcomes

In an attempt to keep pace with the demand for housing and rate of economic growth, authorities have demolished whole urban districts—destroying the cultural, social and built heritage of the area, as is shown in Figure 11.13. Residents are often forced to relocate, and the sense of community that existed is destroyed.

11.13 Historic urban districts and their communities are often destroyed in order to make way for apartments and office towers.

ACTIVITIES

11.1

Knowledge and understanding

1 Describe the pattern of urbanisation throughout Asia.
2 Explain why urbanisation is occurring in Asia.
3 Outline the outcomes of urbanisation in Asia.

Applying and analysing

4 Discuss the likely reasons for Indonesia introducing a law to ban the giving of money to beggars.
5 You have been given the task of assessing the outcomes of urbanisation in Asia. You are to prepare a score sheet that can be used to rate the outcomes for each country. On your score sheet include the outcomes and a way to determine how each outcome will be measured.

Geographical skills

6 Study Table 11.8. Present the following information as a graph or series of bar graphs: urban population; population using improved sanitation; and population living below US$1.25 per day.
7 Study Figure 11.12. Describe the actual and predicted trends in rural and urban populations in Asia.

11.3 Urbanisation in China

With a population of 1.4 billion, China is the world's most populous country. It also has a high rate of urbanisation—a process closely related to the country's emergence as a major economic power.

Growth of urbanisation

At the start of 2011, there were 665.57 million people living in urban areas and the level of urbanisation stood at 49.68 per cent. By 2030, over one billion people will live in China's cities. Meeting the needs of these new urban dwellers has been a major challenge for Chinese authorities.

Chinese urbanisation

In 1949, there were just 132 cities with 57.67 million people, or 10.65 per cent of the population, living in urban areas in China. In 1987, the government adopted a strategy to control urban growth. This involved limiting the size of large cities, developing mid-sized cities (between 200 000 and 500 000 people), and encouraging the growth of small cities and small market centres. The outcomes of these policies are shown in Figures 11.14 and 11.15.

11.14 Levels of urbanisation, China and the world, 1950–2050

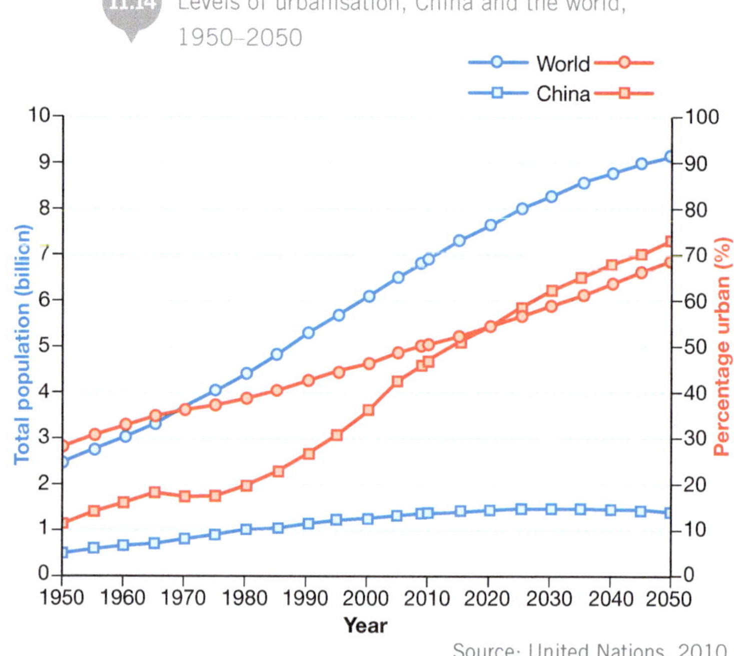

Source: United Nations, 2010

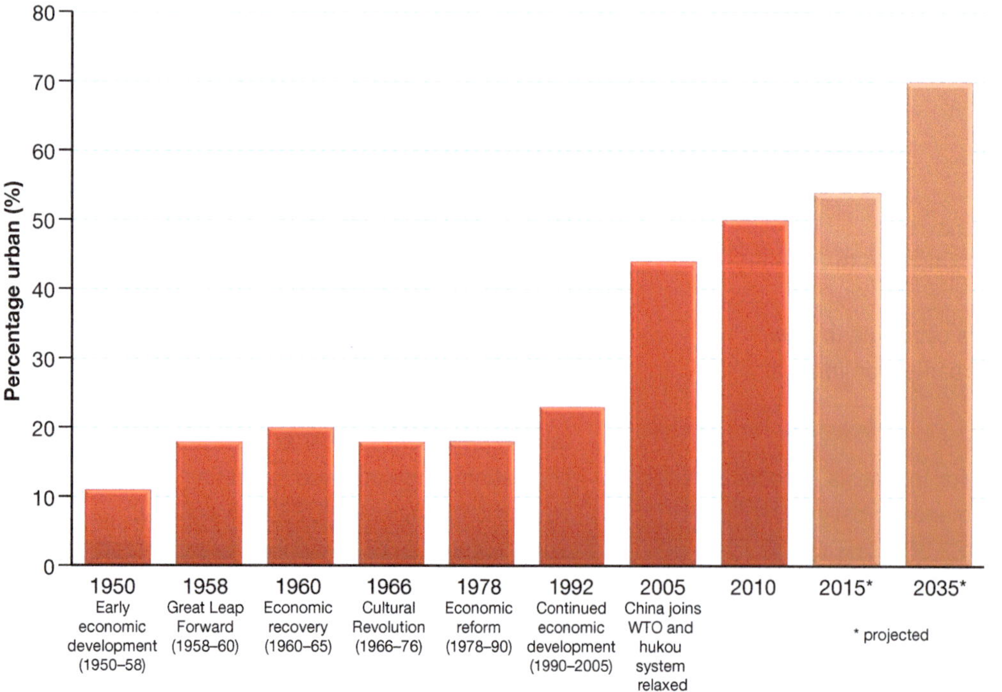

11.15 Economic policy and urbanisation in China, 1950–2015

Source: McKinsey Global Institute (2009) *Preparing for China's Urban Billion*

The early years of the strategy were focused on limiting large cities. The government decentralised economic activity and power. State-owned enterprises were relocated to smaller cities, local authorities were given more power to achieve local economic growth targets, and some cities have been designated to offer incentives to investors.

Urban spatial pattern

The Chinese government has favoured the development of small- and medium-sized cities over the development of very large cities. This has resulted in a very unique spatial pattern of urban places. The urban system consists of large- and medium-sized cities linked to smaller urban centres, as illustrated in Figure 11.16.

In 2013, there were three main metropolitan areas consisting of a number of highly connected cities and towns. These were the Yangzi River delta, the Pearl River delta and the Beijing–Tianjin–Bohai region. Throughout the rest of the country there are a number of urban clusters that feature one or more cities at their centre. These are important zones of economic development, not just for the local region but for the country as a whole.

Outcomes of urbanisation

Economic benefits

Urbanisation has brought economic benefits for China. In 2011, China had the world's second largest economy and was the largest importer and exporter of goods in the world.

Meeting the needs of those moving to the cities also generates economic activity. For example, every additional 10 million urban residents results in an increase in the country's GDP of 4 per cent; 3.6 per cent of this is due to the demand for housing.

The distribution of cities (based on population) and city clusters (based on GDP) in eastern China

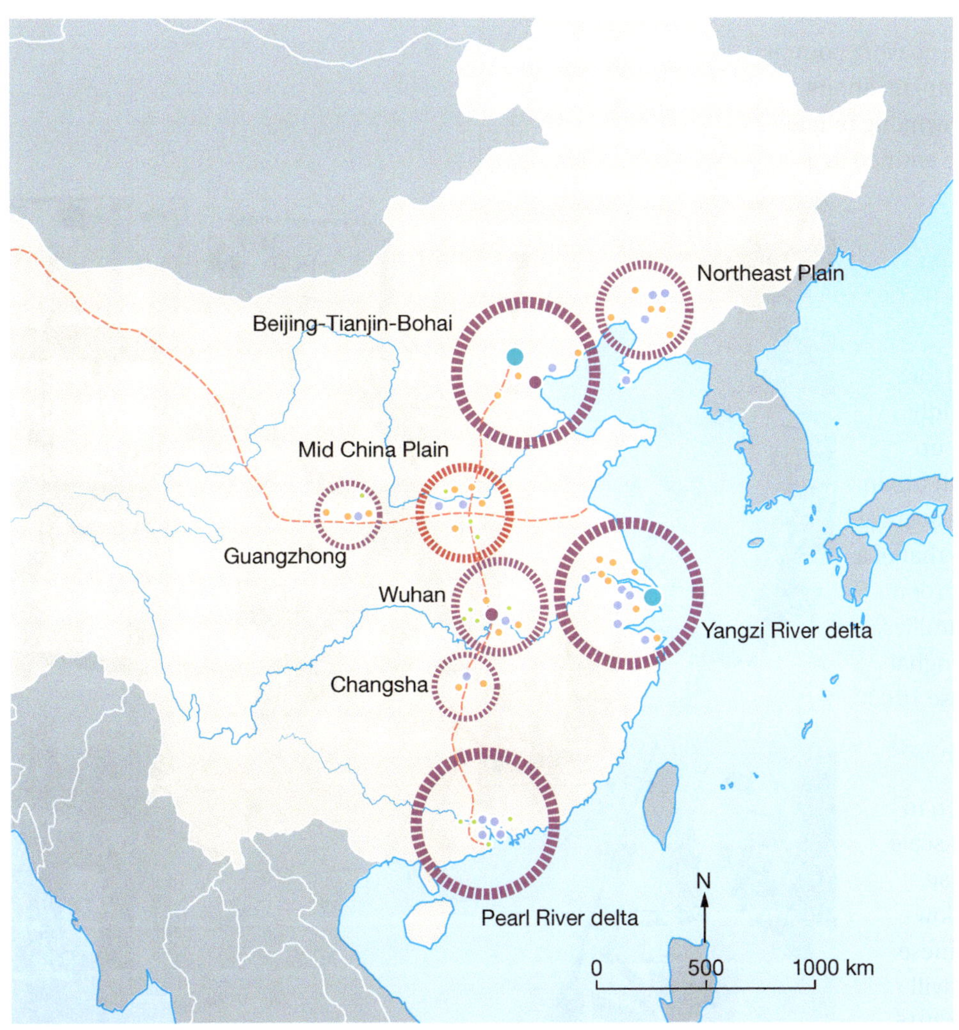

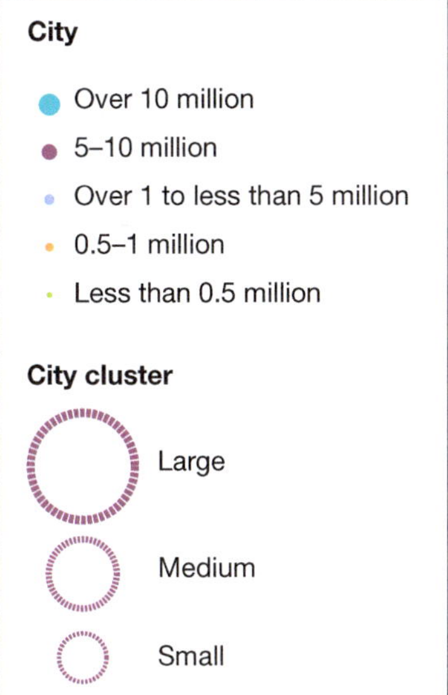

The growth in China's national economy has been accompanied by improvements in people's standard of living. A huge number of people have been lifted out of poverty. In 1981, 85 per cent of people were living in poverty (defined as living on less than US$1.25 per day). By 2005, that figure had been reduced to just 16 per cent. The economic benefits have not benefited everybody, and the gap between the rich and poor has widened.

The plight of rural–urban workers

While urbanisation has been accompanied by greater employment and an improvement in living conditions, some groups still struggle economically. People who travel from rural areas to the cities looking for contract-based work are known as rural–urban workers, or *mingong* ('peasants who became workers'). In 2009, there were 230 million rural–urban workers.

Rural–urban workers face many difficulties.

- Many are employed in low-paid and dangerous jobs.
- They have long working hours—usually six and sometimes seven days a week.
- They live in poor or cramped housing, as shown in Figure 11.17.
- Approximately 40 per cent have no work contracts, meaning they are unable to complain about mistreatment, abuse or unsafe working practices.
- Approximately 21.3 per cent are entitled to a pension
- Only 8.5 per cent are able to get unemployment benefits.

Housing shortages

In China, homes are expensive, and low-cost housing is in short supply. Rapidly increasing urban populations push up prices and rents. An apartment in an urban area in China can cost 8.3 times the average income of a household, or 29 times that of a low-income worker. In many cases, rooms in apartments are sublet to other families, resulting in overcrowding. One Shanghai complex was recently found to house 10 000 people in just 600 apartments.

The government's response has been to allow developers to construct large-scale housing developments. Most of these, however, are still too expensive for low-income earners. As a result, the Chinese government has announced that it will build 36 million low-cost homes by 2015.

Health

Urban living gives people access to modern medical facilities and an income to pay for personal healthcare. However, the health of people in cities is threatened by air pollution, a more sedentary lifestyle, greater work-related stress, social detachment and high-fat diets, all of which contribute to poorer health outcomes.

Pollution

According to the World Health Organization (WHO), seven of the world's ten most polluted cities are found in China. The scale and extent of urban pollution in China is immense. The air, water and soil are all affected and biodiversity has been reduced. In January 2013, north-eastern China recorded its highest ever levels of pollution. WHO considers concentrations of 25 PM2.5 particles (the most dangerous particles that people can inhale) per cubic metre to be unsafe; on 12 January, Beijing recorded a PM2.5 level of 993 per cubic metre. In some areas, visibility was reduced to 200 metres and flights were cancelled. Officials ordered schools to suspend outdoor activities and people were advised to stay indoors.

11.17 Some rural–urban workers bring their whole families to the cities, where they often live in squalid conditions.

China's factories release huge amounts of pollutants, such as sulfur dioxide, into the atmosphere. The air quality of many of China's cities is described as moderately or severely polluted. WHO estimates that there are 300 000 premature deaths each year in Chinese cities because of outdoor air pollution. In addition, many of the pollutants react in the air to produce acid rain, which can affect soil acidity. Soil acidification (increase in the acidity of the soil) is an increasing problem for China's agricultural regions because it affects the fertility of the soil and the ability to grow crops.

Water

Half of China's largest cities are facing water shortages. Three hundred million people have no access to clean water and 700 million people drink water below WHO standards. Eighty per cent of China's rivers are too polluted to fish from. For example, less than 5 per cent of wastewater flowing into the largest river in Shanghai, the Huangpu, is treated, resulting in a river that is essentially dead, with 3.4 million cubic metres of industrial and domestic waste dumped into it every day.

Hukou

Where a person may live in China has been controlled since ancient times by a system of residency permits known as the **hukou system**. Under the system, a person is officially identified as a resident of an area. In 1958, the Chinese government publicised the family register system to control the movement of people between urban and rural areas. Individuals were broadly classified as either 'rural' or 'urban' workers. A worker wanting to move from the country to an urban area to take up non-agricultural work had to apply for permission. The number of workers allowed to make such moves was tightly controlled. Authorities feared a massive movement of people into the cities, causing strain on city government services, damage to rural economies and increases in social unrest and crime. Until 1976, police regularly rounded up those without valid residence permits, detained them in detention centres and expelled them from cities. Today, the hukou system is only partially enforced, as the government is keen in increase the supply of skilled workers for industry.

Figure 11.18 shows the increase in the 'population with urban rights'. Residents who do not have hukou permits are often denied subsidised healthcare and their children are denied access to local schools.

11.18 The growth of China's total urban population and the number of people with permits to live in urban areas, 1970–2010

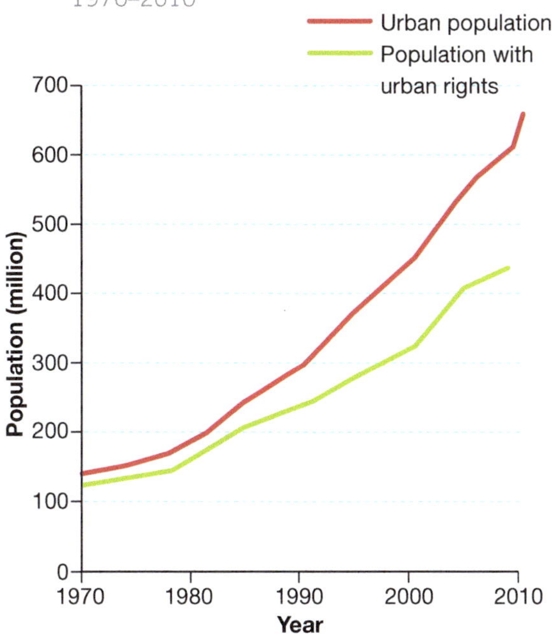

Source: National Bureau of Statistics and Ministry of Public Security, China

ACTIVITIES

Knowledge and understanding

1. Outline how the nature of China's urbanisation has differed from that of other countries since 1949.
2. Explain how the hukou system has been used to control China's rate of urbanisation.
3. Define the term '*mingong*'. Discuss the living and working conditions of these people.
4. Explain why housing is a hotly debated issue in China. What has been the government's response?

Applying and analysing

5. List the outcomes of urbanisation in China and place each outcome in a PMI chart.
6. Imagine you are a Chinese resident living in a rural village. Would you move to the city without a permit to live? Explain.

Geographical skills

7. Study Figure 11.15. Which policy/policies have had the greatest influence on urbanisation in China since 1950?
8. Study Figure 11.16. Describe the distribution of China's urban population.
9. Study Figure 11.18. Describe the trends in China's urban population and the number of Chinese with permission to live in urban areas.

11.4 Impacts of urbanisation

Rates of urbanisation are highest in the countries of the developing world. Poverty, overcrowding, pollution, traffic congestion and squatter settlements are often the result of this process. Providing employment, housing, transport, clean water, electricity and sewerage systems for the new urban dwellers is a major challenge.

Problems of large cities

The problems of large cities in the **developing world** are summarised in Table 11.19.

Quality of life and standard of living

The happiness, wellbeing and satisfaction of a person is referred to as their **quality of life**. Among the many factors that influence a person's quality of life are their family, health, income and access to services. A person's **standard of living** is usually measured in terms of their possessions, such as a house, a car or a computer. While these indicate a high standard of living, they cannot guarantee a high quality of life.

Life in the favelas of São Paulo, Brazil

One hundred years ago, São Paulo was home to some 265 000 people; today there are more than 20.5 million people.

Twenty per cent of the city's population live in its sprawling favelas (slums). Elsewhere, in the richest suburbs, the wealthy live in houses protected by fortress-like walls and security guards. For both rich and poor, daily life focuses on traffic congestion, crime and housing: three issues that highlight the problems that accompany rapid urbanisation. The city's murder rate of more than 9000 a year (compared with fewer than 500 in New York) is the result of slum violence and drug-related gang wars.

São Paulo's favelas are built on unclaimed land. Some have been built on steep hillsides, which are especially dangerous after heavy rain, and low-lying, mosquito-infested wetlands. Elsewhere, favelas have been built under motorway flyovers, in rubbish dumps and next to heavily polluted industrial areas.

11.19 Problems of large cities in the developing world

High unemployment	The number of settlers from rural areas often exceeds the number of jobs available. Many people cope by working in the informal economy—for example as street vendors.
Poverty	Lack of access to employment results in poverty. Many people are forced to make a living by sorting waste for recyclables, begging in the streets or resorting to crime.
Poor-quality, overcrowded housing	Lack of affordable housing forces people to crowd into slums or squatter settlements—illegal, self-built housing on unused land. These areas often lack even the most basic services (water, sewerage systems, electricity).
Inadequate infrastructure	The existing urban infrastructure is unable to cope with increased usage, resulting in failing transport, water supply and sewerage systems.
Too little water; water is often contaminated	The increased demand for water can exceed the capacity of existing sources. Groundwater sources and local waterways are often polluted by untreated sewage and toxic industrial wastes.
Overcrowded education and health facilities	Hospitals and schools find it difficult to cope with the increased demands placed on them by the new urban dwellers. Building new facilities is expensive.
Environmental issues	Rubbish dumped in streets and waterways often remains uncollected. Open space in urban areas is lost to squatter settlements; trees are chopped down for firewood.
Governance	Corruption is common, resulting in the loss of funds for new infrastructure and housing. Laws protecting the environment and public health are not properly enforced.

The quality of life for those living in the favelas is summarised below.

- ***Housing*** Usually this consists of simple, one-roomed shacks made out of scrap materials. The shacks are crammed together, as space is in short supply (see Figure 11.20).
- ***Sanitation*** There is no clean running water and no sewerage system. Streams flowing in and around the favelas serve as a source of drinking water and a place to dump rubbish and human waste. Water-related diseases spread quickly in this way. Dysentery and typhoid are common and infant death rates are high.
- ***Lack of security*** Crime rates are high. Also, because the squatters have no right to be on the land, it is not uncommon for authorities to bulldoze the shacks.
- ***Employment*** Jobs are difficult to find. Wages are low and the journey to and from work can be long.

11.20 São Paulo favela

- ***Education*** There are few schools, so literacy rates are low. Young children are encouraged to earn money by street selling, shoe cleaning and begging on the streets. Many are exploited and subject to abuse.

ACTIVITIES

Knowledge and understanding

1 Explain the difference between quality of life and standard of living.

2 Outline the problems resulting from São Paulo's rapid population growth. What are favelas? Where are they often located?

Applying and analysing

3 Study Table 11.19. Draw up a table with three columns and label the column 'Economic', 'Social' and 'Environmental'. Place each issue in the correct column. Note: a problem may be listed in more than one column.

4 Imagine you are a teenager living in a São Paulo favela. Write a paragraph describing what it is like to live there.

Investigating

5 Investigate the challenges faced by a large city in the developing world. Find out about the solutions to the problems experienced. Develop a multimedia presentation to communicate your research findings.

a Select a city and provide the following information about the people who live in the city:
- population size
- population growth rate
- literacy levels (male and female)
- age levels
- levels of education
- unemployment
- access to medical care and hospitals
- access to clean water and sanitation …

b Copy and complete the following table.

City:		
Issue	**Outline the issue**	**Solutions proposed**
High unemployment		
Poverty		
Poor-quality, overcrowded housing		
Inadequate infrastructure		
Too little water; water is often contaminated		
Overcrowded education and health facilities		
Environmental issues		
Governance		

c In your opinion are the proposed solutions achievable? Discuss.

11.5 Urbanisation in Australia

Australia's small population (23 million in April 2013) is one of the world's most urbanised, with an estimated 86 per cent of people living in towns and cities. Australia has thirty cities with more than 50 000 people.

The growth of the state capitals

In 1910, fewer than 40 per cent of Australians lived in the six state capitals. Today the state capitals account for 64 per cent of the population. A common characteristic of these cities is **urban sprawl**. The population density of the two fastest-growing cities, Brisbane and Perth, is only one-fifth of the average in European cities. Sydney and Melbourne, Australia's two largest cities, sprawl over an area four times larger than European cities with a similar population, consume more than double the amount of fuel for transport and generate three times the amount of greenhouse gases. The continued growth of these cities threatens the quality of the water from their surrounding catchments, the quality of air and the cleanliness of the oceans.

Non-capital cities

Australia's non-capital cities tend to be much smaller than the capital cities. The largest of these are the industrial cities of Newcastle, Wollongong and Geelong, and tourist centres such as the Gold Coast. There are also **regional centres**, groups of large cities and towns, such as Tamworth, Dubbo, Ballarat, Townsville, Alice Springs and Launceston. Each of these cities and centres exists to supply goods and services to its surrounding area, or hinterland, on the edges of towns and cities. They have large shops, educational centres, professional services (such as lawyers and specialist doctors) and regional government and business offices.

Most of these regional centres continue to grow, but often at the expense of smaller rural communities. In smaller rural communities, populations are declining and levels of unemployment and welfare dependency

11.21 Sydney is Australia's largest city.

are rising. Some shops, banks and businesses in these communities have been forced to close and some government services have been withdrawn or are now delivered by new technologies.

Recent trends

The most rapid urban growth is now taking place in Queensland coastal areas (see Table 11.22). Much urban growth is associated with the expansion of tourism and the attractiveness of these locations to retirees. Families leaving the large cities for a **sea change**, or more relaxed lifestyle in the smaller coastal communities, also contribute to this growth.

Another trend is the **tree change**, or movement of people from suburban areas to communities just beyond the fringes of large cities. Many of these people **commute** to the city for work.

Regional centres located close to Australia's major resource projects have experienced rapid population growth during the current mining boom. These include Karratha in Western Australia, Singleton in the Hunter Valley, and Gladstone and Mackay in Queensland.

DID YOU KNOW?

Melbourne's population is expected to be larger than Sydney's by 2028.

Australia's twenty largest urban areas, 2011

Rank	Name	Population	Average annual growth rate 2006–11 (%)
1	Sydney	4 627 345	1.56
2	Melbourne	4 137 432	2.02
3	Brisbane	2 074 222	2.23
4	Perth	1 738 807	2.74
5	Adelaide	1 212 982	1.15
6	Gold Coast	600 475	2.75
7	Newcastle	552 776	1.33
8	Canberra	417 860	1.84
9	Wollongong	293 503	1.09
10	Sunshine Coast	254 650	2.59
11	Hobart	216 656	1.06
12	Geelong	180 805	1.51
13	Townsville	176 993	2.85
14	Cairns	153 075	3.06
15	Toowoomba	132 936	1.75
16	Darwin	128 073	2.29
17	Albury–Wodonga	107 086	1.28
18	Launceston	106 655	0.63
19	Ballarat	97 810	2.03
20	Bendigo	92 935	1.83

Source: Australian Bureau of Statistics, cat. no. 3101.0

ACTIVITIES

Knowledge and understanding

1 Explain why Australia is described as one of the world's most urbanised countries.
2 Identify the factors that distinguish Australian cities from those in Europe.
3 Describe the impact of the continued growth of the capital cities on the environment.

Applying and analysing

4 As regional centres grow, smaller communities decline. What effect does this have on the smaller community?
5 Your advertising firm has been asked to prepare a campaign to attract people and investment to a major regional centre. In groups, develop your ideas for promoting the city. Use the following questions to guide your discussion.
   a What does the city have to offer?
   b Who should the campaign target?
   c What would be the most effective way of promoting the city?
6 Working in groups, develop a concept map featuring the advantages and disadvantages of a sea change or tree change. Share your group's findings with the rest of the class.

Geographical skills

7 Study Table 11.22 and then do to the following tasks.
   a Identify the capital cities with the highest and lowest rates of growth. Can you suggest reasons for this variation?
   b Identify the fastest growing non-capital urban centres. Speculate on the factors responsible for this growth.

11.6 Economics of large cities

Cities that grow in size and importance generally do so because they have economic advantages relative to other cities. As they attract additional economic activities and more people, city growth is stimulated. However, the sheer size of large cities can also create difficulties for businesses and individuals.

Business

Cities are often seen as engines of economic progress and the larger a city is, the more appeal it has for business owners.

Advantages

Large cities have several benefits for business. These include the following.

- ***Professional networks*** Clusters of firms concentrate in big cities. Transnational corporations (TNCs) operate in many countries and need to draw on a range of sophisticated and highly specialised services, such as finance and accounting. As business transactions cross national borders, TNCs also need lawyers who are experts in international commercial law.
- ***Infrastructure*** Efficient transport systems make it easier for people and goods to move around. Large cities are well served by road and rail networks. Ports are especially important for trade. Los Angeles is the second-largest city in the United States of America, and its location on the west coast has contributed to its continued growth, as it links the United States with Asia.
- ***Technological innovations*** Large cities provide an environment in which creativity and innovation thrive, as they attract people with the skills to develop new products. They are where communications and telecommunication technologies are most advanced.
- ***Large workforce*** A big city has more workers with a wider range of skills, from highly specialised professionals to people in low-skilled jobs.
- ***Large market*** Cities have a massive accumulation of wealth with residents earning relatively high incomes. To serve the consumer needs of these people, there is a concentration of retailing in large shopping complexes, such as Chadstone in Melbourne, pictured in Figure 11.23.

Disadvantages

Not all businesses prosper in large cities. Smaller business can struggle to survive for a number of reasons. These include:

- ***High costs*** Businesses can only operate successfully if they are able to make a profit. Keeping costs down is an important part of making a profit. The high demand for space in a large city pushes up the rent of shops and offices. Insurance and labour are also more costly.
- ***Competition*** With so many businesses already established in cities, it can be difficult to start up a new business and be competitive. Unlike smaller towns, where locals are loyal to businesses, with so much choice in a large city, customer loyalty cannot be relied upon.
- ***Traffic congestion*** Time is money for a business, and delays due to heavy traffic can be very costly. Deliveries, going to meetings or even getting to work on time can be a persistent problem.

An individual's perspective

Many people are attracted to cities because they believe they will be better off financially (see Table 11.24), as cities offer more career opportunities and higher wages. There is also the prospect of making contacts and being able to advance a career. This is assisted by the availability of educational institutions such as universities.

Transport can be more convenient and cheaper in a city. Regular bus, train and tram services enable some people to manage without a car. Living in a large city also appeals to consumers, as there is a wide range of commodities to choose from. The biggest disadvantage of living in a large city is the cost of housing. Rents and house prices are generally high in comparison with regional and rural areas, as are council rates for services such as garbage collections.

11.23 Chadstone, Melbourne, is one of the largest shopping complexes in Australia.

11.24 There are significant differences between small towns and large cities.

Town/suburb	Werris Creek	Mosman
Location	Small town in the north-west of NSW	Large suburb on Sydney's North Shore
Total population	1729	27 452
Median weekly household income	$706	$2465
Mean monthly mortgage repayments	$1114	$3033
Median weekly rent	$160	$473
People with a university or tertiary education	5.5%	17.7%

ACTIVITIES

Knowledge and understanding

1 Explain why TNCs are concentrated in large cities.
2 Describe the nature of the infrastructure found in large cities.
3 Explain Los Angeles' continued growth as an urban centre.
4 Outline the difficulties faced by a new business opening up in a large city.

Applying and analysing

5 Construct a T-chart showing the economic advantages and disadvantages of living in a large city.
6 List reasons why it is more economical to use public transport rather than own a car in a large city.
7 Refer to Table 11.24. Use the data to compare the economic advantages and disadvantages of Werris Creek and Mosman.

11.7 Environment and large cities

The environmental impact of urbanisation raises many concerns. Large cities use land and resources, returning their wastes and pollutants to the very environment they depend upon. However, some experts believe that urbanisation is actually advantageous to the environment.

Environmental impacts of large cities

Cities impact upon the land they cover with buildings and roads. Cities also use resources beyond their boundaries. As large cities grow, their interference in the natural world increases, as Figure 11.25 shows.

Loss of habitat and agricultural land

With the pressure of population growth, cities push past their outer limits into the countryside beyond. When this occurs, market gardens just outside the city disappear and any remaining natural vegetation is cleared to make way for development. This outward growth of cities is known as urban sprawl. In Australia, the Australian Farm Institute claims that in the 12 years between 1997 and 2009, Australia's agricultural land area was reduced by 11 per cent, from 462 million hectares to 409 million hectares, due in part to urban sprawl. Approximately two-thirds of Australia's perishable fruit and vegetables are grown on the outskirts of cities, as shown in Figure 11.26, and these areas are under threat from urban sprawl.

Air and pollution

Large cities have an impact on air and water quality. Pollutants from cars and factories produce photochemical smog, which can affect people's health. Waste materials are difficult and expensive to manage, especially those that are toxic, and landfill sites are needed to dispose of household and business waste.

Cities depend on the environment and they also impact heavily upon it.

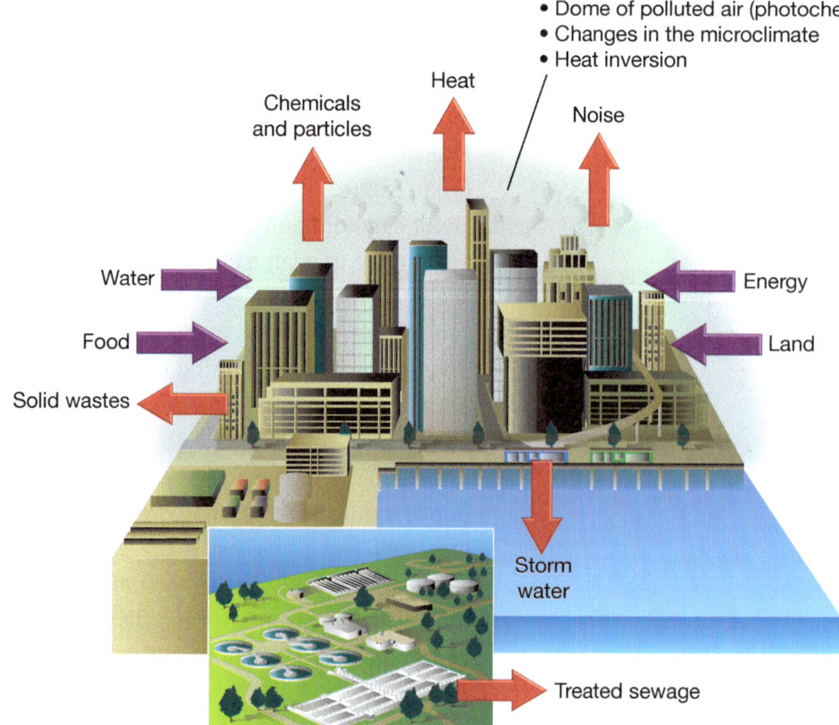

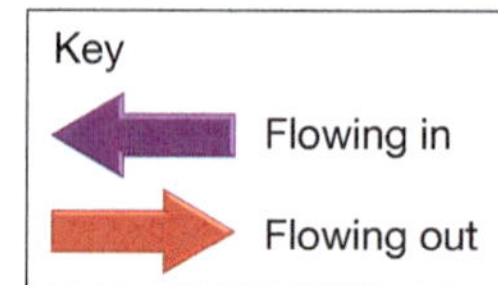

11.26 Market gardens under threat, Werribee, Victoria

Rethinking cities

In China, a 1.3-square kilometre 'Great City' has been proposed. It will be a massive development that is completely sustainable and affordable. The city, planned for 80 000 people, will be centred on a mass transit hub, with all destinations within a few minutes walk, reducing the need for cars.

The design will include a surrounding buffer landscape, which will include forests, valleys and lakes that are integrated into the city. Fifteen per cent of the land within the city will be devoted to parks and green spaces. Sixty per cent will be reserved for construction. The remaining 25 per cent will be used for roads and walkways.

Benefits of concentration

Some experts believe that concentrating growing populations in large cities can actually help save the environment.

Energy efficiencies

If more high-rise buildings were built in urban environments, the resulting higher population densities would improve energy efficiency and lower emissions per capita. New York is one of the most energy-efficient cities in the United States of America, as more than half the households use public transport rather than own a car.

Funding for environmental projects

Cities generate wealth that could be used to fund projects to protect the environment elsewhere. For example, money could be used to save endangered species and conserve habitats. Investment could be made in new technologies or infrastructure to reduce people's environmental impact. One example of this is the increasing number of bike paths being built in Australian cities, which help reduce traffic congestion and emissions from cars by encouraging people to ride their bikes.

11.27 China's pedestrian-focused city of the future

ACTIVITIES

Knowledge and understanding

1 Describe the impacts of urban sprawl.
2 Explain how the growth of cities causes pollution.
3 List the benefits of bigger cities.

Investigating

4 Choose one of the following topics:
- photochemical smog
- changes in the urban microclimate
- heat inversion.

With the aid of a diagram, prepare a PowerPoint presentation, explaining how it occurs and the impact it has on the city environment and its residents.

11.8 People and large cities

In a city, large numbers of people live in a relatively small area. Many people are happy living so close together and thrive on all that is on offer around them. However, the fast pace of city life does not appeal to everyone and some residents find it difficult to cope with the challenges of living in a city.

Social attractions

Those who find city living an enriching experience enjoy their contact with others and take advantage of the opportunities available. These include the following.

- ***Social interaction*** With so many living in cities there are more opportunities to socialise and make friends. People can also take advantage of the ethnic, community and lifestyle diversity that large cities have to offer. Some areas become popular within particular demographic groups, for example young professionals or tourist populations. Other neighbourhoods may become identified with particular ethnicities.

Canal Street, Chinatown, in New York is popular with residents and tourists, who pack the footpaths looking for bargains from street vendors.

- ***Cultural enrichment*** The bright lights and excitement of cities attract people. There is a broader and more diverse range of activities for people to choose from. People are able to go shopping, visit galleries and museums, go to the theatre and attend sporting events.

  Most large cities have experienced increases in their foreign-born populations in the last decade. For example, New York is home to over 200 ethnic communities. Locals can browse in specialty grocery shops or enjoy the varied cuisine. Manhattan's Chinatown is a vibrant district, famous for its wide selection of dim sum restaurants. Canal Street, shown in Figure 11.28, forms the main spine of Chinatown, and separates it from the neighbourhoods of Little Italy and SoHo.

Social disadvantages

Some people find living in the midst of so many others confronting and difficult, for the following reasons.

- ***Stress*** The fast pace of city life is not for everyone. Many people become frustrated when they are held up by traffic or unable to find a parking spot. Much of the stress associated with city living is financial. With house prices and rents so high, people settle for tiny apartments close to where they work, which may not be what they would prefer. Others who choose to live further out face long commutes, with tolls to pay and higher petrol bills.
- ***Crime*** Many people do not feel safe in large cities. The large number of people, properties and retail outlets mean more opportunities for crime. Figure 11.29 shows that most property crime in Australia occurs in retail premises.
- ***Isolation*** People in large cities can be anonymous, and can suffer from isolation and loneliness. People are not always interested in helping those they do not know and contributing to communities they may belong to only briefly. In a very sad illustration of such isolation, in 2011 the remains of an elderly woman who had died eight years earlier were discovered in a terrace house in Sydney's Surry Hills.

11.29 Property crime statistics, 2010–11

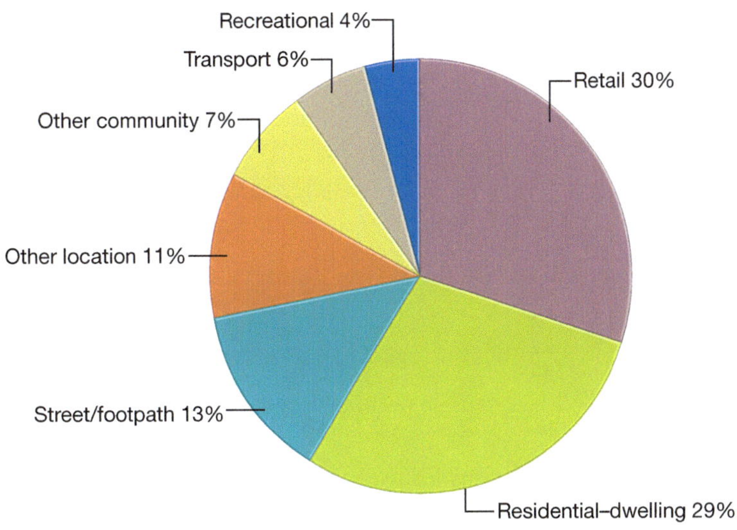

Source: Australian Institute of Criminology

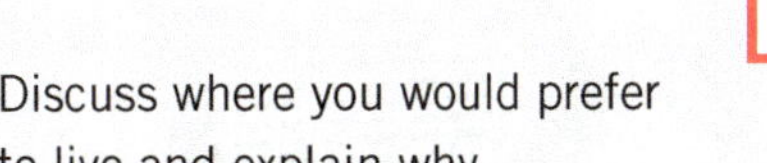
DID YOU KNOW?

Sydney, Melbourne and Perth rank among the world's nineteen most expensive cities, according to the 2012 Mercer Worldwide Cost of Living Survey.

ACTIVITIES

Knowledge and understanding

1 Describe the positive social aspects of city living.
2 Explain how the city environment can make some people happy while others are lonely.

Applying and analysing

3 The advantages of living in a big city outweigh the disadvantages. Discuss.
4 Prepare a T-chart identifying the advantages and disadvantages of living in a small apartment near the centre of a city.
5 Discuss where you would prefer to live and explain why.

Geographical skills

6 Study Figure 11.29 and do to the following tasks.
   a List the two most common locations for property crime.
   b What percentage do they account for?
   c List the least common locations for property crime.
   d What percentage of property crime occurs outside the home?

CHAPTER 12 CITIES IN AUSTRALIA AND THE USA

The world's population is unevenly distributed. This distribution is not random. There is a range of environmental, political, economic and social processes that determine whether populations concentrate or disperse. Australia and the United States of America have among the world's highest levels of urban concentration. There are, however, some important differences. The United States has more large inland cities. In Australia, most of the country's population is concentrated in a few large coastal urban centres, mainly the state capitals.

In this chapter, we look at the reasons for the high urban concentrations in Australia and compare the Australian experience to the settlement pattern in the United States. Central to this study are the geographical concepts of spatial distribution and change. Case studies of Perth, New York and Las Vegas are used to examine the process of urban concentration and its consequences.

KEY IDEAS

- To understand the causes and consequences of the concentration of Australia's urban population in just a few cities
- To describe how the pattern of urban concentration in the United States of America differs from the pattern in Australia

12.0 Miami Beach coast, Florida, United States of America

GLOSSARY

exurban region	a community within two to three hours' drive of a major urban centre
hinterland	the area surrounding a city that is linked to the city by lines of exchange or interaction
infrastructure	physical structures such as buildings, roads, water pipelines, sewers, electricity distribution systems, railways and airports
internal migration	the movement of people within a country
melting pot	a metaphor used to describe a multicultural society
population density	the number of people per square kilometre
population distribution	the arrangement or spread of people living in a given area
urban decay	the deterioration of the built environment. Urban infrastructure falls into a state of disrepair and buildings are left empty for long periods of time
urbanised society	a society in which most of the population lives in towns and cities

12.1 Australia's population distribution

Eighty-five per cent of Australians live in a narrow coastal strip stretching from Brisbane to Adelaide—an area that represents 3 per cent of the continent's land area.

Population density

Population density in Australia is associated with different landuses.

- The closely settled coastal strip contains almost all the major urban centres and most of the large-scale industries and businesses.
- The moderately settled zone is dominated by agriculture and some small-scale industries.
- The sparsely settled zone is dominated by extensive grazing and scattered mining activities.

More than 86 per cent of the population, or 19.6 million people, live in urban centres. Fifty-four per cent of the population, or 12.3 million, live in Brisbane, Sydney, Melbourne and Adelaide. This is why Australia is often described as the most **urbanised society** on earth.

Australia's population density is 2.9 people per square kilometre. This is the lowest of any of the continents, except for Antarctica. The highest population densities are found in the south-east corner of the continent. The uneven distribution of the Australian population is shown in Figure 12.1.

12.1 Most of Australia's population is concentrated in two widely separated coastal regions—the south-east and east, and the south-west.

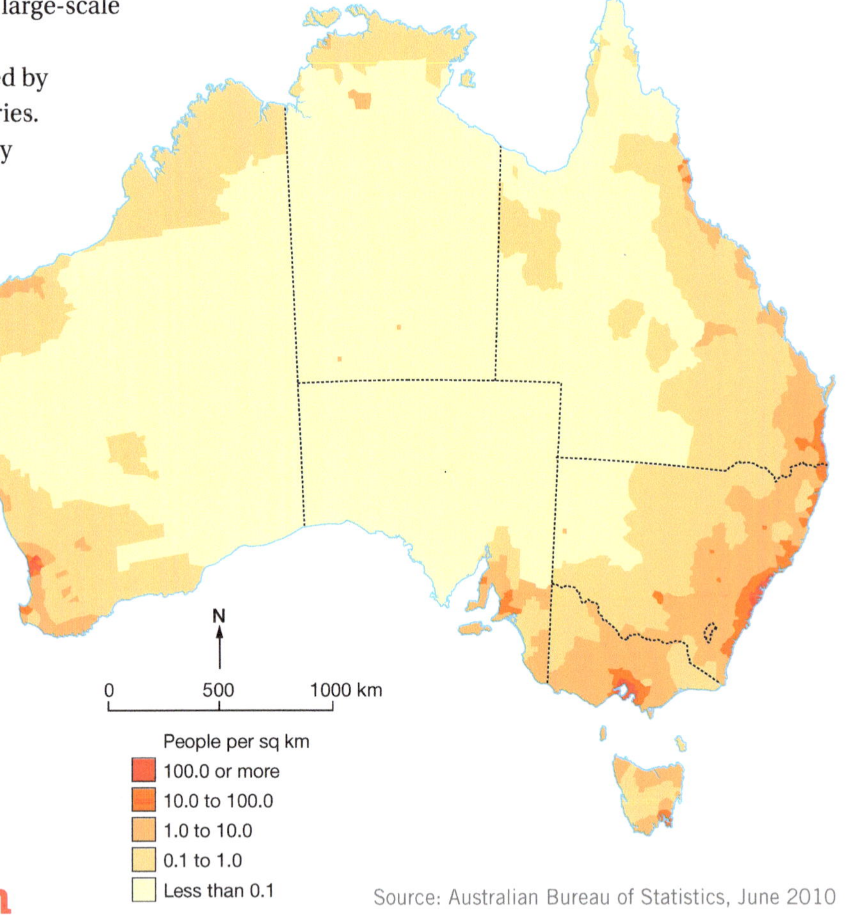

Source: Australian Bureau of Statistics, June 2010

Changing distribution

The distribution of Australia's population is changing. The proportion of the population living in rural areas is declining. As a result, many small rural communities are struggling to survive.

The proportion of people living in the Northern Territory, Western Australia and Queensland has increased, while the proportion living in New South Wales, Victoria, South Australia and Tasmania has declined.

There are regional variations in population growth rates. During the 1950s and 1960s the fastest-growing areas were the state capitals—in particular Sydney and Melbourne—partly due to the expansion of industry, the arrival of immigrants and the decision of governments to expand cities.

The fastest-growing areas of Australia today are the areas surrounding the large cities and, most recently, the capitals of the resource-rich states of Queensland (Brisbane) and Western Australia (Perth). Most of the non-capital-city growth has occurred along the coast of New South Wales and south-east Queensland.

Australia's big cities

Figure 12.2 shows Australia's major urban concentrations.

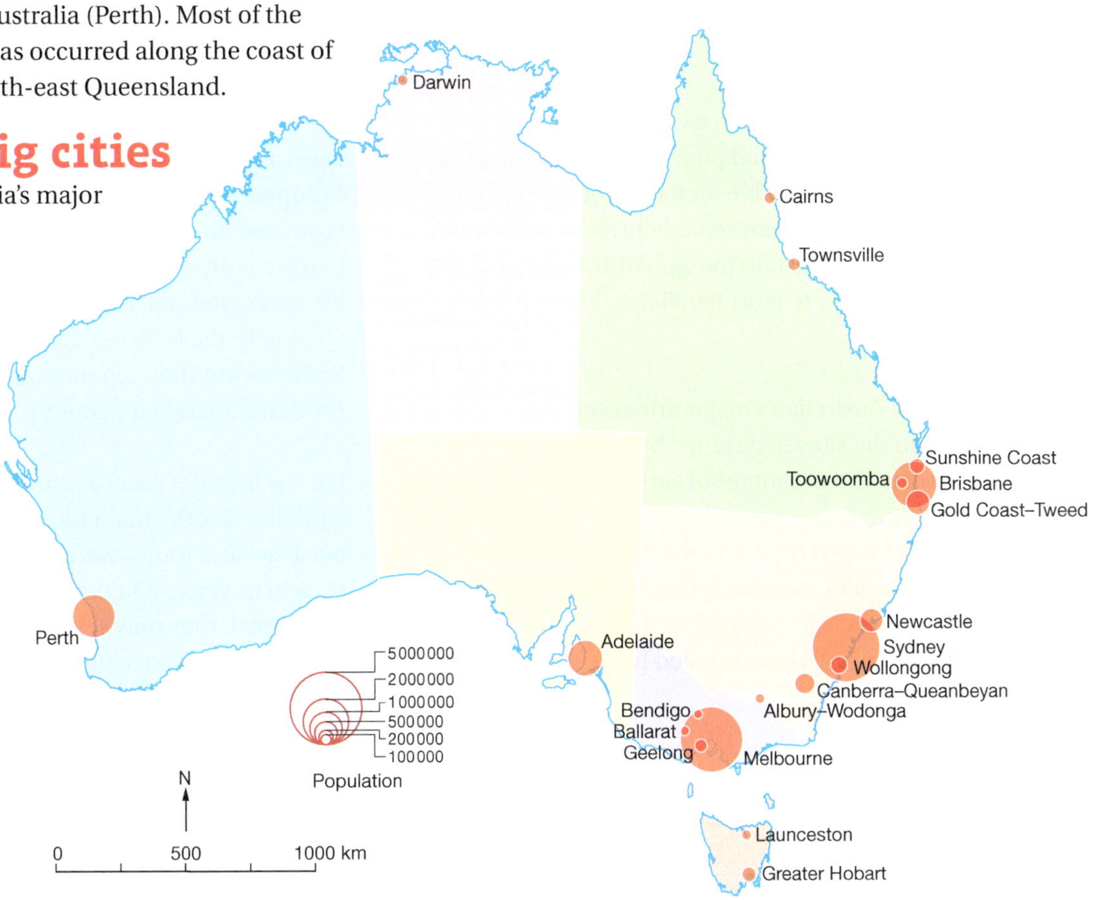

12.2 Proportional circle graph showing the distribution of Australia's twenty largest urban centres

12.3 Hobart is Australia's second-oldest capital city.

Factors affecting distribution

A variety of physical, historical and economic factors have worked together to influence population density in Australia.

Physical factors

Australia's harsh climate, scarce water resources, low rainfall (see Figure 12.4) and poor soils limit the area of the continent that is suitable for food production and settlement. Early European settlement was limited to the areas that could sustain the agricultural systems with which the settlers were most familiar.

Historical factors

In most cases, Australian's major urban centres developed on the sites first settled by European settlers. This gave them the advantage of early growth.

Australia was formed from six separate colonies in 1901. This is significant in explaining the present distribution of population. In each state and territory, the capital city is the largest city and is surrounded by rural settlements.

12.4 Australian rainfall

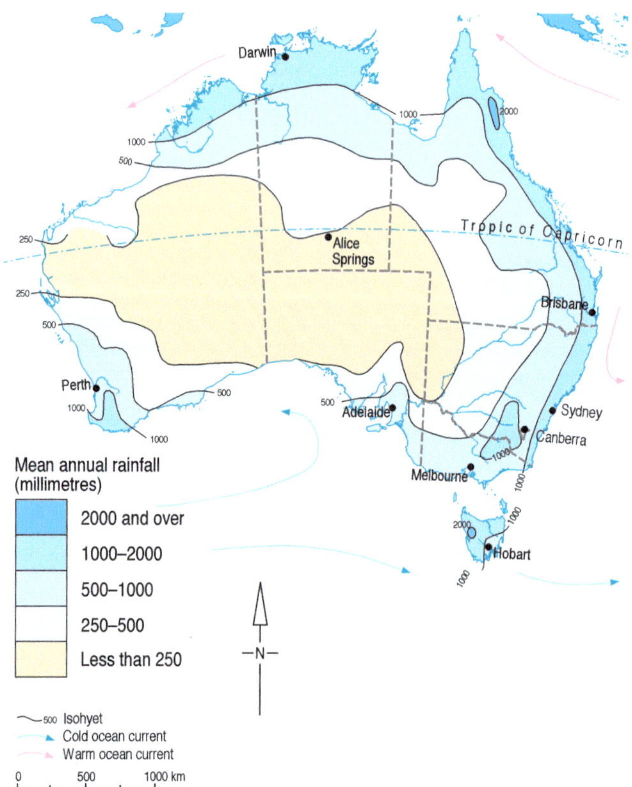

Source: *Heinemann Atlas*, 5th edition

The tendency of state and territory governments to centralise administrative functions in the capitals has reinforced the dominance of the capital cities. For the most part, government departments and agencies are based in the capitals.

Economic factors

Due to their reliance on maritime trade, the first European settlements were located on the coast. In each case, the settlement's port facilities became central to the colony's economic wellbeing. For many years, life was sustained by goods imported from Europe. Gradually, the colonies developed their own economic activities but their reliance on Europe continued; it was the major market for goods produced in the colonies.

For the first 100 years of European settlement, coastal shipping was the main means of communication because land routes were difficult to construct and slow to traverse. When rail and road networks were developed, they only served to reinforce the dominance of the large coastal cities. Usually they radiated out from the capital. Their function, in part, was to transport commodities to the ports, from where they could be exported to the world. Figure 12.5 shows the Port of Fremantle, which was established in 1829 and is still in operation today.

Industry, business and manufacturing

Due to the colonies' reliance on trade, most of the large financial institutions, including the banks, established their headquarters in these large port cities. Industry too was attracted to these centres. They provided easy access to transport, resources and markets, both local and international. Traditionally, employers have always found it easier to attract labour to coastal locations.

The location of manufacturing growth after 1945 served to reinforce the dominance of the large urban centres. There was a ready supply of labour and a market for the goods produced by manufacturers. Locations close to the major ports facilitated access to raw materials and export of the goods produced.

More recently, the ageing of the population has contributed to the development of coastal resort and retirement centres, such as the Gold Coast and Port Macquarie in New South Wales.

 12.5 The Port of Fremantle

Urban concentrations

There is a range of advantages and disadvantages associated with large urban concentrations.

Advantages

Due to the number of people living in a specific area it is less costly for the governments to meet needs in cities. The advantages of urban areas are:

- better **infrastructure** (water, electricity, broadband, sewerage and public transport)
- better services (education, healthcare)
- a more diverse range of cultural and sporting activities
- a wide variety of restaurants, nightclubs and other forms of entertainment
- economic advantages (more businesses supplying all sorts of goods and services)
- more shops and competition
- more jobs and a variety of work.

Disadvantages

The disadvantages of urban areas are:

- traffic congestion
- air pollution
- polluted rivers and creeks
- expensive housing
- a lack of a sense of community
- higher levels of crime.

ACTIVITIES

Knowledge and understanding

1 Outline the relationship between landuse and population distribution.
2 List the changes taking place in the distribution of the Australian population.
3 Outline the factors affecting the distribution of the Australian population.

Applying and analysing

4 Brainstorm and list the advantages and disadvantages of urban concentrations. Construct a mind map summarising the key points.

Geographical skills

5 Study Figure 12.1. With the aid of an atlas, describe the distribution of Australia's population.
6 Study Figures 12.1 and 12.4. Describe the relationship between rainfall and population distribution.
7 Study Figure 12.2. Describe the distribution of Australia's major urban concentrations. What is the relationship between this pattern and Australia's political divisions and history?

12.2 Case study: Perth

Perth is one of the world's most remote large cities. It is more than 2000 kilometres from Adelaide and over 3000 kilometres from Australia's east coast. Despite its isolation, Perth is a modern, vibrant and cosmopolitan city at the forefront of Australia's mining-based economic boom.

Population growth

Since the year 2000, Perth, shown in Figure 12.6, has experienced a rate of population growth greater than Australia's other large cities. This growth closely reflects the changes taking place in the Australian economy. Australia's mining boom and associated development have greatly increased the demand for skilled workers.

12.6 Perth's skyline reflects the conditions associated with the state's mining and development boom.

People have been moving to the west from other Australian states (**internal migration**) and from overseas. More than any other city in Australia, Perth attracts migrants from the United Kingdom and South Africa. People from the United Kingdom now make up 12 per cent of the city's population (compared with 5 per cent Australia-wide), and South Africans make up 1.3 per cent (0.5 per cent Australia-wide).

Geography

Metropolitan Perth extends 150 kilometres north to south along the Western Australian coast—from Two Rocks, 61 kilometres to the north of Perth, to Dawesville (south of Mandurah), 89 kilometres south of Perth. For most of this distance the urban area rarely extends for more than a few kilometres inland. Perth's most easterly suburbs are located in the Shire of Mundaring, only 30 kilometres from the CBD, shown in Figures 12.7 and 12.8.

Perth has a Mediterranean climate with hot, dry summers and mild, wet winters. There is very little rainfall in summer, with most of the rainfall coming in winter. Perth's total annual rainfall is 850 millimetres.

Population

Perth is Australia's fourth largest city. Most of Western Australia's 2.4 million people—1.75 million people, or approximately 74 per cent of the state's total population—live in Perth. This is the highest proportion of any state's population living in its capital city.

In the decade 2001–11, the city's population grew by 25 per cent, or 346 000 people. Much of this growth was driven by the state's mining boom and associated development. At 2.5 per cent, Perth's annual rate of population growth is much higher than Sydney's (1.3 per cent), Melbourne's (1.6 per cent), Brisbane's (1.7 per cent), or Adelaide's (1.1 per cent).

It hasn't always been this way. For many years, Perth was a small, isolated city. Most of the city's growth has occurred since World War II, the era in which private car ownership became common. As a result, the city does not have a dense Victorian-era core like the large urban centres of the eastern states. Only in 1980 did the level of Perth's population overtake that of Adelaide. Today, Perth has more than half a million more people than Adelaide.

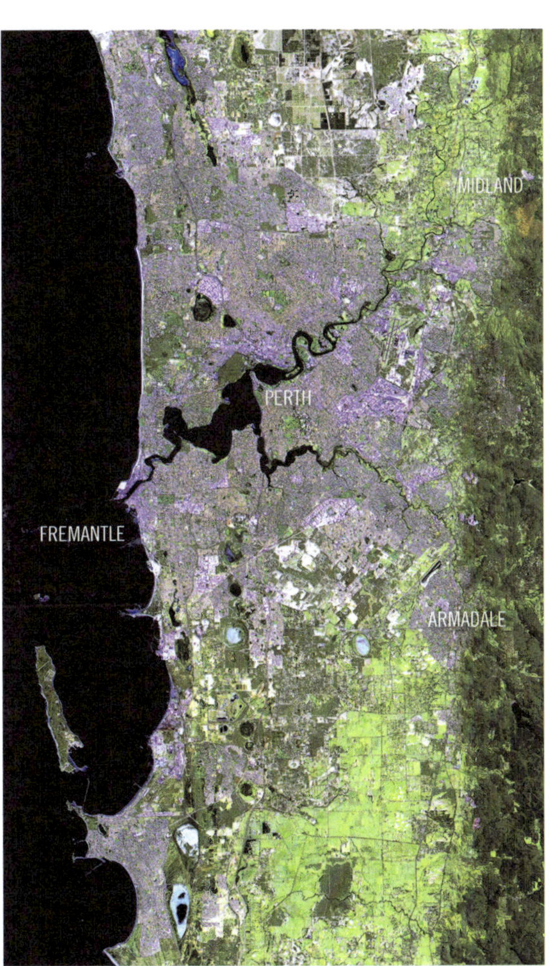

12.7 This satellite image of Perth highlights the importance of the ocean in shaping the geography of the urban area.

Source: *Pearson Atlas*, 2006

12.8 Metropolitan Perth

Where population growth is occurring

Perth's population growth in recent years has been concentrated in the outer suburbs, especially those hugging the coastline. The City of Wanneroo, in the north of the metropolitan area, almost doubled in size between 2001 and 2011—an increase from 84 000 to 156 000. Most of the other high-growth areas were in the coastal suburbs to the south of the city centre (shown in Figure 12.9). These include Rockingham (up 46 per cent), Kwinana (40 per cent) and Cockburn (36 per cent). Mandurah's population increased by 52 per cent and Murray Shire's by 42 per cent. The city's south-eastern corridor also grew strongly. The population of Serpentine–Jarrahdale grew by 57 per cent and Gosnells by 31 per cent. The suburbs in the city's east are also growing. The population of the City of Swan has grown 19.1 per cent. Areas such as Mundaring and Kalamunda are attracting smaller numbers of lifestyle-focused 'tree-changers'.

The lowest rates of population growth are found in Perth's established suburbs, especially those in the affluent inner and western areas such as Claremont (up 6 per cent), Nedlands (6 per cent) and Subiaco (9.8 per cent).

High-rise apartment living in the city's CBD is a relatively recent development.

Perth's ethnic diversity

The ethnic landscape of Perth has changed quite significantly over the past decade and will continue to do so into the future. Perth is clearly on a pathway to becoming a more multicultural city. In 2011, 34.9 per cent of Perth's 573 255 population was born overseas, as shown in Table 12.10 and Figure 12.11.

DID YOU KNOW?

In 2011, 576 731 people living in Perth were born overseas and 27 per cent had arrived in Australia within the previous 5 years.

12.9 Growth of Perth's metropolitan area

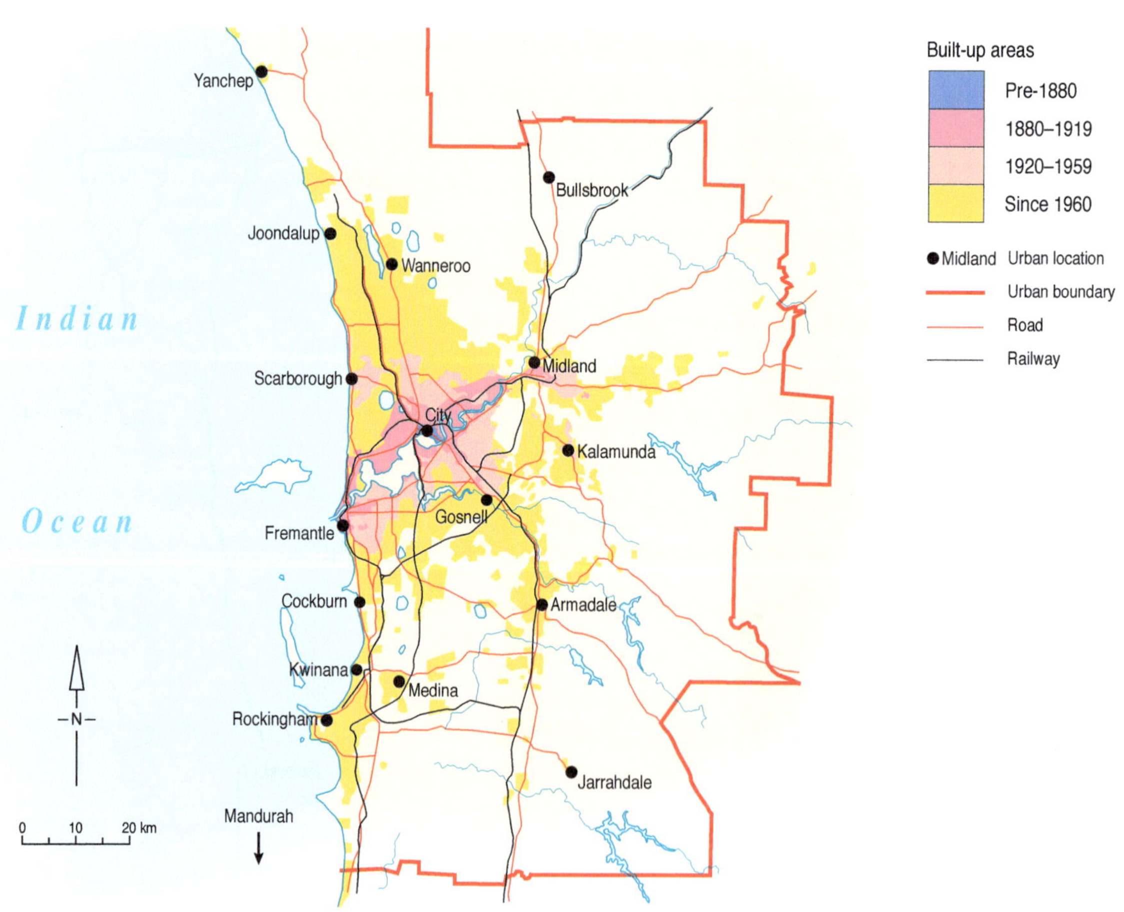

Source: *Heinemann Atlas*, 5th edition

12.10 Origin of overseas-born residents of Perth, 2011

Country of birth	Number of people	Percentage of overseas-born population
United Kingdom	186 546	32.3
New Zealand	49 806	8.6
South Africa	29 160	5.0
India	27 849	4.8
Malaysia	23 685	4.1
Italy	17 489	3.0
China	15 714	2.7
Singapore	13 372	2.3
Philippines	12 987	2.3
Ireland	12 246	2.1
Other	187 876	32.6

Source: Australian Bureau of Statistics, Census 2011

12.11 Change in country of birth of migrants in Perth, 2006 to 2011

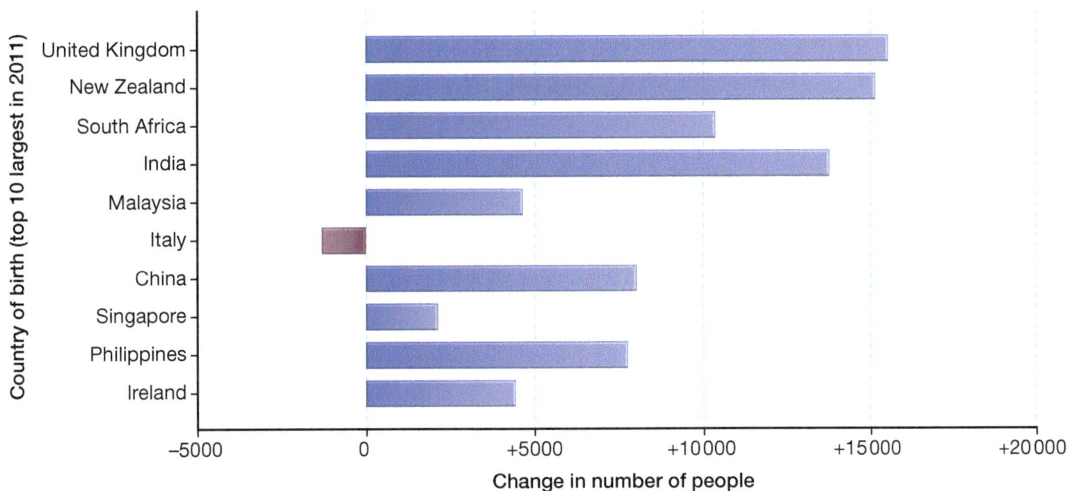

Infrastructure needs

In response to Perth's rapid population growth, a range of infrastructure projects are either in development or being planned. These projects are aimed at improving the liveability of the city. Projects include:

- a desalination plant at Kwinana (built)
- a desalination plant near Inningup (under construction)
- new freeways and highways
- expansion of the metropolitan railway
- removal of roads and barricades to open up the city to the river.

Many people believe that the government still needs to improve the metropolitan railway and connecting bus network. This includes extending the railway network.

ACTIVITIES

Knowledge and understanding

1. Explain why Perth experienced such rapid population growth in the period 2001–11.
2. Describe the geography of Perth.
3. Explain how Perth differs from the large urban concentrations on the east coast.
4. State the parts of Perth in which population growth has been most rapid in the decade 2001–11.

Geographical skills

5. Study Figure 12.9. Suggest reasons for the rapid expansion of metropolitan Perth after 1960.
6. Study Table 12.10. Construct a pie graph showing the country of origin of foreign-born Perth residents.
7. Study Figure 12.11. Using data from the graph, outline the change in country of birth of Perth's foreign-born population in the period 2006–2011. As a class, list reasons why these trends occurred.

12.3 USA's population distribution

The 2010 US Census reported that the population of the United States of America had risen to 308.7 million people, from 281.4 million people in 2000. The distribution of this population is uneven.

Distribution of population

The population of the United States of America is concentrated along the country's coasts (see Figure 12.12). Two-thirds of the people live in states along the three major coasts—38 per cent along the Atlantic Ocean, 16 per cent along the Pacific Ocean, and 12 per cent along the Gulf of Mexico. The smallest numbers live in the vast area between the Mississippi River and the Rocky Mountains, particularly in the central and northern Great Plains. While the Rocky Mountains and Plains states account for about half of the landmass of the country, they account for only 34 per cent of the population.

Overall, the population of the United States is more widely distributed over its landmass than is the case in Australia. The Great Plains, for example, is used quite extensively for agriculture. Much of inland Australia is desert. Figures 12.12 and 12.13 illustrate the relationship between precipitation and population density and distribution in the United States.

The United States has a much larger population than Australia. Australia's population density is close to 2.9 inhabitants per square kilometre compared with 34 inhabitants per square kilometre for the United States. Significantly, however, a higher proportion of Australians (75 per cent) live on or near the coast.

12.12 Population distribution, United States of America 2012

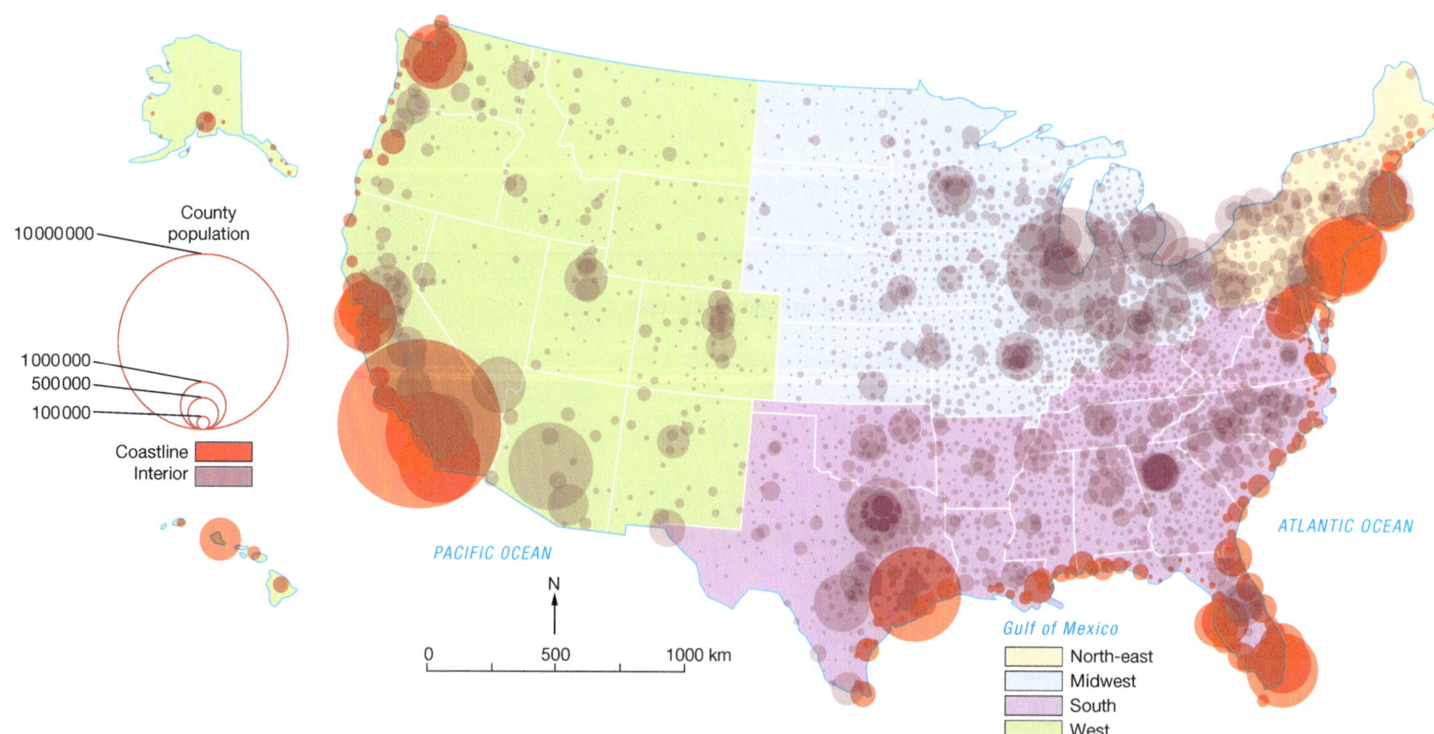

Source: United States Census Bureau, 2012

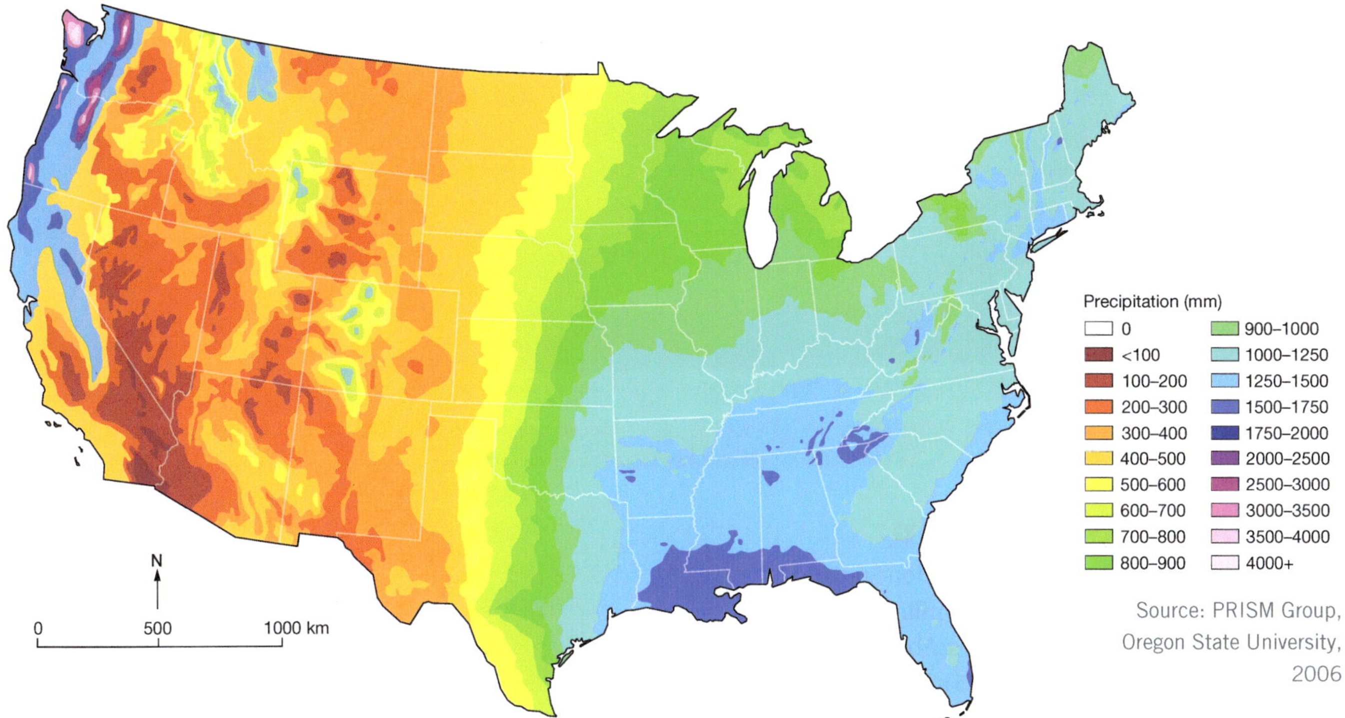

12.13 Annual precipitation, United States of America

Internal migrations

Since the 1960s, Americans have migrated to the south and west to a group of states known as the 'Sun Belt' (includes Florida, Nevada, Arizona, Utah, Colorado and New Mexico), seeking jobs, a warmer climate and sometimes a lower cost of living. Hundreds of thousands of retirees have also settled there.

Internal migration is closely related to economic conditions. People will migrate to areas that have job opportunities. The 2010 census showed that North Dakota had the fastest-growing population, due to the growing oil shale industry.

The flight to the suburbs

Since the 1950s, there has been a shift in population away from America's inner cities to 'greenfield' suburban development well beyond the outskirts of the city. As a result, urban areas have deteriorated. Local governments in the United States are responsible for services such as schooling and policing. When an increasing number of middle-class taxpayers move to the greenfield suburbs, or **exurban regions**, local authorities in urban areas are unable to raise sufficient taxes needed to fund such services. This has resulted in a downward spiral of **urban decay**. As services decline and decay, even more people flee.

This movement also has a racial dimension. The term 'white flight' has been used to describe the large-scale migration from racially mixed inner cities to more racially homogeneous suburban regions.

12.2

ACTIVITIES

Knowledge and understanding

1. Identify the parts of the United States that have experienced the fastest population growth. Which have shown the slowest growth?
2. Account for the uneven pattern of growth.
3. State what the distribution of the US population has in common with that of Australia. In what ways does it differ?
4. Account for the pattern of internal migration in the United States.
5. Describe the relationship between internal migration and economic conditions.
6. Explain how the movement of people in US cities could be described as having a racial dimension.

Geographical skills

7. Study Figures 12.12 and 12.13. Describe the relationship between precipitation and population distribution in the United States. Think of reasons why this relationship exists.

12.4 Migration to the USA

The United States of America is one of the most multicultural countries on earth. Every year since 2001, more than a million migrants have made new homes for themselves in the United States.

Illegal immigration

The United States Office for Immigration Statistics estimates that there are about 11.5 million illegal migrants living in the United States. About half of these people arrived between 1995 and 2004. The great majority of these migrants are Mexicans, as shown in Figure 12.14. Most illegal migrants are looking for work.

Multicultural United States

The United States is often called the world's '**melting pot**'. This means that it is a country to which people from all around the world migrate to live. Figure 12.15 shows the regions from which the migrants to the United States came in both 1999 and 2010.

12.14 The origin of the estimated 11–12 million undocumented migrants living in the United States of America in 2009.

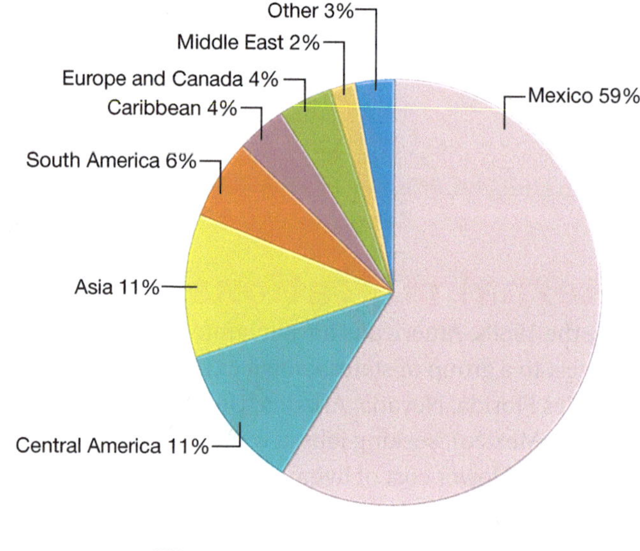

12.15 Source of migrants to the United States of America, 1999 and 2010

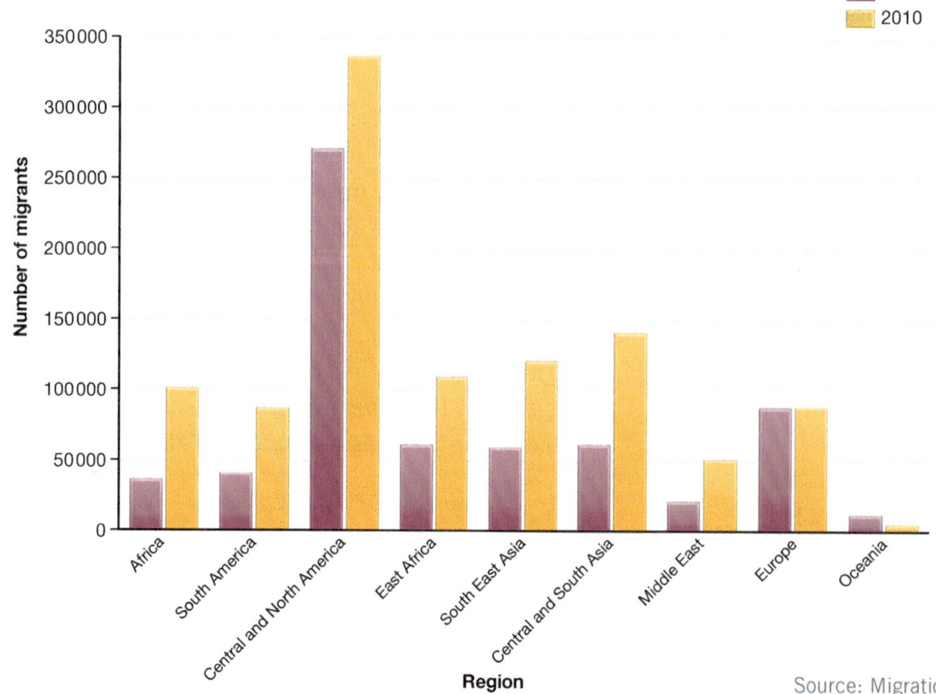

Source: Migration Policy Institute

Mexico remains the most important source country for migrants to the United States. About 30 per cent of all migrants living in the United States today were originally born in Mexico. Given that Mexico and the United States share a common border, this is not surprising. Also, with such a large population of Mexicans in the United States, there are a large number of family reunion migrants. However, the situation is beginning to change, due to the slowing US economy and better job prospects in Mexico.

Destinations

The destinations of migrants once they arrive in the United States vary greatly from place to place. Figure 12.16 shows the US states and cities with the highest proportion of migrant populations. Those states that are close to Mexico, the United States' largest supplier of migrants, have very high proportions. For example, more than 25 per cent of California's population was born elsewhere. Large cities such as New York and Los Angeles, which offer employment opportunities, also have very high migrant populations.

12.16 Migrant populations in the United States of America

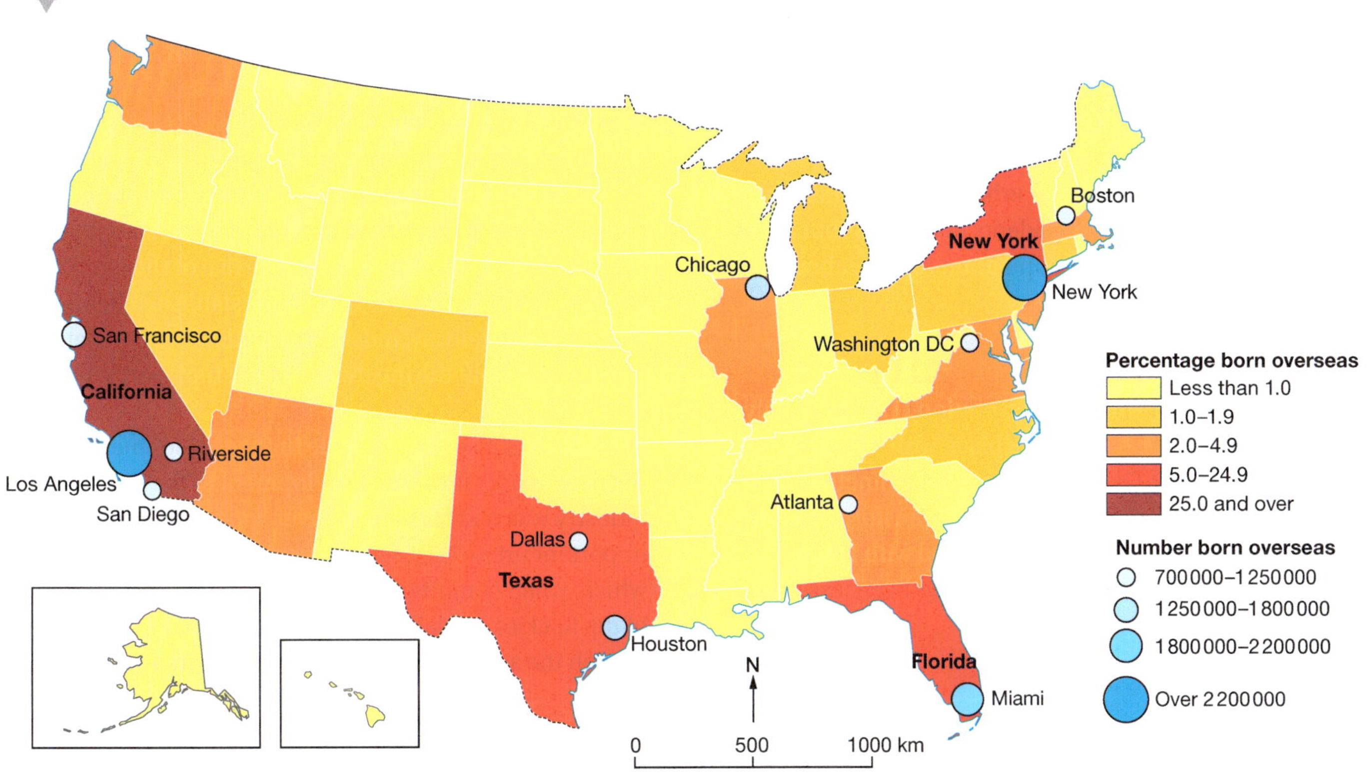

Source: Adapted from United States Census Bureau

ACTIVITIES

Knowledge and understanding

1 Outline the issue of illegal migration in the United States.
2 Explain why the United States is sometimes called a 'melting pot'.
3 Explain why Mexico is such important source of migrants to the United States.
4 Outline why Mexican migration to the United States is beginning to decline.

Geographical skills

5 Study Figure 12.14. Write a paragraph outlining the origin of the estimated 11–12 million undocumented immigrants living in the United States in 2009.
6 Study Figure 12.15. Calculate the difference in the number of migrants from Africa between 1999 and 2010.
7 Study Figure 12.16 and do the following tasks.
   a Identify the regions that have the greatest/least percentage of migrants.
   b Write a paragraph outlining the distribution of migrants in the United States.

12.5 Cities in the USA

The distribution of urban concentrations in the United States is quite different from that in Australia. The United States has a significant number of large, inland cities. Other than Canberra, Australia has none. The reasons for this have to do with the differences in topography and climate.

North America's cities

About 82 per cent of the population of the United States of America lives in urban areas (see Figure 12.17). These areas occupy just 2 per cent of the country's land surface. The majority of urbanised residents are suburbanites; inner-city residents make up about 30 per cent of the urbanised population (that is, about 60 million out of 210 million).

The rise of the US city

At the time of the American Revolution (1775–83), only one person in twenty lived in a place with a population greater than 2500. Over the course of the nineteenth century, however, the landscape was transformed by the process of urbanisation. But compared with Britain (and Australia), the United States has always been less urban.

In contrast to Britain and Europe, the US city system never developed a single, disproportionately large, dominant city. It developed a more spread-out, or dispersed, network of large cities. Many of these new cities were a product of the age of rail. With few exceptions, all the cities west of the Mississippi River were established as the railway spread to the west. While five of the ten largest cities in the United States had been established before the coming of the railway, most of the national urban network of cities was built during the era of rail transportation.

Other factors also played a role in the development of the network of cities. The development of the mining industry was one of these. Denver, for example, provided important urban services to its ranching **hinterland**, which, when combined with mining on the eastern slopes of the Rockies, explains its rapid growth in the 1870s, even though the area was too dry for non-irrigated agriculture.

12.17 Distribution of North America's million plus cities, 2010 and 2025.

Location of cities

In addition to the age of rail, a range of factors account for the existing pattern of urban concentration in North America.

Climate and landuse

More of the United States is suitable for agriculture than in Australia. While the Great Plains—which lie west of the Mississippi River and east of the Rocky Mountains—are classified as semi-arid prairie grasslands, they are capable of supporting ranching and irrigation-based agriculture. The climate is one of extremes. Very cold and harsh winters are followed by very hot and humid summers. East of the Mississippi River, population densities and rainfall increase. In the south the climate is milder, but hot and humid in summer. In the north, winters are harsh. Precipitation is high enough to support more intensive forms of farming.

Geography

The existence of navigable waterways, most notably the St Lawrence River, which links the Great Lakes to the Atlantic Ocean, and the Mississippi River, allow ocean-going vessels to travel some distance inland. In the case of the St Lawrence River, cities such as Detroit and Cleveland thrived as a result of their access to the Atlantic Ocean.

History

The largest concentration of large urban centres is in the north-east of the country—the original focus of European settlement. This region remains the most densely settled part of the United States.

Governance

The United States comprises fifty states, each with its own capital city complete with all the administrative functions associated with government. This approach to government has contributed to the distribution of relatively large urban centres across the United States.

Changing urban populations

The urban population pattern of the United States is undergoing change. The country's industrial cities are losing population while the service-based cities are expanding. Figure 12.18 shows that while cities such as New York and Los Angeles are growing, others, including Chicago, Philadelphia and Baltimore, are declining.

12.18 While US coastal cities have continued to grow, inland cities, especially those traditionally associated with manufacturing, have experienced declining populations.

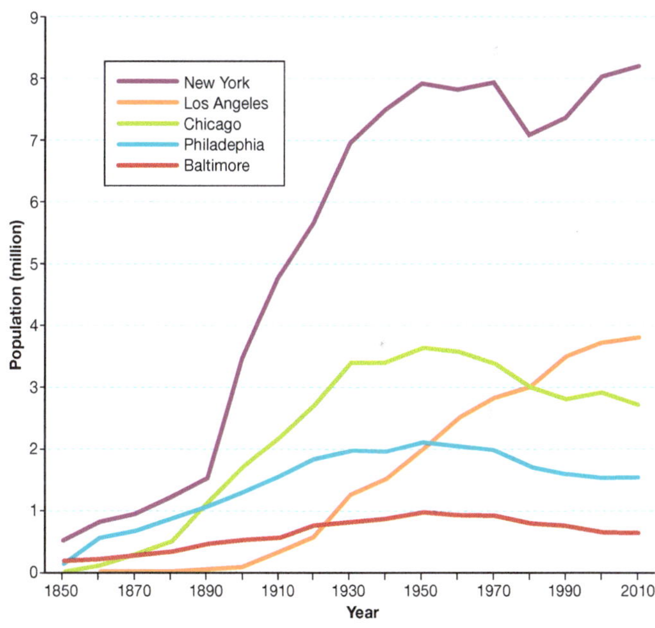

Source: United States Census Bureau, 2012

ACTIVITIES

Knowledge and understanding

1. State how the pattern of urban concentration in the USA differs from that of Australia.
2. Outline how the urbanisation process in the USA differed from the European experience.
3. Explain the role that railways played in developing the USA's pattern of urban concentration.
4. Outline how climate, landuse and geography have influenced the distribution of urban concentration in the USA.
5. Explain how the US system of governance has affected the urban pattern of the country.

Geographical skills

6. Study Figure 12.17 and do the following tasks.
   a. Identify the overall settlement pattern.
   b. List the possible reasons for this distribution.
7. Study Figure 12.18. Using data from the graph, outline the trends in urban populations for New York, Los Angeles, Chicago, Philadelphia and Baltimore.

12.6 Case study: New York City

New York City is at the centre of the largest urban concentration on earth. It is also a world city—a place that influences global commerce, finance, media, fashion, art, entertainment, research and technological innovation. It is the home of the United Nations and as such plays a central role in world affairs. New York is also a magnet for tourists. Fifty million tourists visit the city each year.

Geography

New York City is located in the north-east of the United States of America, approximately halfway between Washington, DC and Boston. The city straddles three islands at the mouth of the Hudson River—Manhattan, Staten Island and Long Island.

Manhattan—a 59.5 square kilometre island—is generally divided into Downtown (Lower Manhattan), Midtown (Midtown Manhattan) and Uptown (Upper Manhattan), with Fifth Avenue dividing Manhattan's east and west sides. Manhattan is bounded by the Hudson River to the west, the East River to the east and the Harlem River in the north. These water bodies are clearly visible in Figure 12.19, as is Central Park.

New York's summers are hot and humid, while winters are cold and damp. Outbreaks of very cold air flowing south from Canada occur from time to time, bringing snow to the city.

History

New York was founded as a trading post in 1624 by Dutch colonists and was named New Amsterdam. It fell to British forces in 1664 and was renamed New York—a tribute to King Charles II's brother, the Duke of York. The city served as the capital of the United States from 1785 to 1790. It has been the United States' largest city since 1790.

In the nineteenth century, New York grew rapidly and became a major port (see Figure 12.20). This, in turn, encouraged the growth of manufacturing and commerce. Great fortunes were made. Fortunes are still made, and lost, in New York City. The city, especially the financial district focused on Wall Street, is now the centre of global finance and a control centre of the global economy (see Figure 12.21).

12.19 Satellite image of New York City

Population

More than 8.2 million people live in New York City. The New York metropolitan area has 18.9 million people. When adjacent metropolitan areas are taken into account, the total rises to 22 million, nearly the size of the entire Australian population. New York City is the world's most culturally diverse city. Throughout its history it has been a major entry point for immigrants seeking a better life.

The term 'melting pot' is used to describe the densely populated immigrant neighbourhoods. At the beginning of the nineteenth century, overpopulation meant that many new arrivals at first lived in the city's crowded slums. Today, the mix of cultures is the city's defining quality. The city's population in 2010 was 44 per cent white, 25.5 per cent black and 12.7 per cent Asian. New York is the world's most linguistically diverse city with more than 100 languages spoken.

Manhattan's distinctive skyline took shape as the city grew skyward in an effort to accommodate its ever-increasing population. Today, New York is the most densely settled city in the United States with 26 924 people per square kilometre.

DID YOU KNOW?

The Statue of Liberty (a gift of the French people) greeted millions of immigrants arriving by ship in the late nineteenth and early twentieth centuries. Today it is the globally recognised symbol of the United States of America and its democratic tradition and the concept of freedom.

12.20 Manhattan and New York City, 1873. The famous Brooklyn Bridge is under construction.

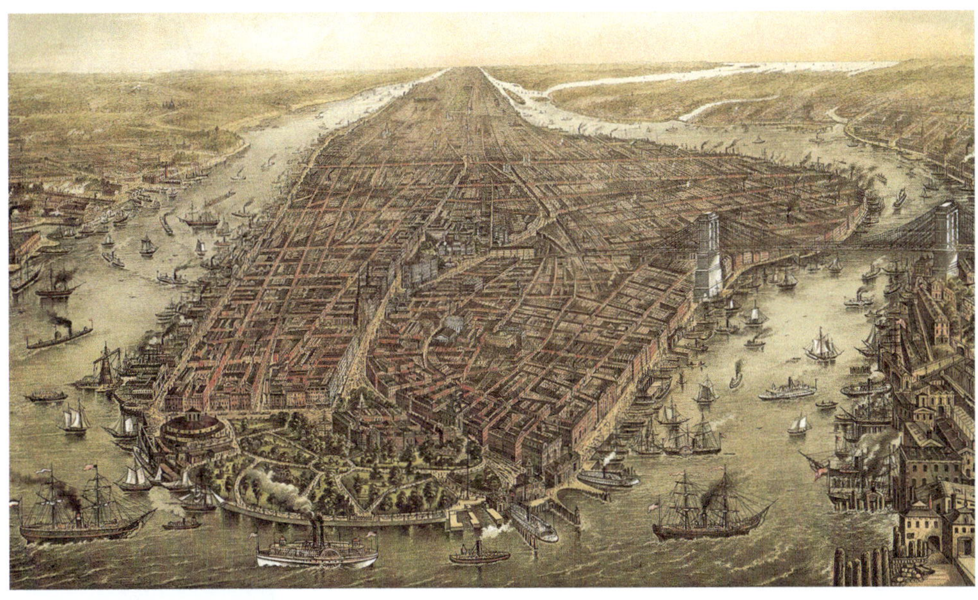

12.21 Manhattan and New York City, 2012. The famous Brooklyn Bridge is still a prominent landmark.

Neighbourhoods and their characteristics

Over time, New York's various neighbourhoods have developed their own distinctive character or 'sense of place'. Figure 12.23 highlights this urban diversity.

SPOTLIGHT

Central Park

Central Park is a massive, 3.4 square kilometres of parkland in Upper Manhattan. It is shown in Figure 12.22. It was first opened to the public on its completion in 1873.

Today, the park is one of the world's largest urban public parks. While the landforms and planting appear natural, the park is in fact almost entirely landscaped. Features include several natural-looking lakes and ponds. The park also accommodates the world-famous Central Park Zoo. An estimated 35 million people visit Central Park each year. It features in countless movies and television shows.

12.22 Central Park viewed from the Rockefeller Center in Midtown

1 Lower Manhattan

The birthplace of New York. Today the area is dominated by commerce. It was also in this area that the World Trade Center stood before it was attacked by terrorists on 11 September 2001. Also found in the area are Battery Park, Wall Street, the New York Stock Exchange and the September 11 Memorial. Ferries leave from here for the Statue of Liberty and Ellis Island.

2 Seaport and the Civic Centre

The administrative core of New York City. Home to the Civic Centre, police headquarters, and state and federal court buildings. The docks that dominated the East River shoreline during the age of sail have been transformed into a museum, shops and restaurants. Landmarks include City Hall and the Brooklyn Bridge.

3 SoHo and TriBeCa

A neighbourhood famous for its cast-iron architecture. Popular with artists, the area is home to trendy galleries, cafes, shops and boutiques.

4 Lower East Side

For many years the destination of immigrants arriving in the USA. Here Italians, Chinese and Jews established distinct neighbourhoods, preserving their language, foods, traditions and religions. Many ethnic restaurants are found in the Lower East Side. Landmarks include Chinatown, Little Italy and the New Museum of Contemporary Art.

5 East Village

Another area popular with immigrants. From about 1900 the Irish, Germans, Jews, Poles, Ukrainians and Puerto Ricans left their mark. More recently, the area has been popular with those attracted to alternative lifestyles and the arts. The area is home to many experimental music clubs and theatres.

6 Greenwich Village

A neighbourhood popular with celebrities, artists and writers. An expensive district popular with the gay community. Student housing is found around the campus of New York University. The Meatpacking District is now dominated by smart boutiques and restaurants.

7 Chelsea and the Garment District

Once a warehouse district until it was gentrified. Today, it is filled with art galleries and antique shops. It has a large gay community. It is also home to the enormous Macy's Department Store. Other landmarks include the Empire State Building and the General Post Office.

8 Gramercy and Flatiron District

A largely residential area dotted with trendy cafes and boutiques.

12.23 New York City and its neighbourhoods

DID YOU KNOW?

The New York Stock Exchange is the world's largest, with an estimated annual trading volume of US$5.5 trillion.

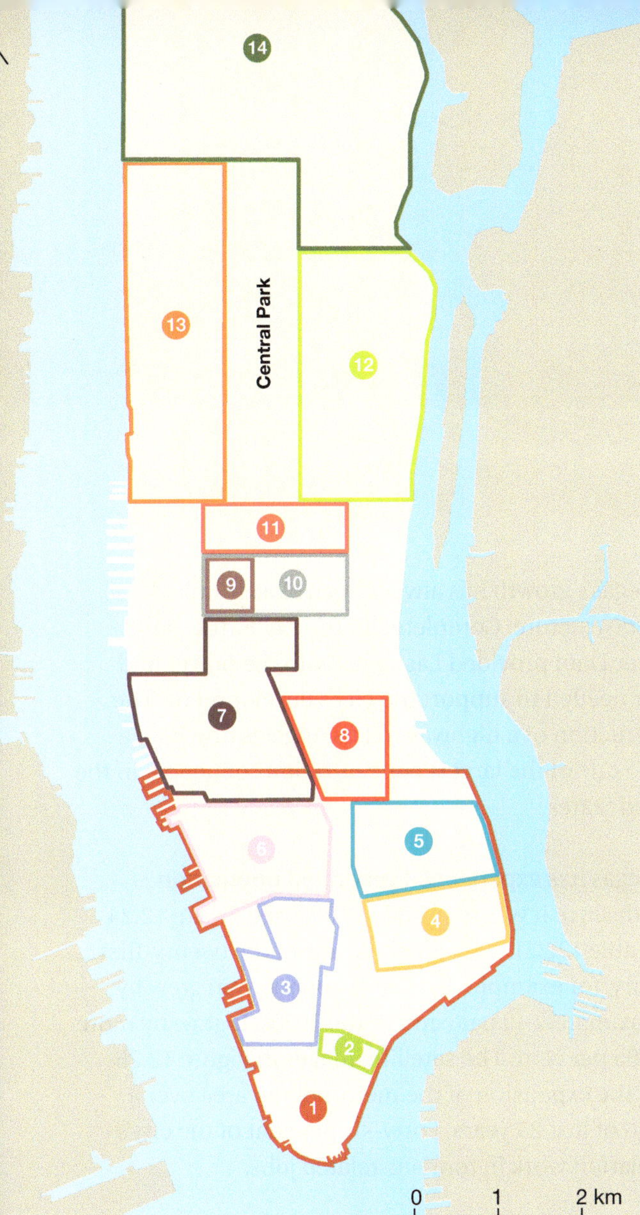

9 **Theatre District**
The opening of the Metropolitan Opera House in 1883 resulted in the development of numerous theatres and restaurants in the area. Landmarks include Times Square, the Rockefeller Center and Carnegie Hall.

10 **Lower Midtown**
The largely commercial and residential area houses some of the city's greatest architecture. Landmarks include the Chrysler, Metlife and Hemsley Buildings, Grand Central Terminal and the United Nations.

11 **Upper Midtown**
Upmarket residential and retail areas, with grand hotels and world-famous department stores. Landmarks include the Fifth Avenue department stores and boutiques, Trump Tower, IBM Building, St Patrick's Cathedral, the Museum of Modern Art (MoMA) and the Plaza and Waldorf–Astoria hotels.

12 **Upper East Side**
New York's wealthy occupy the area's grand apartment buildings on Fifth and Park avenues. Upmarket shops and restaurants are found throughout the area. Landmarks include the Metropolitan Museum of Art and the Frick and Guggenheim museums.

13 **Upper West Side**
An upmarket residential area, especially in the blocks lining Central Park. The famous Dakota Building, the city's first luxury apartment block, was built in 1884. Landmarks include the Museum of Natural History and the Lincoln Center.

14 **Morningside Heights and Harlem**
Home of Columbus University and America's most famous black community.

ACTIVITIES

Knowledge and understanding

1 Explain why New York City is referred to as a 'world city'.
2 Explain the origins of the term 'melting pot'.
3 Explain why New York City reached skyward.
4 Outline the history of New York City.

Geographical skills

5 Study Figures 12.20 and 12.21. Compare and describe the transformation that has taken place. What elements of the landscape have remained fairly constant?

Investigating

6 Study Figure 12.23. Select one of the neighbourhoods highlighted on the map. Undertake research using the internet to answer the following questions.
  a What is special or unique about the neighbourhood? Describe its character.
  b Is it associated with any particular function, social group or ethnicity?
  c What are the major tourist attractions of the area?
  d Does the neighbourhood play a role in terms of popular culture? For example, does it feature in a film or television series?
7 Select one of the New York City tourist attractions featured in Figure 12.23. Using the internet, prepare a short written report explaining why this attraction is special and why a visitor to New York City would want to visit it.

12.7 Case study: Las Vegas

With a population of just under two million, Las Vegas is one of the largest inland urban centres of the United States of America. The city is an internationally renowned tourist destination, famous for its vast casino-hotel-resort complexes and associated entertainment. The city's main entertainment precinct is concentrated along a 6.8 kilometre stretch of South Las Vegas Boulevard known as 'The Strip'. Las Vegas is also a major convention centre.

Geography

Las Vegas is located in an arid desert basin surrounded by mountains. Much of the landscape is rocky and dusty. The environment is dominated by desert-like vegetation, and the area is subject to torrential flash floods. The dry heat is an attraction, especially for tourists from the colder parts of North America.

Origins and growth

Las Vegas was founded as a stopover for the pioneers travelling to the west, and became a railroad town in the early twentieth century. In 1931, the state of Nevada legalised gambling. This led to the development of casinos. Major developments occurred in the 1940s, following the influx of scientists and staff working on the Manhattan Project—the World War II research that led to the development of the atomic bomb.

Las Vegas's growth has always been closely linked to infrastructure. Completed in the 1930s, the giant Hoover Dam provided Las Vegas with the power and water needed to support the city's development. The construction of a highway to Los Angeles provided a link to one of the largest concentrations of people in the United States.

Las Vegas has experienced very rapid population growth since it was founded in 1905 (see Figure 12.24 and Table 12.25). In the first decade of the twenty-first century, for example, Las Vegas's population grew by approximately 40 per cent. In the 1990s, it grew by more than 85 per cent. The satellite images in Figure 12.26 show the expansion of the metropolitan area over a period of just 25 years. Forty-six per cent of the city's population work in tourism-related jobs.

12.24 Las Vegas's famous strip

12.25 Population growth of the Las Vegas metropolitan region, Clark County, Nevada

Year	Clark County population	Change (%)
1910	3321	
1920	4859	46.30
1930	8532	75.60
1940	16414	92.40
1950	48289	194.20
1960	127016	163.00
1970	273288	115.20
1980	463087	69.50
1990	741459	60.10
2000	1375765	85.50
2009	1902834	38.31

Source: United States Census Bureau

12.26 The growth of Las Vegas over the last quarter of a century is shown here in false-colour Landsat. The dark purple grid of city streets and the green of irrigated vegetation extend in every direction to the surrounding desert.

1984

2009

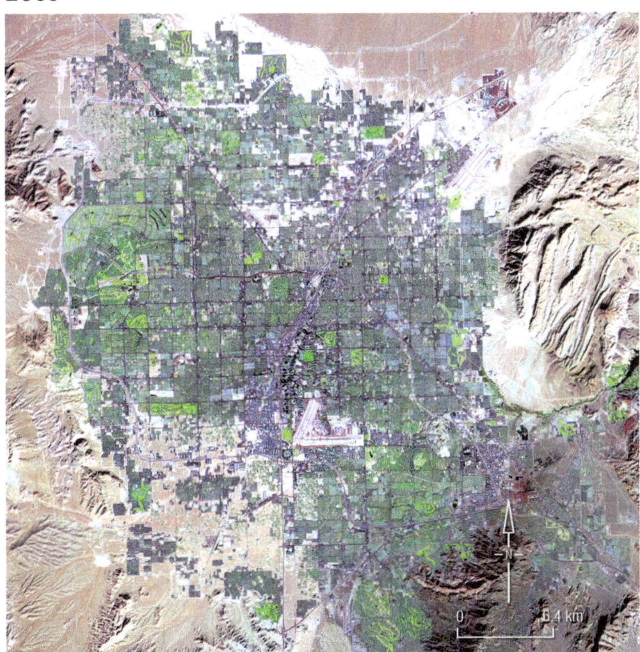

Tourism

Nearly 40 million people visit Las Vegas each year. Five million of these visitors travel there to attend the 19 000 conventions held in the city annually. Sixteen per cent of the city's visitors are international tourists.

The casino and entertainment complex constructed in Las Vegas by 1970 resulted in a rapid growth of airline connections and the development of the convention business. By 2010, McCarran Airport ranked twenty-second in the world for passenger traffic, with 39 757 359 passengers passing through the terminal. The airport ranked ninth in the world for aircraft movement, with over half a million take-offs and landings.

The other advantage of Las Vegas is lots of hotel rooms. Nine of the top ten largest hotels in the world can be found in Las Vegas. The availability of so many hotel rooms has resulted in the emergence of the nation's largest convention business.

Gambling

Las Vegas is also a leading centre of gaming. Revenue from gambling now exceeds US$9.2 billion per year. Las Vegas's initial advantage came from being home to the first large-scale gaming industry. Due to the increase in gambling around the world, especially in Macau, China, Las Vegas-based firms are now major suppliers of gambling expertise worldwide.

DID YOU KNOW?

Seventeen of the twenty biggest hotels in the US are in Las Vegas.

ACTIVITIES

Knowledge and understanding

1. State why Las Vegas is so well known internationally.
2. Outline the scale of the tourism and convention industry in Las Vegas.
3. Describe the geographical setting of Las Vegas.
4. Draw a timeline outlining the origins and stages of development of Las Vegas.

Geographical skills

5. Study Table 12.25. Construct a line graph illustrating the population growth of Las Vegas from 1910 to 2009. In which decade was growth most rapid?
6. Study Figure 12.26. Estimate the increase in the area of Las Vegas between 1984 and 2009.

CHAPTER 13

MIGRATION TO AUSTRALIA

Since early 1945, more than seven million people have come to Australia as new settlers. Today, nearly one in four Australians was born overseas. These new arrivals have reshaped the nation. Australia is now truly a multicultural society.

The vast majority of new settlers choose to settle in Australia's large cities, especially Sydney and Melbourne. These cities are now extremely diverse places, and our lifestyle has been transformed and enriched.

In this chapter we will identify and explain the main types of international migration, the patterns of international migration in Australia, where international migrants settle in Australia, and how this reinforces urban concentration. We will also examine how international migration affects urban lifestyles.

KEY IDEAS

- To understand the main types of international migration
- To investigate the main origins and destinations of Australia's international migrants and how these reinforce urban concentration in Australia
- To describe how international migration has affected Australian urban lifestyles

13.0 Visitors viewing the Welcome Wall at the Australian National Maritime Museum, Sydney

GLOSSARY

cosmopolitan lifestyle	a way of life drawn from aspects of the lifestyles of a range of cultures across the world
guest worker	a person who moves temporarily to another country for work
internal migration	the movement of people within a country
international migration	the movement of people between countries on a permanent or semi-permanent basis
involuntary migration	the forced movement of people due to war, civil unrest, drought and famine
multicultural society	a community made up of people from different cultural backgrounds
urban consolidation	increased population densities in existing urban areas
urban decay	the deterioration of the built environment. Urban infrastructure falls into a state of disrepair and buildings are left empty for long periods of time
urban renewal	the redevelopment of an urban area so that it better meets the needs of people. Also known as urban redevelopment
White Australia Policy	various historical federal government policies aimed at restricting non-white immigration to Australia from 1901 to 1973

13.1 International migration

Migration can be defined as simply the movement of people. Movement within a country, for example people moving from rural areas to the city, is called internal migration. Movement between countries is called international migration.

Who migrates

Each migrant undertakes the move for their own reasons. For some people the choice is voluntary and is made for lifestyle or economic reasons. For others there is no choice: natural disaster, war or a range of other concerns force them to move.

There are five broad categories of migrants: settlers, contract workers, professionals, undocumented workers, and refugees and asylum seekers.

Settlers

Settlers choose to move permanently to a new country. These people need to apply and then be accepted into their new country, usually by passing some type of test.

Contract workers

Contract workers are accepted for a short time into a country, usually for a set type of employment. For example, ski instructors may be permitted into Canada for the ski season.

Professionals

As the world economy grows and companies become more global, the need for company employees to travel increases. Many of these transnational corporations move their staff around the world. These migrants are often referred to as expatriates and are usually short-term (2–3 years) migrants.

Undocumented workers

Undocumented workers are usually referred to as illegal immigrants. Many people in this category are smuggled into host nations. For example, there are an estimated six million illegal immigrants in the United States of America, mostly from Mexico. Other illegal immigrants may arrive in a country legally, usually as tourists, but then overstay.

Refugees and asylum seekers

Refugees and asylum seekers flee their homes because of fear. If the authorities in the countries to which they flee confirm that they are in danger they are entitled to protection and perhaps even resettling in a new country. Asylum seekers are people who seek to have a government accept them. This type of migration is often referred to as **involuntary migration** because the migrants are forced to move.

Why migrate

The reasons for migration are complex and differ from person to person. However, there are three main reasons why a person would want to move to another location. They can be summarised as follows.

- *Pull factors* These are often referred to as demand factors. They are those things that make a person want to go to a new location. Better employment opportunities and higher wages are typical pull factors. Social factors, such as the desire to be reunited with family members, are also pull factors.
- *Push factors* These are often referred to as supply factors. They are those things that make a person want to leave a place. Unemployment, low wages, natural disasters—for example drought or famine—and war are common push factors. When people flee conflict they try to bring as many of their belongings as possible, as shown in Figure 13.1.
- *Network factors* These factors include the desire for a new experience or moving as a result of a company restructure.

Migration has taken place for thousands of years. Globalisation, however, has made it easier for people to move. Work and education have become more international and many people spend some of their lives in countries other than their own.

Many migration patterns reflect traditional links. For example, many migrants to Great Britain come from Pakistan and India. Pakistan and India are former dominions (colonies) of Britain and so have historical links with Britain and British influences in their cultures. Changes in government policies—such as the dismantling of Australia's **White Australia Policy**—have enabled more global migration patterns. Figure 13.2 shows the ten countries with the largest migrant populations.

13.1 Bangladeshi men who have been working in Libya flee the unrest there, heading to a refugee camp after crossing the Tunisian–Libyan border.

13.2 Top 10 countries with the largest migrant populations, 2010

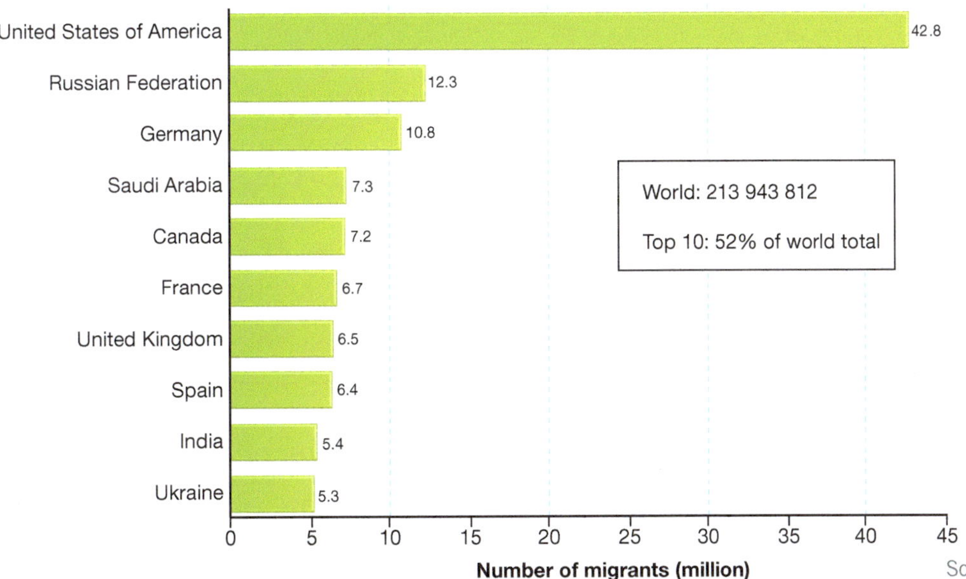

Source: Migration Policy Institute

Migrant destinations

There are many countries around the world that have large overseas-born populations. These countries often offer opportunities for work, for example the United States of America, Canada and Germany.

Some countries have a large population of temporary migrants. For example, Saudi Arabia has a very large number of **guest workers**. These are people who temporarily move to a country for work. These guest workers do not have the full rights of other residents and usually do work that local people do not want to do. In Saudi Arabia, guest workers commonly do construction work, domestic work and other manual labour. In Hong Kong, an estimated 100 000 Filipino guest workers work as maids. These guest workers—all women—send the money they earn back to the Philippines, to help their families. Figure 13.3 shows Filipino guest workers sending money back to the Philippines.

Figure 13.4 shows some of the main flows of people for work around the world. This map shows that the most common pattern of movement is from less developed countries to more developed economies. For example, there are significant flows of people from Mexico and South America to the United States. There are also large numbers of people moving from North Africa and Eastern Europe to the more prosperous and developed Western Europe.

Guest workers are not common in Australia, although in recent years there has been some movement of people to fill skilled vacancies in the mining industry. It is more common in Australia that people migrate permanently, or at least for long periods of time. Australia also has a large number of migrants from New Zealand. Australia and New Zealand have a special arrangement that allows the citizens of one country to work freely in the other for an unlimited period of time.

Filipino guest workers sending money back to the Philippines

13.4 Major movements of people for work

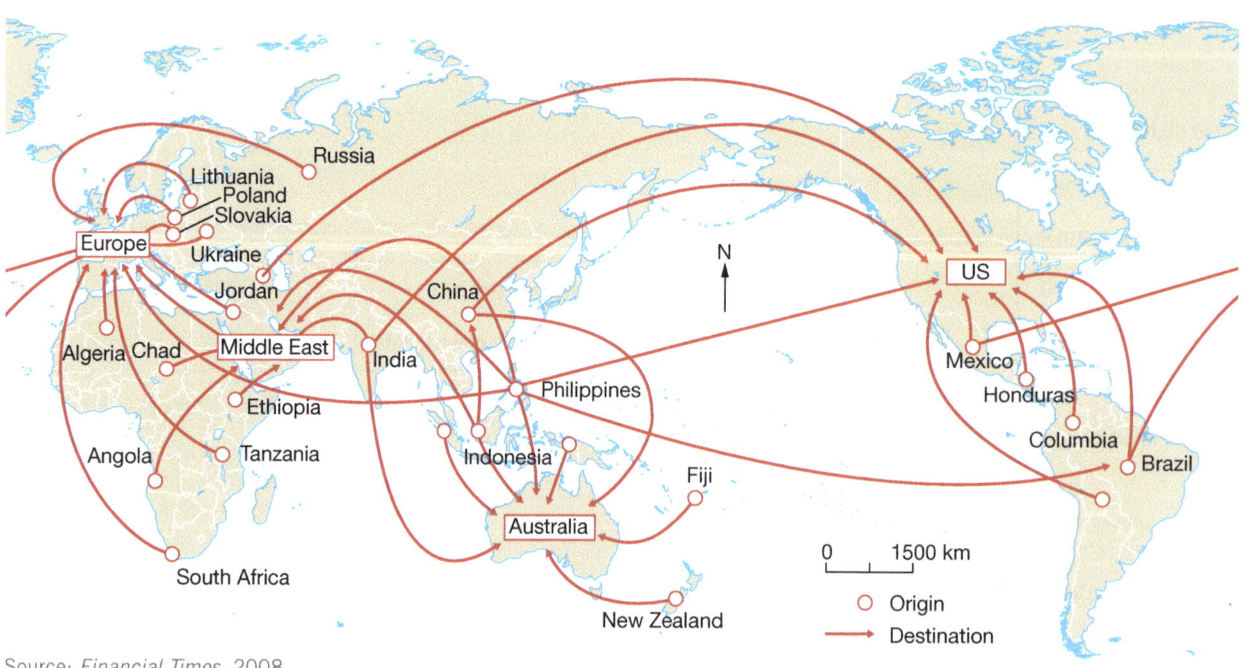

Source: *Financial Times*, 2008

SPOTLIGHT

People trafficking

Unfortunately, not all migration is undertaken willingly. Human trafficking occurs when people are taken against their will from one country to another. In most cases this involves people being kidnapped or tricked into travelling to other countries for work, in many cases in the illegal sex industry. The United Nations estimates that up to 800 000 people are trafficked across borders illegally each year.

Research indicates that there are an estimated 700 000 victims of trafficking currently in the United States. This includes a large number of labourers who are forced into working in terrible conditions on farms and in small factories. Other countries in which there are large numbers of people-trafficking victims are Britain, Nigeria, Brazil, France, Saudi Arabia, the Netherlands and Japan.

Here in Australia there have been 184 victims of trafficking identified by police and assisted between 2004 and 2008, mostly young women from South-East Asia who were trafficked into Australia to work in the sex industry. Thirteen people were convicted under Australian law in this period.

People-trafficking victims nearly always come from developing nations. Common source countries are in central Africa, South-East Asia and Eastern Europe.

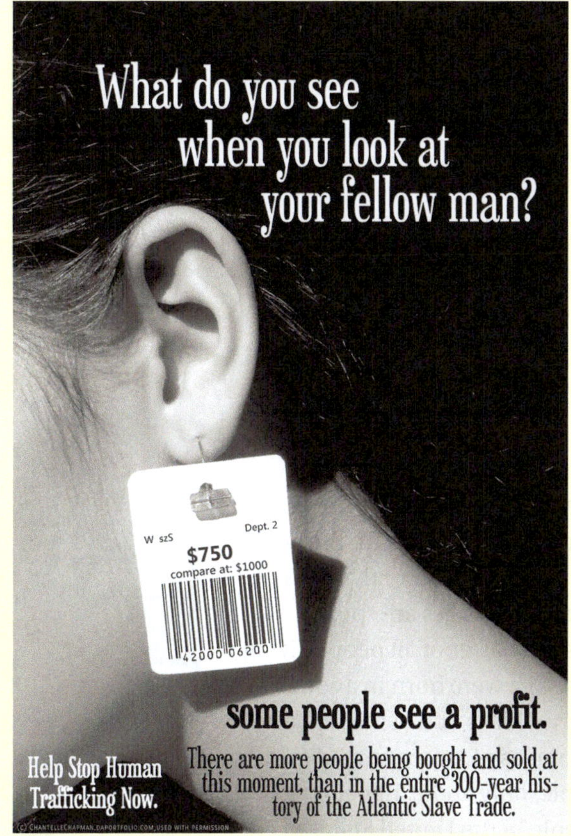

13.5 This advertising campaign was run in the USA to highlight the problem of human trafficking.

ACTIVITIES

Knowledge and understanding

1 Distinguish between internal and international migration.
2 Explain what a refugee is. How does a refugee differ from other types of migrants?
3 Outline the pull factors for migration.
4 Outline the push factors for migration.
5 Explain what a guest worker is.
6 Describe the meaning of the term 'people trafficking'.
7 Explain the extent of the problem of people trafficking around the world.

Applying and analysing

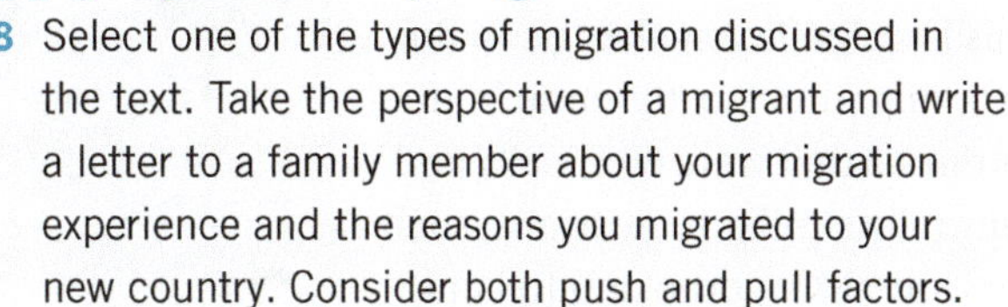

8 Select one of the types of migration discussed in the text. Take the perspective of a migrant and write a letter to a family member about your migration experience and the reasons you migrated to your new country. Consider both push and pull factors.
9 In many countries, guest workers are open to exploitation. In a small group, discuss why you think this is the case. Share your ideas with the class.

Geographical skills

10 Study Figure 13.2. What percentage of the world's migrants live in the United States?
11 Study Figure 13.4. Write a short report outlining some of the major flows of workers shown on the map. Explain why you think these flows exist.

13.2 Australian migration

In the first half of the twentieth century the White Australia Policy ensured that migrants were white Europeans. After World War II, this policy was gradually relaxed. More and more migrants, initially from southern Europe, particularly Greece and Italy, and then from Asia and finally the Middle East, came to Australia.

Multicultural Australia

Australia is now one of the most multicultural countries in the world (see Figure 13.6). More than 200 different languages are spoken. Almost a quarter of all people living in Australia were born in another country. This is one of the highest percentages of any country. For example, only about 10 per cent of people in the United States of America were born outside the United States of America.

13.6 Australia is now one of the most culturally diverse countries in the world.

Origin of settlers

Australia accepts migrants from around the world. The top ten source countries for migrants to Australia in 2010–11 are shown in Figure 13.7. As can been seen in the graph, most migrants to Australia now come from less developed countries. This is a significant change. In the past, migrants mostly came from countries that were ethnically and culturally similar to Australia, such as the United Kingdom and Ireland.

Types of migrants

The largest category of new settlers is skilled migrants. For the 12 months between July 2011 and June 2012, 125 755 people migrated to Australia under the skilled migration program. This special program is designed to encourage migrants who have skills that Australia needs. For example, 6914 accountants, 4836 cooks and chefs and 2688 computer programmers from around the world were allowed to migrate to Australia in 2011–12. These are all areas in which Australia has a skills shortage.

Destination of settlers

Australia's large capital cities are the destination of most migrants when they arrive in Australia. More than 80 per cent of newly arrived migrants first settle in Sydney, Melbourne, Brisbane, Perth or Adelaide. More than 30 per cent of migrants settle in Sydney and about 25 per cent make their home in Melbourne. This reflects the fact that these large cities provide better employment prospects, especially for skilled migrants. The large cities also tend to have better access to support networks for migrants, such as intensive English language schools.

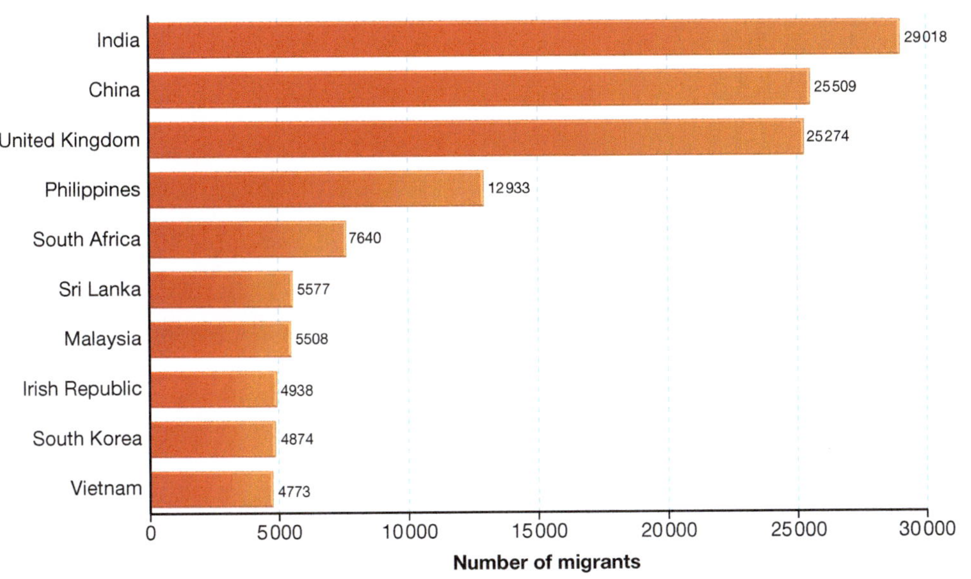

13.7 Migration currently accounts for much of Australia's population growth

Source: Department of Immigration and Citizenship, 2011–12 Migration Program Report

SPOTLIGHT

Humanitarian migration

An important part of Australian migration is the humanitarian program. This program supports refugees. Refugees are people who need the protection of another country as they can no longer live in safety in their own country. In 2010–11, more than 13000 people were accepted into Australia under this program. Typically, these people were fleeing war and persecution in their homelands, such as Burma, Afghanistan, Iraq and Sri Lanka (see Table 13.8). In 2012, the Australian government announced that it would be increasing the humanitarian migration program to allow up to 20000 people a year to migrate to Australia.

13.8 Top 10 countries of birth of refugees accepted into Australia in 2010–11 under the offshore humanitarian program

Source	Number
Iraq	2151
Burma	1443
Afghanistan	1027
Bhutan	1001
Democratic Republic of Congo	565
Ethiopia	381
Sri Lanka	289
Iran	271
Sudan	243
Somalia	190

Source: Australian Bureau of Statistics

ACTIVITIES

13.2

Knowledge and understanding

1 Describe how Australia's approach to migration changed throughout the twentieth century.
2 What is a skilled migrant?
3 Why do you think Australia has a focus on skilled migration?
4 Outline Australia's humanitarian migration program.
5 List the main destinations of migrants once they arrive in Australia.

Applying and analysing

6 Study Figure 13.7. Write a short report outlining the major sources of migrants to Australia.
7 Study the data in Table 13.8 and do the following tasks.
   a Describe the region/s from which the majority of refugees come.
   b Construct a column graph to summarise this data.

Investigating

8 Select one of the countries shown in Figure 13.8 and use the internet and library resources to conduct research and prepare a short digital presentation on why refugees would be leaving this country. Present your presentation to the class. Include in your presentation:
   - map of the country and an overview of the population (for example size and ethnic diversity)
   - history behind the reason people are fleeing and the current situation.

13.3 Australia's cultural diversity

Building on its rich Indigenous heritage, Australia was settled by people of many different cultures. Many of these arrived following World War II, encouraged by the Australian government's policy of 'populate or perish'. A massive immigration program resulted in millions of people coming to start a new life in Australia.

A nation of migrants

Immigration has played a key role in Australia's population growth and accounts for much of the cultural diversity of the country. The first arrivals were the ancestors of Indigenous Australians, who moved down from the islands of the Indonesian archipelago and New Guinea over 60 000 years ago.

European settlement

From the 1600s, during the great European voyages of discovery, there were a number of coastal landings, but mainly on the arid western side of the Australian continent. The British first sailed along the well-watered east coast in 1770, and the first permanent European settlement, the British Crown Colony of New South Wales, was established in 1788. Initially a convict colony, free settlers started moving in from the early 1790s. The gold rush attracted many people, including the Chinese. Following Federation in 1901, the government implemented the White Australia Policy, which excluded non-white immigrants. By 1945, the Australian population had reached seven million and was mainly Anglo-Celtic.

After World War II

After World War II (1939–45), Australia began a huge immigration program, based on the belief that it must 'populate or perish', following the near invasion by the Japanese during the war. As a result of the program, hundreds of thousands of people left war-torn Europe and migrated to Australia.

Between 1945 and 1972, over a million British people immigrated under the 'Ten Pound Poms' scheme, whereby British people—favoured by the Australian government—could migrate to Australia cheaply.

From the 1970s, millions of migrants and refugees arrived in Australia, looking for a life free from war, persecution and poverty. Waves of migrants came from the Asia–Pacific region, Africa and the Middle East.

In total, more than 6.5 million migrants and 675 000 refugees have settled in Australia since 1945.

Multiculturalism

Multiculturalism is a social policy based on the ideas of inclusion, recognition and tolerance. It recognises the right of all Australians to enjoy their cultural heritage and to receive equal treatment and opportunities regardless of their background. It carries with it a responsibility to respect Australian values.

Top 10 countries of birth for the overseas-born population, 2011

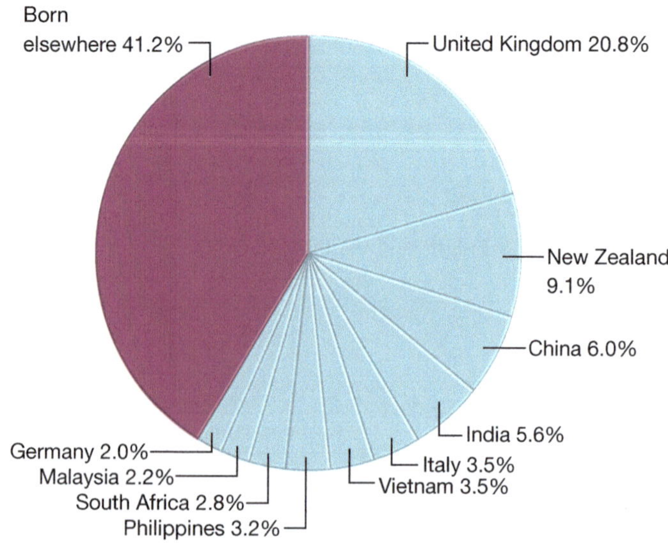

Source: Australian Bureau of Statistics

13.10 Many different cultures are recognised and celebrated in Australia.

People and culture

Australia's linguistic and cultural diversity reflects the long history of migration to the continent. The cultural traditions of Indigenous Australians are among the oldest in the world, and Australia's migrants have brought their cultural traditions from over 200 countries.

Migrants have enriched the Australian way of life in many ways, as is illustrated in Figure 13.10. They have, for example:

- introduced new cultural celebrations such as Chinese New Year and St Patrick's Day
- contributed to the creative and performing arts, for example the Borovansky Ballet Company
- added colour and interest to Australian cities, for example in Lygon Street in Melbourne and Chinatown in Sydney
- introduced new sports and recreational pastimes, such as futsal and taekwondo
- introduced architectural styles, for example with joss houses, mosques and synagogues
- enriched school curricula through the addition of languages such as Mandarin and Arabic
- greatly expanded the range of foods eaten by Australians, for example pizza, sushi and burritos.

ACTIVITIES

Knowledge and understanding

1 Draw a timeline of Australian migration starting in 1788.
2 Define the term 'multiculturalism'.

Geographical skills

3 Study Figure 13.9 and do the following tasks.
   a In 2011, what percentage of Australia's population had been born in the United Kingdom?
   b What is the next largest population born overseas after people from the United Kingdom?
   c Prepare a mind map demonstrating the ways in which migrants have enriched the Australian way of life. Use examples different from those listed above to illustrate this.

13.4 International migration and urban lifestyles

Immigration has helped to shape the urban lifestyles of Australia since the first European settlement was established at Sydney Cove in 1788. Today, the lifestyles of those living in Australian cities reflect the growing ethnic and cultural diversity of the Australian population.

Increasing urban concentrations

In the period 1947–2010 in Australia, there was an increase in the number of people living in cities. This pattern was most evident among the migrant population, as shown in Figure 13.11. The proportion of the Australian-born population living in large cities grew from 50 to 61 per cent. The proportion of Australian-born living in rural areas decreased from 32 per cent to 14 per cent, while the proportion of overseas-born Australians in rural areas fell from 25 to just 6 per cent in the same period.

The majority of migrants to Australia have settled in the capital cities, particularly migrants from non-English speaking countries.

13.11 Australia's overseas-born population living in major cities, 1947 and 2010

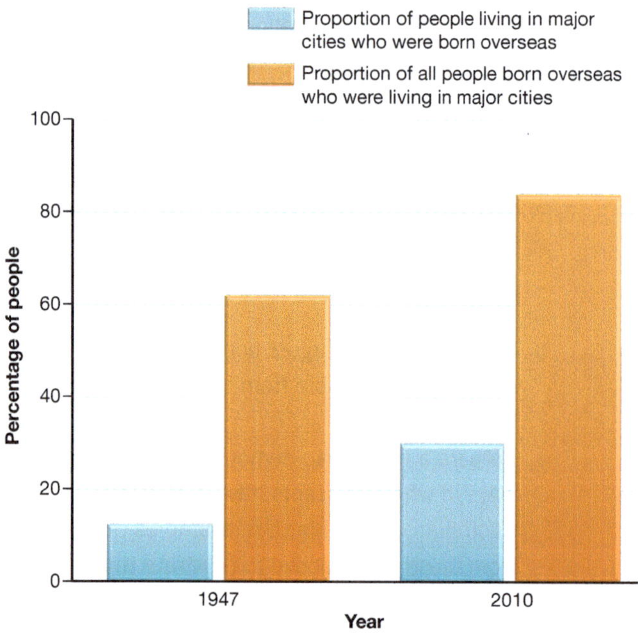

Sydney and Melbourne

The majority of post World War II migrants decided to settle in Sydney and Melbourne. As a result, these cities have the most international or **cosmopolitan lifestyles**.

In 2013, Sydney and Melbourne's share of the nation's overseas-born population was almost 55 per cent. While Sydney remains the most important centre of immigrant settlement, there is some evidence of a shift, as people move from Sydney to the other capitals. There is also evidence of increased settlement beyond the capital cities.

Ethnic and cultural influences

The diversity of Australia's cities has been influenced by waves of migration, each characterised by a different mix of birthplace groups. These waves reflect changes in Australia's immigration policy and the changing national and global economic and political situation.

Migrant settlement patterns

Patterns of settlement vary among migrant groups. New South Wales, for example, has 75 per cent of Australia's Lebanese-born population, 63 per cent of the Iraqi-born, 63 per cent of the South Korean-born, 60 per cent of the Fijian-born and 60 per cent of the Chinese-born. Victoria has 51 per cent of Australia's Sri-Lankan-born population, 50 per cent of the Turkish-born, 49 per cent of the Greek-born and 42 per cent of the Italian-born. Western Australia, the state with the highest foreign-born proportion of population, attracts 30 per cent of all Singapore-born Australian residents, and is narrowly behind New South Wales in having the largest population of British-born. Queensland has attracted the greatest proportion of people born in Papua New Guinea (52 per cent) and New Zealand (38 per cent).

13.12 Melbourne's cafe culture is a lifestyle-related outcome of Australia's ethnic diversity.

Impact on urban lifestyles

The cosmopolitan lifestyle enjoyed by residents of cities such as Sydney and Melbourne owes much to the cultural and ethnic diversity of their migrant populations. Some urban neighbourhoods have, over time, become associated with particular ethnic groups. These ethnic precincts are essentially clusters of immigrant entrepreneurs and service providers. These, in turn, attract the people who rely on these providers.

One of the most obvious outcomes of this process is the concentration of restaurants representative of different cultures. European, Middle Eastern and Asian cuisines are now an important part of our urban lifestyle. There are clearly defined restaurant and cafe areas in both Sydney and Melbourne. These are good examples of how different ethnic groups can enrich a culture. Residents of both cities have also embraced the European-inspired tradition of outdoor eating. Most cafes now have tables and chairs on the footpath, as shown in Figure 13.12, and some have outdoor areas to match those seen in Italy and France. The practice of eating out at restaurants and relaxing at pavement cafes is now common everywhere, and Australian's obsession with espresso coffee (introduced into Australia by southern Europeans) is central to the country's cafe culture.

ACTIVITIES

Knowledge and understanding

1. Explain why Sydney and Melbourne have the most international, or cosmopolitan, lifestyles.
2. Outline how the pattern of settlement arrivals has changed since World War II.
3. Identify some of the important differences in migrant settlement patterns.
4. Outline the impacts of migration on urban lifestyles.

Applying and analysing

5. List the ways in which your town or neighbourhood has been influenced by migration. Use the information collected to produce a whiteboard or pinboard-mounted mind map.
6. Draw on the ethnic diversity of your community to explore reasons why people chose to settle in Australia. Identify the ways in which different cultures have enriched your community.

13.5 Case study: Melbourne

More than four million people live in Melbourne, making it home to almost 75 per cent of all Victorians and 17 per cent of Australians. In recent years, Melbourne has been the fastest-growing capital city in Australia. During the first decade of the twenty-first century, 1500 people moved to the city every week. The Australian Bureau of Statistics (ABS) estimates that by 2056 Melbourne's population will grow to 6.8 million.

Melbourne's urban sprawl

Most of Melbourne's growth—about 1000 people a week— has taken place in the outer suburbs, where vast new housing estates have been developed over the last decade. For example, the population of Wyndham, shown in Figure 13.13, in the south-west of Melbourne, increased by 12 200 (7.8 per cent) from 2010 to 2011. Much of the growth in Melbourne's outer suburbs is driven by new arrivals to the city. About 60 per cent of new residents move to the outer suburbs.

Impacts of population growth

No other city in Australian history has experienced the rate of growth Melbourne has undergone in recent years. As the population has exploded, the city's infrastructure and services have become strained. Melbourne residents complain about increasing traffic congestion and overcrowding on public transport. There are also concerns that important services such as hospitals and schools are being put under pressure as the city grows.

13.13 Werribee–Wyndham region, located on the outer fringe, is one the fastest-growing suburbs in Melbourne.

In 2012, 68 per cent of Melbourne's population lived more than 20 kilometres from the city centre. As the city has grown outwards, the pressure on infrastructure has increased. Melbourne's tram system does not extend to the outer suburbs, which means that people must commute by car or travel in the overcrowded trains.

Inner city growth

While most of Melbourne's growth has taken place in the outer suburbs, the inner city has also been growing. More than 74 000 people moved to inner Melbourne in the first decade of the 2000s. This represented a growth of about 30 per cent. The inner city is now dominated by large high-rise apartment buildings. This process of increasing population density in established urban areas is called **urban consolidation**.

The Docklands

Melbourne's Docklands, shown in Figure 13.14, is one of the major redevelopments that has taken place in the inner part of Melbourne. Docklands was once the city's port—an area dominated by wharves, railway marshalling yards and storage facilities. With new cargo-handling technologies (especially the introduction of shipping containers) and bigger ships, the docks fell into disuse and **urban decay** accelerated. Urban decay occurs when an urban area becomes derelict and unused.

Beginning at the start of the twenty-first century, a major **urban renewal** program started to transform the area. Large apartment buildings, office towers, parklands and recreational facilities were built. In 2001, just 658 people lived in the area and 600 people worked there. By the end of 2012, 8000 people lived in Docklands and there were 30 000 working there. By 2025, this is expected to increase to 25 000 residents and more than 60 000 workers.

13.14 The Docklands redevelopment has created new housing areas close to Melbourne's CBD.

Melbourne's changing demographics

At the 2006 Census, Melbourne had a population of 3 592 591. By the 2011 Census, the population had increased to 3 999 982, an increase of 11.3 per cent. The number of dwellings (houses and apartments) increased from 1 351 532 to 1 430 665. Table 13.15 shows the changing ethnicity of Melbourne between the two census counts. For example, in 2006, Indian people were not a major category but by 2011 had become the third-largest group. This reflected a big increase in migration from India, especially students coming to study in Melbourne.

13.15 Percentage of Melbourne's population by birthplace, 2006 and 2011

Country of birth	2006 Census (%)	2011 Census (%)
Australia	64.20	63.30
England	3.50	3.40
India	N/A*	2.70
China	1.50	2.30
New Zealand	1.50	1.70
Italy	2.10	1.70

* changes in Census data collection

Source: Australian Bureau of Statistics

ACTIVITIES

Knowledge and understanding

1 Describe the growth of Melbourne's population in recent years.
2 State where most of the growth of Melbourne has taken place.
3 Outline some of the problems associated with the rapid growth of Melbourne's population.
4 Describe the growth of Melbourne's inner city.
5 Define the term 'urban renewal'.
6 Outline how urban renewal has transformed Melbourne's Docklands.

Applying and analysing

7 Prepare a list of the advantages and disadvantages of Melbourne's population growth. Examine your list and decide whether population growth is good or bad overall for Melbourne. Justify your decision.

Geographical skills

8 Study Figure 13.15 and the accompanying text and do the following tasks.
   a Which country of birth has experienced the most rapid increase in population?
   b Which country of birth has declined the most as a percentage of population?
   c Construct a column graph comparing the data from 2006 and 2011.

13.6 Case study: Darwin

The capital of the Northern Territory, Darwin, is the smallest of Australia's state and territory capitals, with a population of just 130 000 people (less than 1 per cent of the Australian population). While Darwin's population is small, it is ethnically diverse. More than eighty ethnic groups are found in the city. It also has the largest Indigenous population of any Australian city or town in Australia, as well as large populations of Greek, Chinese, Italian, Vietnamese, Indonesian, Filipino and Timorese people.

13.16 Darwin, the capital of the Northern Territory, is the smallest of Australia's capital cities.

A growing city

According to the Census, Darwin's population increased from 114 362 in 2006 to 129 024 in 2011, or nearly 13 per cent. The Australian Bureau of Statistics (ABS) predicts a population of 185 000 (an increase of 43 per cent on 2011) in 2021 and close to 250 000 (an increase of 93 per cent on 2011) in 2056. Much of this increase will be due to increasing employment opportunities, for example in the rapidly expanding Northern Territory natural gas industry. The Australian military also has a very big presence in Darwin and this is expected to increase in the future.

New urban areas

As the population of Darwin continued to grow in the first decade of the twenty-first century, the Northern Territory government began planning for a whole new urban area to the south of Darwin. The new city, called Weddell, will ultimately be home to about 50 000 people, and 20 000 new dwellings will be built there.

Weddell lies about 40 kilometres from the centre of Darwin, or about 20 kilometres by water across Darwin Harbour. It will be a satellite city. This type of urban development is attached to a major city but is also self-sufficient. One of the key elements of planning for Weddell is to make it as sustainable as possible. Therefore, it will have a focus on renewable energy, especially solar energy. Planners will also try to minimise car usage by locating residential, recreational and work areas close to each other, as well as establishing a well-developed public transport system link with the city. A plan for the city is shown in Figure 13.17.

There are many challenges in building Weddell. Planners will have to consider everything from the crocodiles that inhabit the creeks and waterways around the site to cyclones, which could cause flooding in low-lying areas. There are also important environmental habitats, Indigenous sacred sites and areas of archaeological importance to protect. By 2021, it is expected that Weddell will have a population of between 3000 and 10 000 people.

SPOTLIGHT

Darwin's Portuguese–Timorese community

The Timorese people have a strong connection with Darwin going back many decades. They originally come from the former Portuguese colony of East Timor—now known as Timor-Leste. This small country lies about 700 kilometres across the Timor Sea to the west of Darwin.

During World War II, East Timor was invaded by the Japanese and many of the Portuguese fled to Darwin. At the end of the war, political control was returned to Portugal until 1975, when East Timor declared its independence. Soon afterwards it was again invaded, this time by Indonesia. A long and bitter civil war began in East Timor, and many refugees fled to Darwin.

In 1999, East Timor won its independence from Indonesia but again a bloody civil war broke out and thousands of refugees were again housed in Darwin. Australian troops were sent to East Timor to bring about an end to the violence.

Today, there are about 5000 Portuguese–Timorese people living in Darwin, nearly 4 per cent of Darwin's population. About 1000 of these people were born in Timor-Leste and the rest are Australian-born with a Timorese heritage.

 Car sharing/market carpark/PV farm/ waste management/Communications masts

 Community facilities

 Commercial

Residential

13.17 Weddell, a new city, is being planned for a site near Darwin.

ACTIVITIES

Knowledge and understanding

1. Describe Darwin's population.
2. Outline the importance of Darwin to the Portuguese–Timorese people.
3. Outline the growth of Darwin's population in recent years.
4. List the factors driving Darwin's population increase.
5. Explain why the city of Weddell is being developed. What issues need to be overcome for this new city to be successful?

Applying and analysing

6. You have been appointed by the Northern Territory government to help design the new city of Weddell. Develop a concept plan for Weddell. Think about features that would help to make the city sustainable and also make it an attractive place for residents to move to.
7. With the aid of an atlas, write a short report describing the location of Darwin in relation to Australia's other capital cities. What issues do you think Darwin's isolation from other cities might create?

CHAPTER 14

INTERNAL MIGRATION IN AUSTRALIA

Australia's economy is being transformed. The demand for the country's mineral and energy resources by the emerging economies of South and East Asia has initiated the biggest mining boom in Australian history. Billions of dollars are being spent on new mines and related transport and port infrastructure. Thousands of new jobs are being created. This transformation, together with factors such as demographic change and people's changing lifestyle choices, has led to changes in the distribution of the Australian population.

In this chapter, we look at the nature and causes of Australia's economic transformation, internal migrations, Australia's mining boom, and its impacts on Australia's regional economies. Case studies focus on coalmining in Queensland's Bowen Basin and Western Australia's North West Shelf. This development illustrates how such developments influence the distribution of Australia's population. It also illustrates Australia's expanding links with the rapidly growing economies of South and East Asia.

KEY IDEAS

- To describe the factors that influence population movements within Australia
- To understand how the mining boom has influenced the distribution of the Australian population
- To investigate two mining locations in Australia

14.0 Modern open-cut and shaft mine system in the upper Hunter Valley, New South Wales

GLOSSARY

basin	a bowl-shaped depression in the earth's crust that formed when layers of rocks buckled under pressure
coal seam	a thin layer of coal found between layers of rock
counter-urbanisation	the process whereby people move away from urban areas to settle in rural areas
exurbanisation	the movement of people into communities within two or three hours' drive of major urban centres
gross domestic product (GDP)	the total monetary value of goods and services produced in a country during one year
Industrial Revolution	the rapid development of industry in the early nineteenth century, brought about by the introduction of steam-driven machines
nomadic	a way of life that involves moving across a territory seasonally in search of food and water
productivity	(of land) the amount of food or fibre produced from a unit of land
regional Australia	the non-metropolitan areas of Australia outside the major state capital cities and their surrounding suburbs
sea changer	a person who relocates from the city to the coast
service sector	the sector of the economy that provides services, such as shops, banks, hotels etc.
songlines	paths that are recorded in traditional Aboriginal and Torres Strait Islander songs, stories and dance
tree changer	person who relocates from the city to a rural or regional area
weightless economy	the part of the economy that produces, exchanges and consumes knowledge-based services rather than goods

14.1 Australia's economic transformation

Australia's economy is being transformed. The resources boom, driven by demand for the country's minerals and energy resources, has influenced internal migration in Australia.

Economic restructuring

In the 1800s, about one-third of Australia's total output of goods and services was agricultural products. But by the early 1930s, agriculture's share of Australia's economic output had started to decline. Today agriculture is just 3 per cent. This does not mean that Australia is producing fewer agricultural products. The country is actually producing more than ever, but other parts of the economy have grown at a much faster rate than agriculture.

14.1 Australia exports wheat around the world.

As agriculture's share of economic activity started to decline, manufacturing's share began to increase. In 1900, manufacturing was about 15 per cent of total output. In the 1960s, it peaked at 25 per cent. Since then its share has also declined and today it is less than 12 per cent.

Australia's service industries have grown from about half of all economic activity in the early 1900s to 60 per cent in the 1960s and 80 per cent today.

Economic exports

The make-up of Australia's exports has also changed. In the 1970s, 62 per cent of total exports were agricultural products. Today, agriculture is 10.8 per cent. In 2012, the only agricultural product in Australia's top ten exports was wheat. Figure 14.1 shows wheat being prepared for export. In the 1960s, mining's share was 15 per cent; in 2012 it was 50.6 per cent. Australia is a commodity-exporting economy.

Manufacturing's share of total exports is 13.3 per cent, and services' share is 16 per cent. Service-based exports include tourism and education.

The global economy

The process of industrialisation, which began in Europe in 1750 with the **Industrial Revolution**, has finally reached the world's developing countries. The world's two most populous nations, China and India, are rapidly industrialising, as shown in Figure 14.2. Within the next 20 or 30 years they will regain the economic dominance they once had. This process of industrialisation will require enormous amounts of iron ore, coal and other raw materials. These raw materials are likely to come from Australian mines.

A key feature of the re-emergence of China and India as major economic powers is the global shift of all but the most sophisticated types of manufacturing from developed countries to South and East Asia. This is a process affecting all the developed economies, not just Australia. Developed countries are having to find other things to do as their manufacturing moves to developing countries.

Australia's response

Australia has responded by adapting to the changed circumstances. It is promoting those economic activities best suited to a wealthy, well-educated, highly paid economy. While the poor countries are becoming manufacturing economies, the wealthy countries, such as Australia, are becoming knowledge-based economies.

At the centre of the knowledge-based economy are highly educated and skilled workers selling the products of their knowledge and skills. This economy includes all the professions: medicine, teaching, research, law, accounting, engineering, architecture, design, computing, consulting and management.

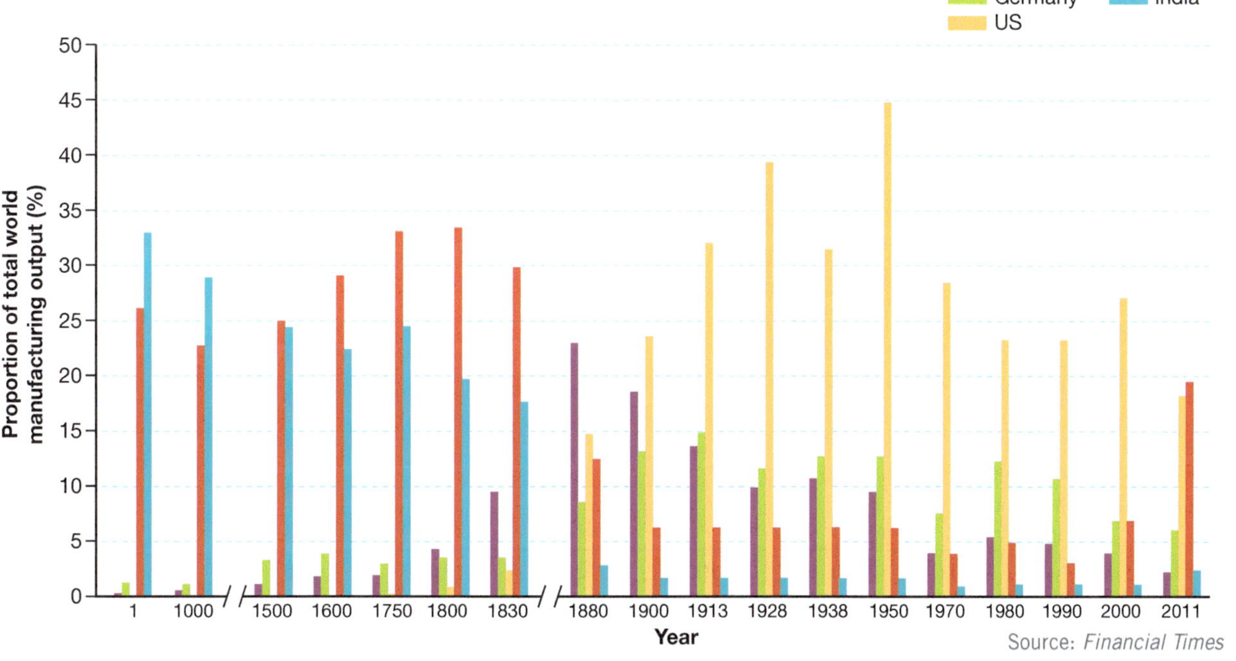

14.2 Trends in global manufacturing over the past 2000 years

A 'weightless' economy

An economy that produces, exchanges and consumes knowledge-based services rather than goods is known as '**weightless**', because nothing is created. While some of these services are fairly basic (for example car washing and lawn mowing), the fastest-growing categories involve a high level of knowledge and skill.

Employment in Australia's manufacturing sector has been falling since the 1980s, but since the mid-1980s total employment has grown by three-quarters to 11.5 million. Figure 14.3 shows that the employment share of the 'weightless' services sector has steadily increased since 1960.

Changing structure of employment

The economic transformation of Australia has been accompanied by a change in the structure of Australian employment. Manufacturing employment has declined while employment in the services sector has surged, as shown in Figure 14.4. Mining employment has also increased significantly but is only 1.6 per cent of the Australian workforce, or just over 200 000 people.

Australian employment by sector, 1910–2010

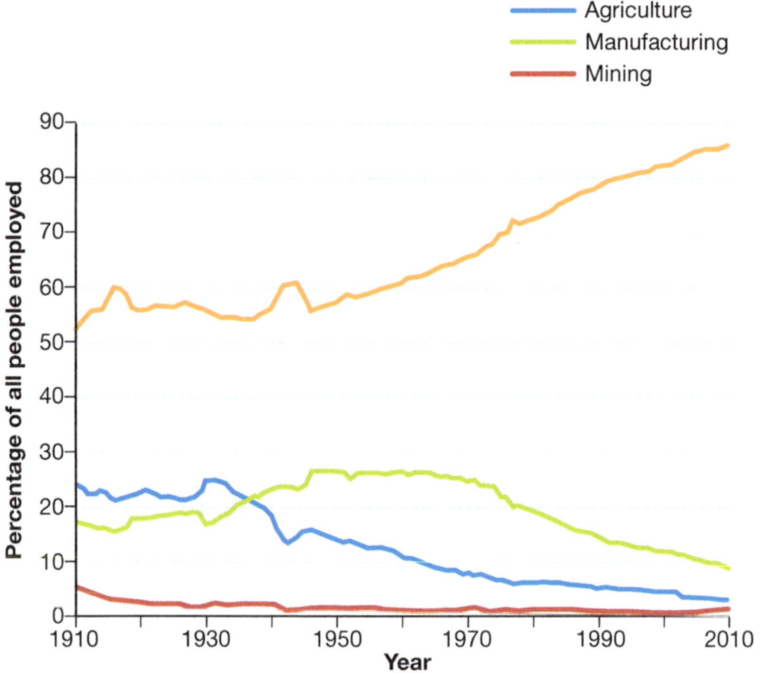

Source: Australian Bureau of Statistics; Reserve Bank of Australia

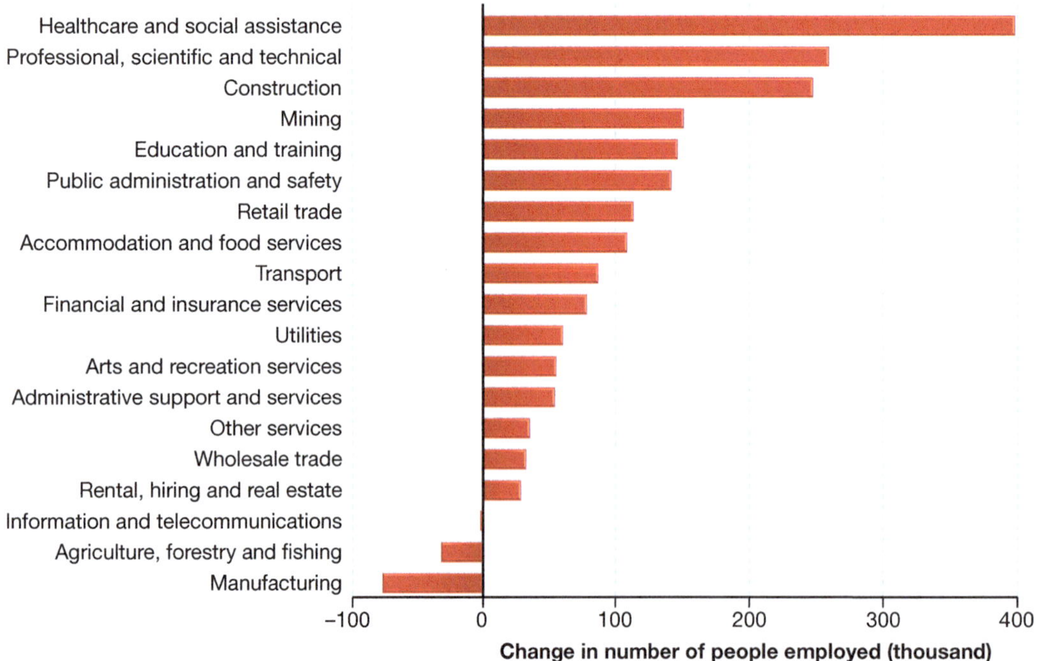

Source: Australian Bureau of Statistics

Population movements

Australia's economic transformation has contributed to a change in the distribution of Australia's population. The decline in manufacturing, combined with the growth in mining, has resulted in a shift in population from those states traditionally associated with manufacturing (New South Wales, Victoria and South Australia) towards the resource-rich states of Queensland and Western Australia, as shown in Figure 14.5. Employment is only one of the reasons people move within Australia. Lifestyle and demographic factors are also major considerations.

DID YOU KNOW?

Mines occupy 0.02 per cent of the Australian landmass.

14.5 Population distribution by state and territory, 1859–2009

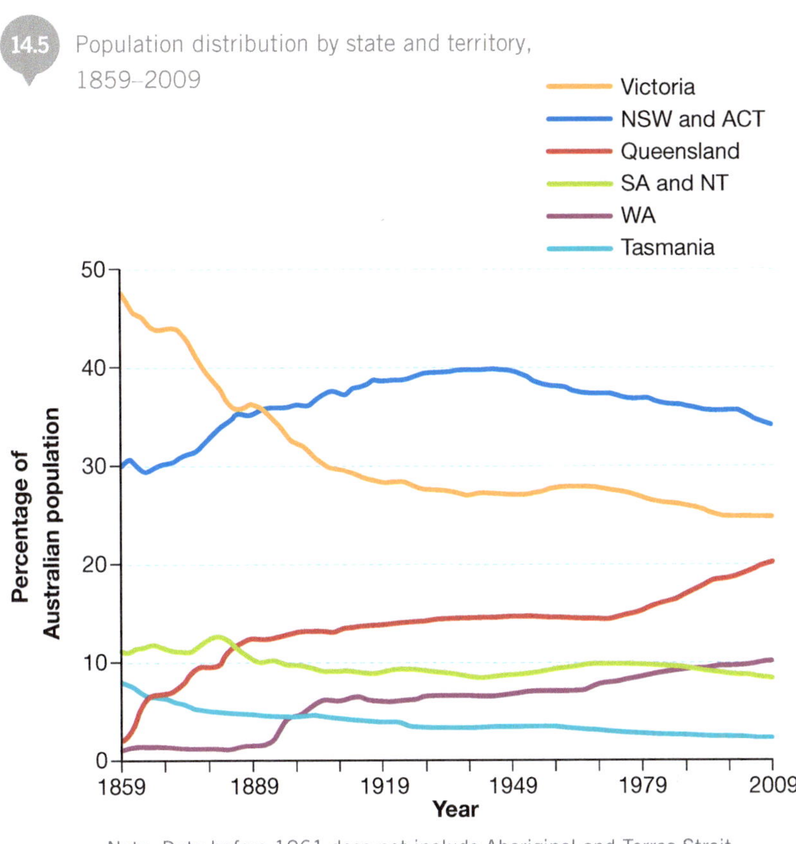

Note: Data before 1961 does not include Aboriginal and Torres Strait Islander people

Source: Australian Bureau of Statistics, Reserve Bank of Australia

ACTIVITIES

Knowledge and understanding

1 Draw a timeline outlining Australia's economic transformation, starting in the 1800s.

2 Explain what is meant by the term 'service-based economy'.

3 Outline how the composition of Australia's exports has changed over time.

4 Outline the changes taking place in the global economy.

5 Explain what is meant by the 'weightless' economy. Give examples.

6 Outline the impact of Australia's economic transformation on the distribution of the country's population.

Geographical skills

7 Study Figure 14.2 and answer the following questions.

a When did the USA's share of global manufacturing peak? What has been the trend since? Suggest why its share of manufacturing has declined.

b When did the UK emerge as a major manufacturing power? What has been the trend since then?

c Outline the trends in German manufacturing.

d Using information from the graph, evaluate the following statement: *The recent growth of manufacturing in China represents a re-emergence of China as a major economic power.*

8 Study Figure 14.3. Using data from the graph, outline the trends in employment by industry in Australia.

9 Study Figure 14.4. Outline the areas of employment growth and employment decline.

10 Study Figure 14.5. Which states and territories have a declining share of Australia's population? Which have an increasing share?

14.2 Australia: internal migrations

Internal migration occurs when people move to other places within their own country. In Australia, Queensland and Western Australia receive the highest number of internal migrants. In 2010, more than 86 400 people moved to Queensland from other states and territories.

Movement of people

Four of the five fastest-growing regions in Australia are now found in Queensland. These are Brisbane, the Gold Coast, Moreton Bay and the Sunshine Coast. All are located in the south-east corner of Queensland. Figure 14.6 shows average net interstate migration, 2001–02 to 2010–11.

Economic migration

The reasons for internal migration are varied. Employment opportunities can attract people to certain areas. For example, there has been significant migration to North Queensland and parts of Western Australia due to those states' booming mining industries, where people can earn higher incomes.

14.6 Net interstate migration, annual average movement 2001–02 to 2010–11

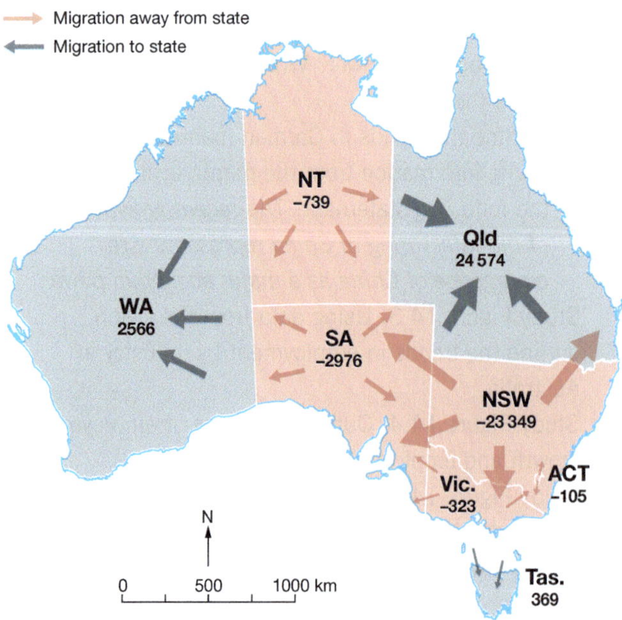

Source: Australian Bureau of Statistics

Economic migration can bring about very significant change within communities. The increase in the population can be so sudden that local communities can be overwhelmed. Important infrastructure, such as roads, utilities (water and electricity supply), schools and hospitals, and a range of community services, are quickly overstretched. There can be many economic benefits to local communities of internal migration, but in some cases, where the population grows very quickly, the impact on local economies can be negative. For example, housing shortages are very common and this has meant that house prices and rents have skyrocketed.

Lifestyle and retirement migration

Another group of internal migrants are the retirees and those who choose to move out of the cities for lifestyle reasons. Moving to be closer to family is another key factor in internal migrations.

There is also a significant movement of people within states. For decades, populations have been declining in many of Australia's smaller country towns. The people who once populated these towns have moved to the large cities on the coasts or to the larger rural centres.

Tree and sea changes

Over the last 15 to 20 years there has been an increase in the number of people moving away from cities to a rural or coastal area.

The terms '**counter-urbanisation**' and '**exurbanisation**' are used to describe the process of people moving away from cities but maintaining important links with the city, such as work. For example, in South Australia many people have moved from Adelaide to the Adelaide Hills. Here they have access to a rural lifestyle but, because the Adelaide Hills are relatively close to Adelaide, they are able to continue to work in the city.

Tree changers are those people who move from the city to inland areas, such as from Melbourne to Ballarat and Bendigo in Victoria. Such regions offer larger blocks of land suitable for small farms and large gardens, smaller communities and a cleaner and greener environment.

Sea changers make their move from the city to the coast. The most popular coastal areas are those relatively close to the big cities.

SPOTLIGHT

Mandurah, Western Australia

Located just 73 kilometres from Perth, Mandurah has become a popular sea change destination for Western Australians. It has a population of more than 70 000 people, and is expected to double in the next decade (see Figure 14.7). This makes Mandurah Western Australia's fastest-growing region. Mandurah is linked to Perth by a freeway and a railway, which opened in 2007. The opening of the railway made Mandurah a more popular destination. A significant proportion of the population, around 20–25 per cent, are expected to be aged over 65. This presents some challenges for the city, especially in terms of providing the health and other services needed by an older population.

As Mandurah has grown, it has ceased to be the small, quiet fishing village it once was. Large shopping malls and housing estates now dominate much of the city. Some long-term residents fear that their town has lost some of its unique character. However, others welcome the increased population and the services and opportunities that this brings.

There are also concerns about the local environment. For example, there are currently 22 kilometres of canals in the city. These canals are carved out of the natural intertidal wetland and mud flats that form along the Mandurah Estuary, shown in Figure 14.8.

14.7 The projected population increase in Mandurah

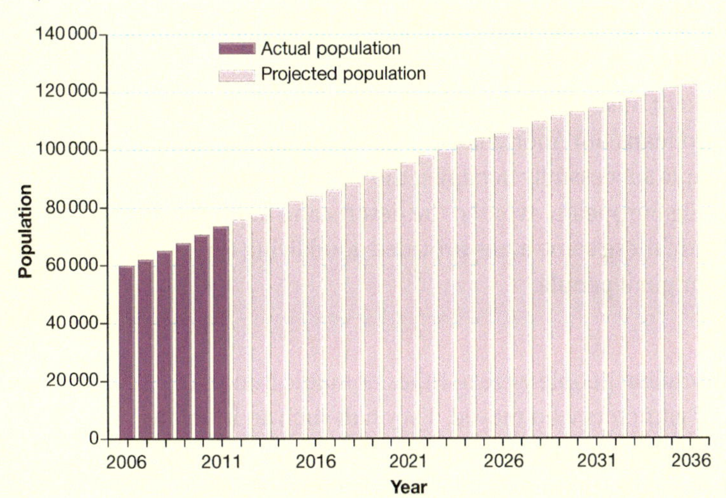

Source: Forecast.id® website

14.8 Canal housing development in Mandurah

ACTIVITIES

Knowledge and understanding

1 Write your own definition of the term 'internal migration'.

2 Outline some of the main reasons for internal migration.

3 Describe how internal migration can become a problem for local communities.

Applying and analysing

4 If you could migrate to anywhere in Australia, where would you go? Why?

Geographical skills

5 Study Figure 14.6 and do the following tasks.

   a List the states that have experienced population gains and losses from internal migration.

   b Write a paragraph describing the impact that internal migration has had on Mandurah.

14.3 Case study: The mobility of Indigenous Australians

Indigenous Australians have been described as a highly mobile people. In particular, Indigenous Australians living in remote and rural Australia move frequently between places. Movement was, and still is, the means by which Indigenous people maintain their relationships with others and the places that are special to them.

The traditional Indigenous way of life

Most of Australia appeared uninhabitable to the first Europeans but Aboriginal and Torres Strait Islander people successfully occupied the entire continent. While Europeans avoided the deserts and hot tropical north, these same areas provided good living for Aboriginal people.

Movement across the land

Aboriginal people were at times nomadic hunters and gatherers who moved. They had such an intimate knowledge of their environment that, in most areas, they could obtain sufficient food and raw materials for shelter, clothing and ceremonies in just a few hours. They were keen observers of the condition of their country and would move on when resources showed signs of declining, enabling the resources to replenish.

The extent to which Indigenous people moved across the land was determined by the **productivity** of the land. This was reflected in population densities over the continent. In the coastal areas, where there was plenty of food and water, they might spend months in one location and rarely move very far. In the arid areas, they had to walk long distances between waterholes and food sources, which were spread over a much wider range.

Indigenous people also moved across the land to exploit seasonal foods and resources. They followed the breeding and movement of animals and fish, as well the cycles of flowering and fruiting plants. Their knowledge of such regular patterns was very deep and built up over thousands of years of careful observation.

By singing songs in an appropriate sequence, Aboriginal People could navigate their way over hundreds of kilometres.

DID YOU KNOW?

Songlines, or Dreaming tracks, are paths that are recorded in traditional songs, stories and dance. By singing the songs, Aboriginal people were able recognise landmarks and find their way over vast distances.

Ceremonies, social exchange and trade

There were also annual migrations over great distances for spiritual purposes. The Snowy Mountains, for example, were of special significance. Aboriginal people were summer visitors to the highest peaks, coming from many directions and walking hundreds of kilometres to gather peacefully for trade, ceremonies and marriages. While in the mountains, they feasted on the bogong moths that were readily available and considered a great delicacy.

The tracks used by Aboriginal people were well-known paths through their country. Songlines joined places of ceremonial significance (see Figure 14.9).

14.10 Most common destinations for Indigenous people in Australia

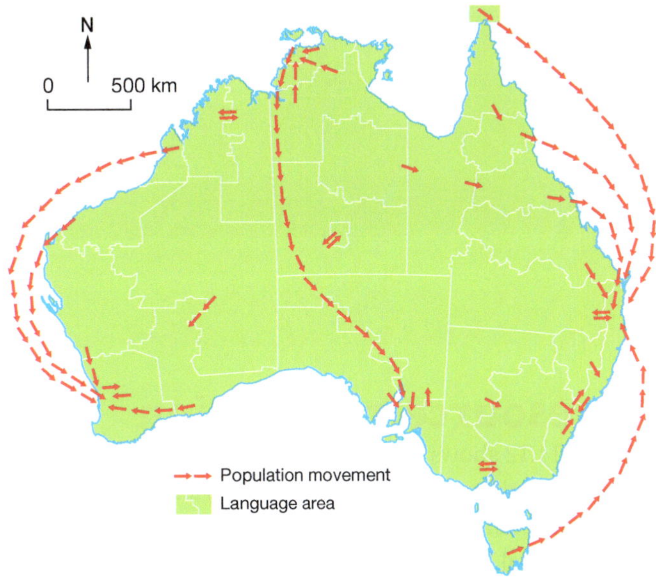

Mobility of Indigenous people today

Aboriginal people are highly mobile. They travel for the usual reasons of education, employment, health services and sport. There has, however, been an overall pattern of migration from remote areas to cities. Despite this mobility there remains a strong attachment to place and kin.

The significance of place

Indigenous people are very attached to the place in which they or their ancestors were born. Many are very conscious of their own local territory, possessing a detailed knowledge of its geography and sacred histories. They retain family stories and memories of their country, which make it significant to them. If they move away, they feel that they must periodically return to their birthplace because of their attachment to place.

Maintaining kin relationships

The kinship system is central to Aboriginal social organisation. Family relationships by blood ties and marriage define the kinship system. 'Family visits' are not just about enjoying social occasions. They fulfil obligations that are essential in the Aboriginal society. This pattern of movement is more common in rural and remote areas, particularly in northern Australia.

> 'Aboriginal people will, at no notice, join a vehicle travelling hundreds of kilometres away, taking with them no money and few provisions, and will have no idea of when or how they will return ...' Such travel is made possible by knowing that the vehicle is in the charge of a relative and secondly that relatives can be found at the end-destination and intermediate stops along the way who would accept unannounced visits and provide support for visiting kin.

14.11 A reflection on Aboriginal people in Central Australia, from A Hamilton, 'Coming and Going: Aboriginal Mobility in North-west South Australia, 1970-71' in *Records of the South Australia Museum*, Vol 20, pp 47–57

ACTIVITIES

Knowledge and understanding

1 Explain how Aboriginal people were able to live off the land over much of the Australian continent.
2 Describe how songlines assisted Aboriginal people to find their way.
3 Explain why Indigenous people are so mobile today.

Applying and analysing

4 Study Figure 14.10. Describe the pattern of movements between states.

Investigating

5 Investigate the Aboriginal trade in ochre that existed prior to European settlement.
  a Using a map of Australia, label and annotate the sources of ochre and the trade routes.
  b Explain how the ochre was gathered.
  c What did the Aboriginal people use to trade for the ochre?
  d Outline how the ochre was used.
  e Were there any other main trade items? Explain.

14.4 Australia's mining boom

The Australian mining industry experienced remarkable growth during the 2000s. The richest resources boom in Australia's history boosted the nation's income. It was driven by the rapid growth of Asian economies, especially China, which need minerals for steel production and energy generation. In response, investments in mining projects in Australia rose to record levels.

A resource-rich nation

Figure 14.12 shows that Australia's mineral wealth has grown greatly since 1996. This wealth is a result of Australia being one of the world's leading exporters of black coal, iron ore, lead and zinc, which are the products of the continent's long geological history. Water moving through ancient igneous rocks dissolved minerals and re-deposited them in concentrated ore bodies. Large areas of swamps were buried millions of years ago and, over time, compressed to form extensive coal deposits.

Western Australia and Queensland are Australia's resource-rich states. Of the 340 mines operating in Australia, almost half are in Western Australia. The state produces almost all of Australia's iron ore exports, mainly from the vast deposits in the Pilbara region. Both New South Wales and Queensland have large coal basins.

The resources boom

Australia's prosperity over the last decade is a result of trade with China. During this period China experienced a sustained period of rapid industrialisation and urbanisation. China turned to Australia for high-quality mineral and energy resources to fuel this growth. In 2008, China became Australia's most important trading partner.

In 1996, China emerged as the largest steel producer in the world, accounting for nearly half the global output. China used its steel to build its massive cities and rail networks. The Reserve Bank has estimated that constructing a typical Chinese apartment requires 6 tonnes of steel, and 10 kilometres of a subway requires 75 000 tonnes. For each tonne of this steel, 1.7 tonnes of iron ore and over half a tonne of coking coal are required. China sources 40 per cent of its iron ore imports from Australia, as well as coking coal for its steel mills.

14.12 Australia's major commodity export earnings

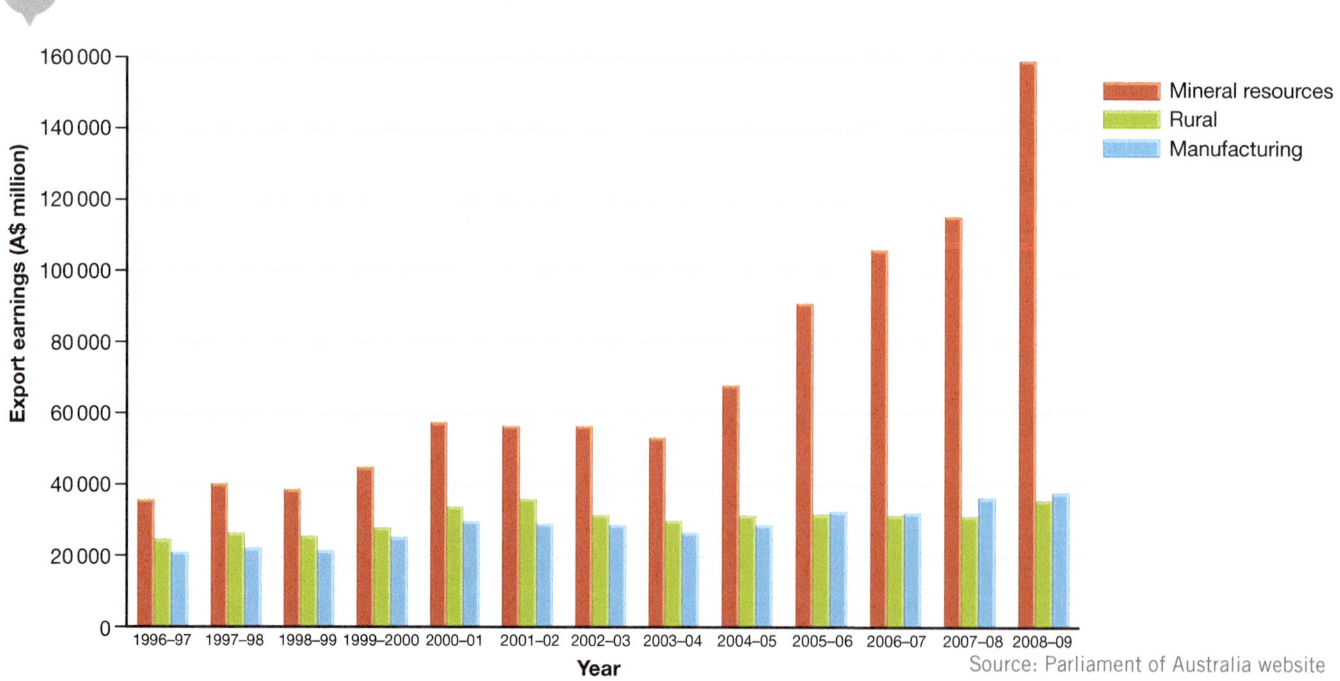

Source: Parliament of Australia website

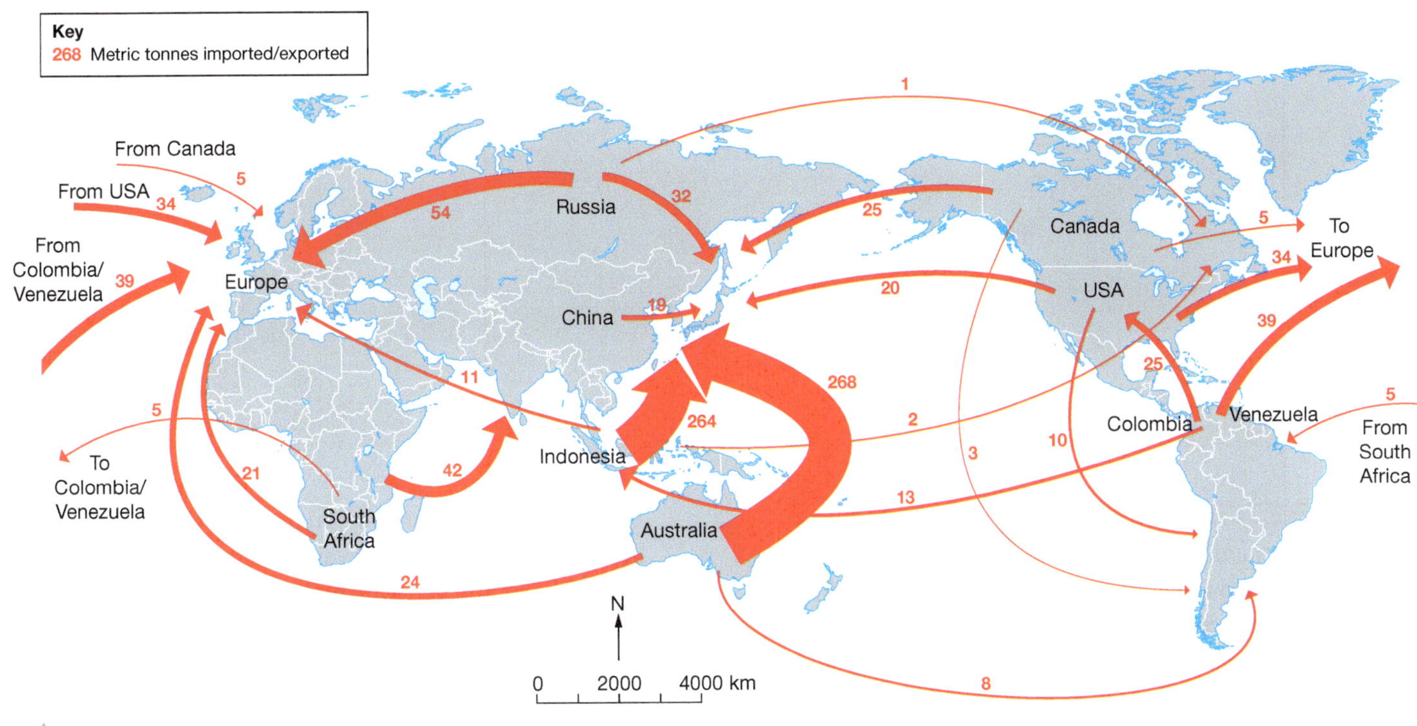

14.13 The global movements of coal

China's urban boom

By 2025, China's urban population will have grown by 350 million people. Two hundred and twenty-one cities will have more than a million people (up from 35 in 2012). To meet the needs of its rapidly growing urban population, China will need to build five million buildings—50 000 of which will be skyscrapers—and 170 mass-transit systems. This construction will drive the demand for the resources needed to make steel, especially iron ore and coal.

Steel production

Without coal, you can't produce steel. There is no alternative for metallurgical coal, which is high in energy and produces the very high temperatures essential to the steel-making process. Coal is baked in furnaces to produce the coke that is needed to smelt iron ore for use in the production of steel. The movement of coal around the world is shown in Figure 14.13.

Impact of the boom

With its extensive reserves and direct shipping links to China, Australia has been well placed to benefit from the increased demand for resources. The effects were evident in the record investments in mining projects.

The prices of key resources used in steel and energy production have increased sharply over the last decade. The global price for Australia's resource exports rose by more than 300 per cent from 2003 to 2011. However, as other miners around the world raised their production, prices declined in 2012.

While Chinese demand for Australian resources may slow down, it is projected to remain at a very high level over the next decade. After China, growth in India will generate further demand for Australian resources.

ACTIVITIES

Knowledge and understanding

1 Describe how Australia has benefited from the mining boom.
2 Explain why Australia is so rich in minerals.
3 Outline the changes that have occurred in China over the last decade and explain why these have driven demand for Australian minerals.

Geographical skills

4 Study Figure 14.12 and do the following tasks:
   a What was the value of mineral exports in 2008–09?
   b Calculate the increase in the value of these exports from 2003–04 to 2008–09.
5 Study 14.13. Describe the pattern of world trade in coal. Suggest why this pattern has developed.

14.5 Impacts on regional Australia

Since minerals were first discovered in Australia, large areas of the country have been opened up for resource development. Infrastructure such as roads, railways and ports have been built and towns established to serve mining operations. The mining boom has continued to impact on regional Australia.

Regional development

The extraction, processing and transportation of minerals to market requires a substantial investment in infrastructure. Historically, Australian railways and ports were largely developed, funded and owned by governments. Towns were established to support mining operations and many large regional centres are a direct result of mining, as outlined in Table 14.14.

14.14 Main mining regions of Australia

State	Region	Mineral
Western Australia	Pilbara	Iron ore
	Kalgoorlie	Gold
Northern Territory	Kakadu	Uranium
	Tanami	Gold
South Australia	Iron Triangle (Port Augusta, Port Pirie, Whyalla)	Copper, silver, lead, iron ore
	Roxby Downs	Copper, uranium, gold
Queensland	Mt Isa	Copper, silver, lead, zinc
	Weipa	Bauxite
	Bowen Basin, Galilee Basin, Surat Basin	Coal
New South Wales	Broken Hill	Silver, lead, zinc
	Hunter Valley, Illawarra, Gunnedah, Lithgow, Ulan	Coal
Victoria	Latrobe Valley	Brown coal
Tasmania	West Coast	Gold, silver, lead, zinc, copper

The prosperity of many regional areas of Australia has been based on resource development. Mines provide employment and generate business in the communities supporting them. As regional centres grow, schools, hospitals and other community services are provided to meet the needs of residents. Port Hedland, Gladstone, Roxby Downs, shown in Figure 14.15, and Muswellbrook are all mining centres.

Recent developments

In the past two decades, rapid expansion of mining has created substantial infrastructure deficiencies and an urgent need for new rail links from the mines to ports. Many of the major resource companies have the money to establish and operate their own infrastructure. Rio Tinto and BHP Billiton have built rail lines to move iron ore in the Pilbara region. Several projects for rail links and ports are planned for Queensland.

Due to the remoteness and large scale of many modern mines, there has been a shift towards mining operations based on long-distance commuting, or 'fly-in-fly-out' (FIFO). Workers have been able commute from coastal communities where they prefer to live.

Regional economic growth

The mining industry offers generous wages and, without such incomes, many regional communities would struggle. Mining towns, particularly those in the Pilbara, have become the highest earning regions in Australia, with residents earning over $2000 a week.

The future viability of mining communities depends on how long the mining boom continues.

Regional population growth

In the last decade, both the Gold Coast and the Sunshine Coast have experienced big population increases. This growth has, in part, been driven by their popularity

14.15 Roxby Downs is a purpose-built mining town. It was first occupied in 1987 and later developed to service BHP Billiton's Olympic Dam mine in South Australia.

with FIFO workers, who live near the coast and 'fly in' to work in the mines. In Western Australia, centres such as Karratha have also grown quickly. The population of the East Pilbara region has had the largest percentage growth in Australia, jumping by 226 per cent to 11 950 between 2001 and 2011.

It is projected that people will continue to move to Western Australia and Queensland. With the promise of employment, the growth in Queensland is expected to be in the south-east corner (where FIFO workers prefer to be based) and regional centres (where investment is continuing in resource-sector development). Gladstone, Isaac and the Central Highlands are expected to be among the fastest-growing regional areas.

Impact of FIFO workers

Despite the economic gains in regional Australia, prosperity has come at a cost. Some towns have come under pressure from the large increase in the number of FIFO workers, with implications for local community sustainability. Such towns:

- are unable to meet the infrastructure and service demands—roads become extremely busy and accidents more common
- face increased housing costs and decreased housing availability
- lose a sense of community, as FIFO workers tend to be less involved in clubs and volunteer groups
- experience an increase in social problems.

ACTIVITIES

Knowledge and understanding

1 List the issues that are created by rapid population growth.
2 Explain how mining has led to the development of parts of regional Australia.
3 Describe the regions most affected by the mining boom.
4 Explain why coastal areas in Queensland have experienced population growth in the last decade.

Applying and analysing

5 Create a mind map of the costs and benefits to regional communities from FIFO workers.
6 Study Figure 14.15. Describe the natural and built environment you observe in the image.

14.6 Case study: Bowen Basin

In 2010–11, Queensland produced 179.8 million tonnes (Mt) of coal. Approximately 162.5 Mt (valued at A$29.05 billion) was exported, mainly to Japan, South Korea, India and China. Mining in Queensland is concentrated in the Bowen Basin of Central Queensland. The basin contains the largest reserve of high-quality coal in Australia.

Australia's major coal basins

The largest deposits of coal (see Figure 14.16) are found in **basins**—bowl-shaped depressions in the earth's crust that formed when layers of rocks buckled under pressure. The coal is found in **seams** trapped between layers of rocks (see Figure 14.17). Coal seams range from a thickness of just a few centimetres up to 30 metres. Geologists estimate that the layer of plants needed to produce a 1-metre seam of coal would have been approximately 20 metres thick.

14.16 Important coal-bearing basins of Australia

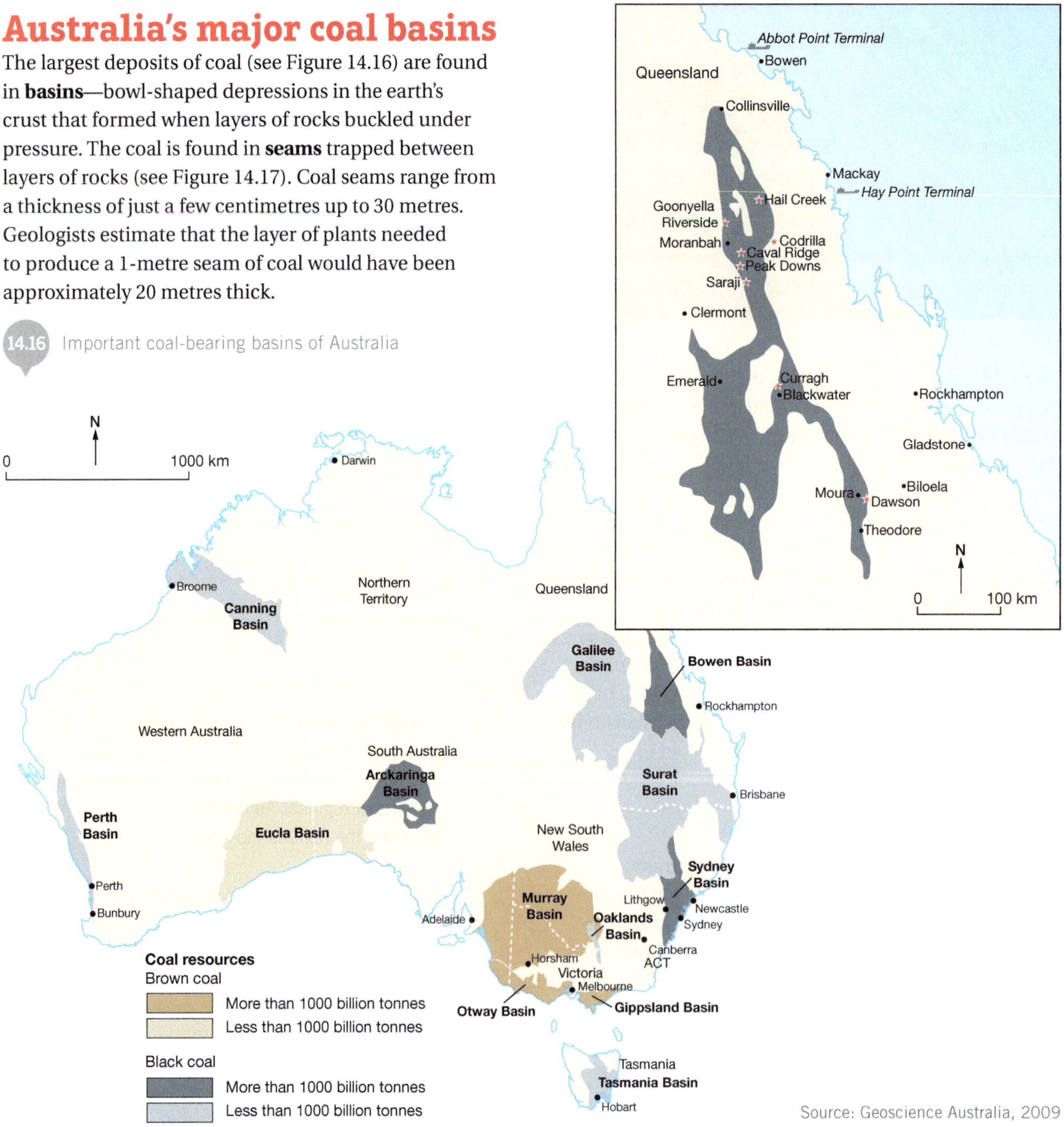

Abbot Point Terminal
Bowen
Queensland
Collinsville
Mackay
Hay Point Terminal
Goonyella Riverside
Hail Creek
Moranbah
Codrilla
Caval Ridge
Peak Downs
Saraji
Clermont
Emerald
Curragh
Blackwater
Rockhampton
Gladstone
Moura
Biloela
Dawson
Theodore
N
0 100 km

N
0 1000 km
Darwin
Broome
Canning Basin
Northern Territory
Queensland
Galilee Basin
Bowen Basin
Rockhampton
Western Australia
South Australia
Arckaringa Basin
Surat Basin
Brisbane
Perth Basin
Eucla Basin
New South Wales
Perth
Bunbury
Sydney Basin
Murray Basin
Lithgow
Newcastle
Sydney
Adelaide
Oaklands Basin
Canberra
ACT
Horsham
Victoria
Melbourne
Otway Basin
Gippsland Basin
Tasmania
Tasmania Basin
Hobart

Coal resources
Brown coal
More than 1000 billion tonnes
Less than 1000 billion tonnes
Black coal
More than 1000 billion tonnes
Less than 1000 billion tonnes

Source: Geoscience Australia, 2009

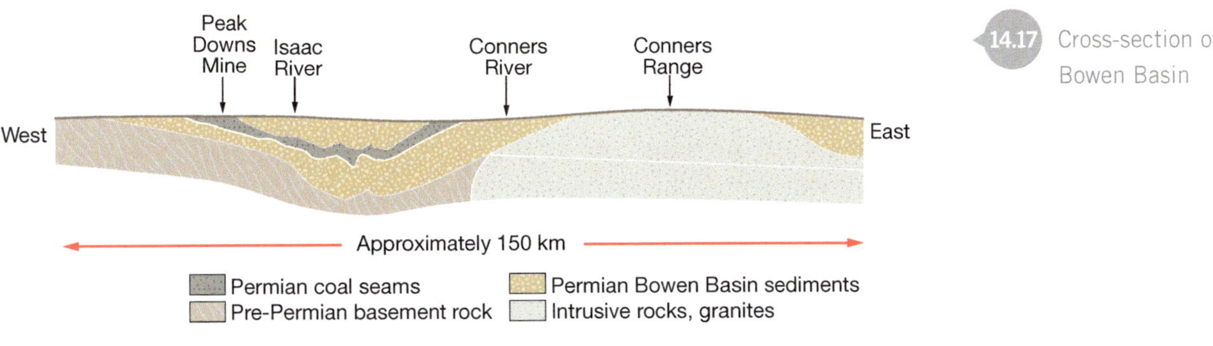

Source: Barry Cook, *Coal in a Sustainable Future*, Osmond Earth Sciences, 2003

14.17 Cross-section of the Bowen Basin

History

Coal was first discovered in the Bowen Basin in 1866 and, although extensive deposits were known to exist, coalmining did not begin until the 1950s because New South Wales had been the focus for Australian coalmining. When a large contract to supply coal to Japan was secured in 1965, many new mines opened up. The population increased rapidly and many new towns were established, including Moranbah (1970) and Glenden (1983).

Coalmining today

There are now forty-seven coalmines operating in the Basin. Together they produced 180 million tonnes of high-grade black coal in 2009–10 and have generated a total of 29 550 jobs. The region is now recognised as one of the most significant black coal producers and exporters in the world.

Queensland's infrastructure has been developed to serve the coal industry. Electrified rail links allow efficient transport of the coal to the ports at Gladstone and Mackay. A 220-kilometre water pipeline has also been built to provide a reliable water supply for the Bowen Basin. This has enabled the mines to produce even more coal.

The future

The mining industry is expected to expand in northern Bowen Basin over the next 20 years, mainly to meet the demand for coal from India and China. Among the new projects are twenty-three coalmines that will tap into five billion tonnes of yet-to-be-developed resources. It is expected that an additional 10 000 jobs will be created in the mines in the near future. There are fifteen communities in the Basin with an estimated population of over 50 000 people. Despite 50 per cent of mine workers in the Bowen Basin being fly-in-fly-out workers, the number of people living in the local communities is expected to increase as coalmining expands in the Basin.

14.18 Giant mine in the Bowen Basin

ACTIVITIES

Knowledge and understanding

1. Define the term 'basin'.
2. Describe how coalmining developed in the Bowen Basin.
3. Assess the importance of coalmining to Queensland.

Geographical skills

4. Study Figure 14.16.
   a. Describe the locations of Australia's coal basins.
   b. Describe the relative location of the Bowen Basin.
   c. Name the principal coal export ports serving the Bowen basin.
5. Study Figure 14.17.
   a. Describe the rock types.
   b. What feature is associated with the coal seams?
   c. What natural feature is located east of the Conners River?

14.7 Case study: North West Shelf

The North West Shelf is a region of large oil and gas reserves off the north-western coast of Australia. These reserves were discovered in the 1970s and during the 1980s construction began on offshore platforms and a natural gas plant at Karratha.

Natural gas production

Australian gas fields, shown in Figure 14.19, produce a large amount of natural gas, either as compressed natural gas (CNG) or liquefied natural gas (LNG). There is an estimated 4000 billion cubic metres of natural gas in proven gas fields throughout Australia. This is enough to power a city of one million people for 8000 years. In Australia, 8.6 per cent of electricity presently comes from natural gas. The transport industry also makes use of natural gas, particularly trucks and commercial vehicles. The largest source of natural gas can be found in the Carnarvon and Browse Basins in the North West Shelf (see Figure 14.20). Australia's exports of liquid petroleum gas (LPG) were valued at $9.5 billion in 2010.

The North West Shelf Joint Venture

The North West Shelf Joint Venture (NWSJV)—Australia's largest resource development project—extracts natural gas at offshore platforms, processes it onshore and

14.19 Australian gas production

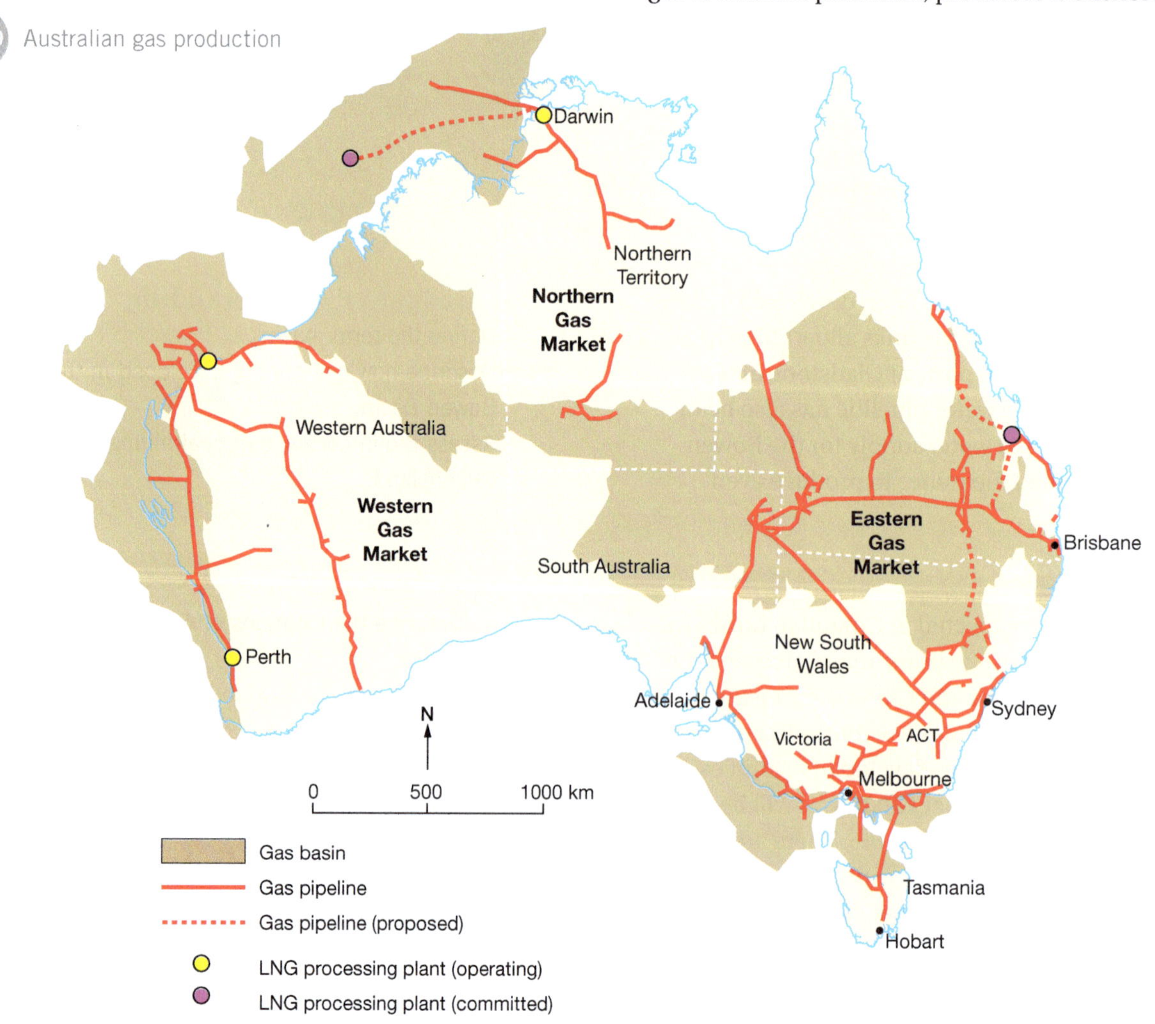

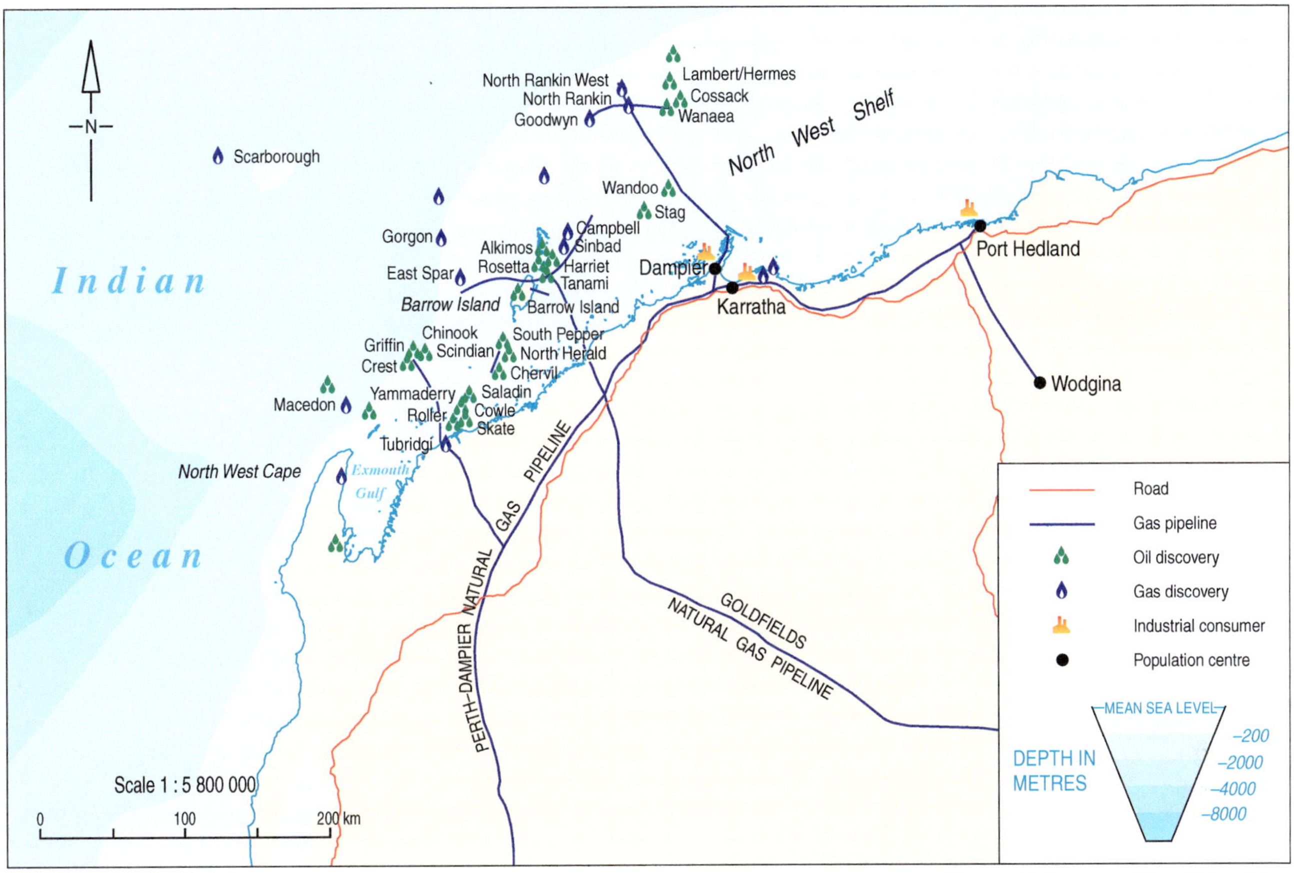

Source: Heinemann Atlas, 5th edition

The location of Australia's North West Shelf natural gas basins and infrastructure

exports natural gas products. The NWSJV delivers LNG to countries all around the world, representing 1 per cent of Australia's **gross domestic product** and providing over $5 billion to state and Commonwealth funds each year. In addition, its operation provides $800 million worth of revenue to associated Australian businesses. The NWSJV is Australia's largest resource development project.

Australian gas production 2000, 2008 and 2009

Australian gas	2000 production (billion cubic feet)	2008 production (billion cubic feet)	2009 production (billion cubic feet)
Domestic gas sales	725.5	1060.4	1000.5
Export gas (LNG)	345.1	727.1	835.3
Total gas	1070.6	1787.5	1835.8

Local communities

Karratha and Port Hedland are two towns that have experienced rapid population growth. They also have a large number of fly-in-fly-out workers, who generally fly in from Perth.

ACTIVITIES

Knowledge and understanding

1 Describe the North West Shelf Joint Venture.
2 List the infrastructure needed to exploit the North West Shelf's natural gas reserves.

Applying and analysing

3 Apply your knowledge by suggesting some possible new natural gas markets for the North West Shelf Joint Venture.

Geographical skills

4 Study Figure 14.19. Describe the location of Australia's gas reserves.
5 Study Table 14.21. Draw three pie graphs to illustrate the information presented in the table.

CHAPTER 15 CHINA IN TRANSITION

The pace of economic change in China has been extremely rapid since the start of economic reforms just over 25 years ago. Over the past two decades China's annual economic growth has averaged 9.5 per cent and seems likely to continue at that pace for some time. National income has been doubling every eight years. Such an increase in output represents one of the most sustained and rapid economic transformations the world has ever seen. Hundreds of millions of people have been lifted out of poverty, cities transformed, vast industrial complexes constructed and a sophisticated transport infrastructure built.

China's economic transformation has initiated one of the most significant redistributions of the human population ever seen. Millions of people have abandoned life in the rural villages and towns in a quest to share in the benefits of the country's economic transformation. Vast industrial complexes, both Chinese and foreign owned, provide employment, which by Chinese standards is reasonably well paid.

In this chapter we examine China's economic transformation, the internal migrations it initiated and the impacts of this redistribution on cities and rural villages and towns.

KEY IDEAS

- To describe the causes and consequences of China's economic transformation
- To explain the impact of China's rapid economic growth on China's population distribution
- To discuss the impact of this redistribution on cities, towns and rural villages.

15.0 Container terminals at the port in Shanghai, China, 2011

GLOSSARY

hukou system	a Chinese household registration system that entitles the holder to a range of services such as healthcare and education, used to regulate the movement of people in China
gross domestic product (GDP)	the total monetary value of goods and services produced in a country during one year
middle class	a class in society between upper and lower
One Child policy	policy introduced by the Chinese government to reduce its population growth—people were permitted to have only one child
per capita	per person; often used with the term GDP
rural–urban migration	the process by which an increasing proportion of a population moves from rural areas to live in towns and cities; also known as urbanisation
special economic zone	a geographical region of China in which economic and other laws are more accommodating of capitalism (free market) than the national laws
urban sprawl	the outward spread of a city and its suburbs as they grow

15.1 China: An emerging economic giant

The global economy is changing as China claims an increasing share of the world's manufacturing. Since the 1970s, China has been transformed from a relatively poor and very inaccessible nation into a giant of the international economy. The extent and speed of this change are staggering and look set to continue.

Biggest economy

China will become the biggest economy in the world within the next decade. This has huge implications, as it will be the first time in centuries that the title of world's most powerful economic power will be claimed by a non-Western nation.

A tumultuous history

China was the world's leading civilisation for centuries, due to its vast wealth, sophisticated culture, art and inventions. In the nineteenth and early twentieth centuries, China became caught up in civil unrest. The country was torn apart by war and its people lived in terrible poverty. In 1949, the Communist Party under Mao Zedong took control and the People's Republic of China was formed. Under communist rule, all business enterprises and private land were seized and became the property of the state. There were heavy restrictions on people's everyday lives. Men and women were given jobs on farms or in factories and issued with food rations. Life was very hard and millions of people starved.

For decades China had a 'closed door policy' in its dealings with the rest of the world. The communist government discouraged any economic, political or cultural links with other countries.

Reforms and open policy

Reforms in China started in 1978 in agriculture. Eventually people were allowed to own their own land and businesses. State-owned industries were assisted and supported by the government and China's economy started to grow.

DID YOU KNOW?

There were 7905 multimillionaires in China at the end of 2011, an increase of 41 per cent since 2007.

Opening the economy

The most visible economic reform was the creation of **special economic zones**. Foreign countries were permitted to establish factories in these zones and were exempted from the taxes and controls imposed elsewhere in China.

Billions of dollars of foreign investment flowed into China, as it became the preferred destination for manufacturing. Exports drove the expansion of the Chinese economy and the 'Made in China' label is now commonplace on consumer goods right across the world. As the wealth of the country increased, China also invested heavily overseas to secure the energy and commodity resources needed by its industries.

China joined the World Trade Organization in 2001 after removing trade barriers.

Rapid economic growth

China has undergone a major economic transformation in recent decades. China became involved in the free market system, and exports generated considerable wealth. Both foreign investment and international trade soared, generating an extraordinary growth in **gross domestic product** (GDP). China has had an average economic growth rate of over 9.5 per cent for over two decades. This growth has raised the living standards of millions of Chinese. In 2010, China displaced Japan as the world's second largest economy, as shown in Table 15.1.

China will continue to grow. The size of its population and economy will drive regional and global growth into the future. The changes in the structure of the economy are reflected by changes in the patterns of employment. As China has developed, employment in primary industry has declined, while employment in tertiary industries has increased, as shown in Figure 15.2.

15.1 GDP growth rates for selected nations, 2005 to 2010. China and India maintained high growth rates while the rest of the world suffered in the global financial crisis (2007).

Country or region	2010 GDP (100 million US$)	GDP growth rate (%)				
		2005	2007	2008	2009	2010
World	629 093	4.57	5.40	2.87	-0.52	5.01
Australia	12 355	3.13	4.59	2.59	1.33	2.75
China	58 783	11.30	14.20	9.60	9.20	10.30
India	15 380	9.17	9.88	6.18	6.76	10.37
Japan	54 589	1.93	2.36	–1.16	–6.238	3.94
UK	22 475	2.17	2.69	–0.06	–4.87	1.25
US	146 578	3.05	1.95	–0.4	–2.63	2.83

Source : International Monetary Fund, China Statistical Yearbook 2011

15.2 China's changing employment structure

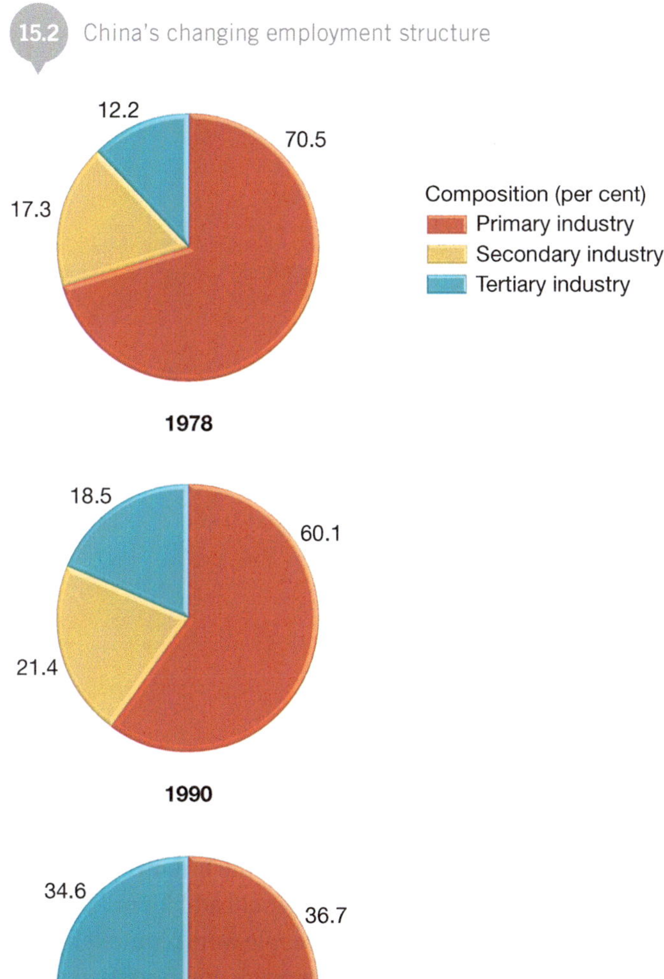

Source: China Statistical Yearbook 2011

ACTIVITIES

Knowledge and understanding

1 Describe the nature of communist government control in China before 1978.

2 Outline the changes in government policy that have occurred since 1978.

3 Describe the transformation of the Chinese economy.

4 Explain why China's economy will continue to grow.

Geographical skills

5 Study Table 15.1 and do the following tasks.
  a Construct a multiple line graph of the GDP growth rates for the countries shown.
  b In what year did China experience its highest growth rate?
  c Which countries experienced negative growth rates? Give reasons for your answer.
  d Explain why Australia did not experience a negative growth rate.

6 Study Figure 15.2 and do the following tasks.
  a Write a paragraph describing the changing structure of Chinese employment.
  b Suggest reasons why so many Chinese people are still employed in primary industries.

15.2 China's internal migrations

China's economic transformation has triggered one of the largest population movements in human history. The country's economic success has led to a rapidly rising standard of living. The main driver of urban growth has been, and will continue to be, rural–urban migration.

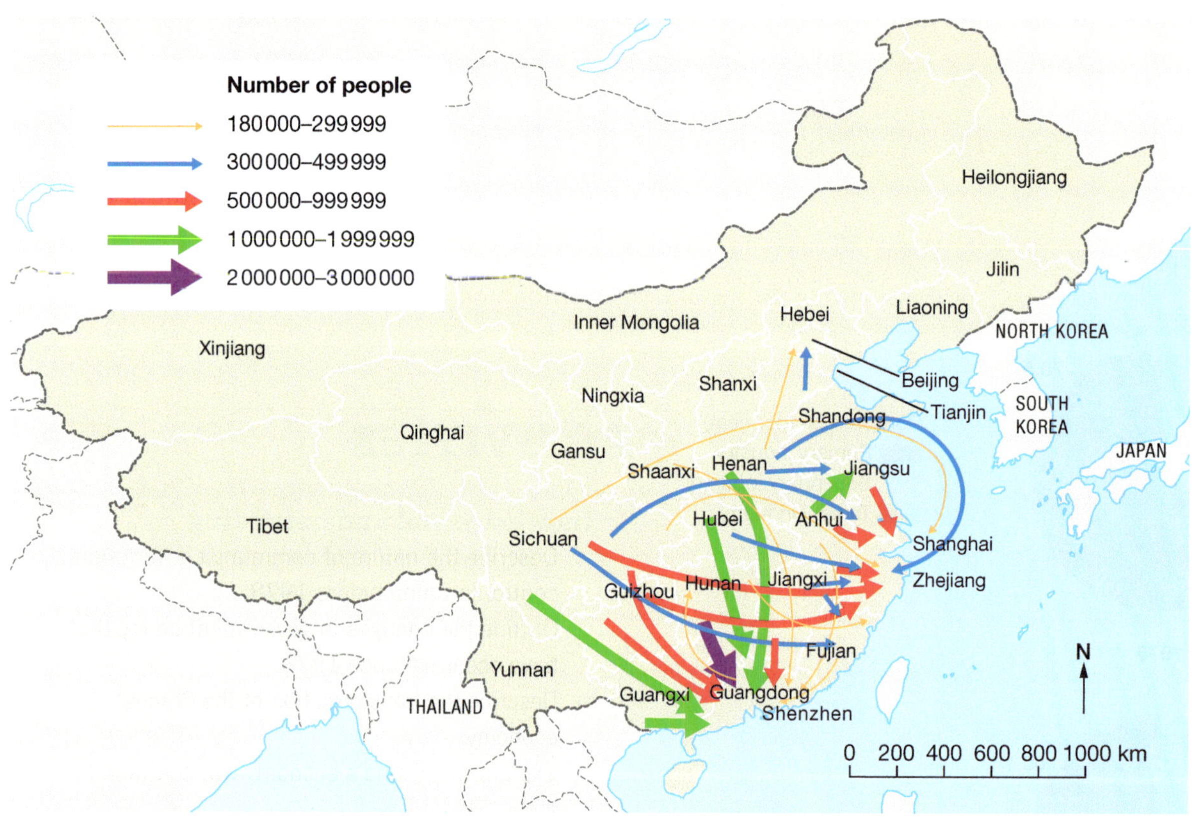

15.3 Internal migration in China

The facts

- Between 2010 and 2025 the population of China's cities will grow by 350 million people, which is equal to today's population of the United States of America.
- One billion people will live in Chinese cities by 2030.
- The number of cities with more than one million people will reach 221 by 2030. Europe has only thirty-five today.
- Five billion square metres of roads will need to be paved.
- Forty billion square metres of floor space will need to be built in five million buildings. Fifty thousand of these buildings will be skyscrapers.
- One hundred and seventy-five new mass-transit systems (public transport networks) will need to be built.

Population mobility

Until the end of the 1970s, 80 per cent of the population were rural dwellers and it was almost impossible for them to leave their villages. Under the **hukou** system, a person was registered in a household and identified as a resident of a particular area. People were required by law to stay in their district and they were not permitted to move elsewhere looking for work. People could qualify for coupons for food or receive medical treatment only in their own designated hukou and so the government services that people needed to survive were tied to where they were registered. This kept the people in the countryside for decades.

15.4 Migrants on the move, seeking jobs in the cities

A shift in government policy

As the post-1978 economic reforms took hold, the rules of hukou system were progressively relaxed. Most significantly, coupons were no longer required to buy food, and a worker without an urban hukou could work in the city, as the government no longer assigned jobs to people. However, in doing so, workers were still ineligible for services such as healthcare, free education for their children and subsidised housing, all of which people who were registered in urban areas received. Despite these remaining limitations, the relaxing of the hukou system was enough for urbanisation to get underway. China's policy shift to accept and promote migration to urban areas was designed to create the pool of labour necessary for economic growth.

The hukou system today

The hukou remains one of the most valuable documents in China today. Migrant workers without an urban hukou still have trouble finding decent housing and education for their children. They are also denied a range of government benefits. It is especially hard to gain admission to university without being an urban hukou holder.

There has been a push to reform the hukou system. Shanghai, Guangzhou and Shenzhen have relaxed the rules. The local hukou can be obtained if the migrant invests in the city or buys property. However, in crowded cities such as Beijing, where one-third of the residents are without the hukou, the government is hesitant to abolish the system, fearing that it would result in a surge of migrants to the city. They fear that such an influx would overwhelm the city's infrastructure and lead to even more overcrowding.

Internal migration

China's economic reforms involved its integration into the global economy. China quickly became the world's largest exporter of manufactured goods and a major importer of resources. Once-sleepy coastal fishing villages were suddenly transformed into major ports and manufacturing centres. The most rapid growth occurred in large cities in the coastal provinces, where the export industries boomed, as shown in Figure 15.3. Guangdong was known as the 'world's factory' due to the high number of factories and factory jobs there.

As the economy grew, it created new jobs and there was a flood of people leaving the countryside for the towns and cities in search of jobs and higher wages. Migrants on the move are shown in Figure 15.4. Migrants were employed in factories, on construction sites and in services (security guards, cleaners, couriers and domestic workers). Most of the migrants were young people who did not want to stay in the rural villages, which offered little in the way of career options.

Lives of internal migrants

Migrant workers from rural areas make up to half of China's urban workforce and half of the country's gross domestic product (GDP), but suffer from discrimination. The migrants are denied access to public services in the cities, due to the hukou system, and often experience exclusion and abuse.

The future

In 2012, the number of people living in urban areas exceeded the rural population for the first time, after 21 million people were added to the population of urban areas in just one year, as shown in Figure 15.5. The sheer numbers involved are staggering. Between 1990 and 2005, 103 million people migrated from rural areas to urban areas and in the following 20 years another 243 million are expected to do the same (see Figure 15.6).

China's urban economy is expected to generate over 90 per cent of GDP. China's economic goals are closely tied to continued urbanisation. The government is committed to quadrupling **per capita** GDP by 2020 (from the 2000 level). This goal implies the continued growth of towns and cities, where the wealth is generated. This growth will create employment—a major pull factor—and internal migrations will continue to impact on China's cities and rural areas.

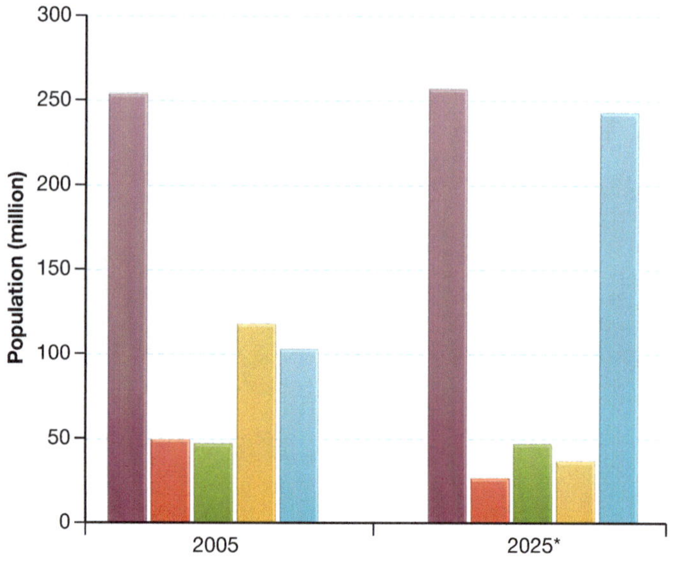

Source: McKinsey Global Institute, China All City Model

15.5 China's rural and urban population growth, 1950–2030

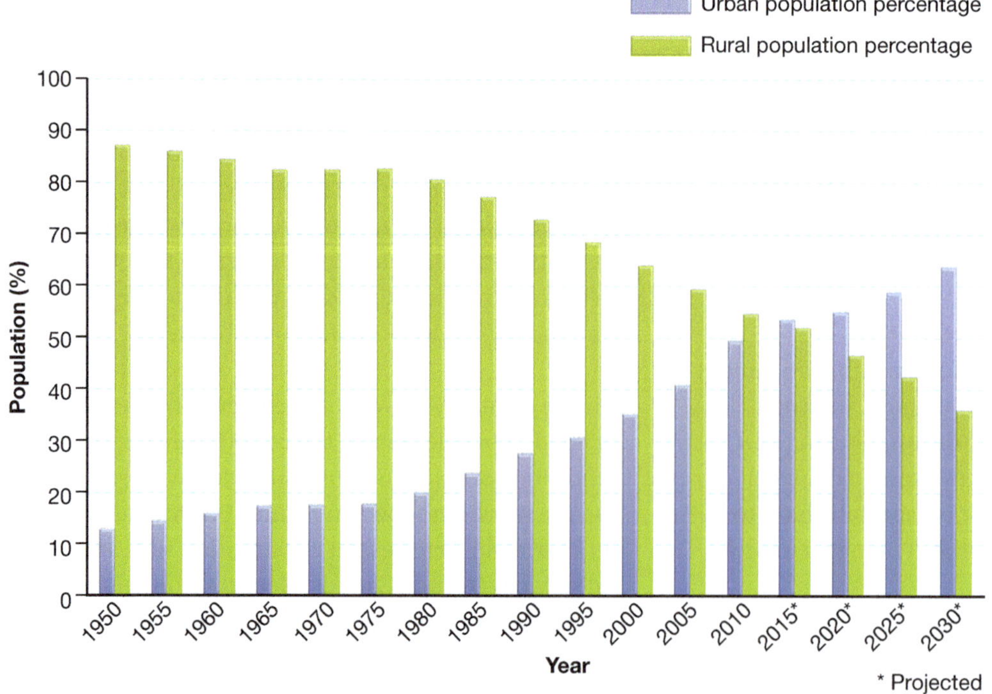

15.7 A densely built residential apartment development near Ordos City in China. With an investment of over US$161 billion by local government and revenue from the region's rich coal deposits, buildings to accommodate at least 300 000 residents, complete with modern facilities and grand plazas, were built on the site of an old desert village.

SPOTLIGHT

Cha Guoqun's story

Cha Guoqun left his village in search of work in the city of Hangzhou, in eastern China. When a cut on his leg became infected in November 2006 and prevented him from working, he visited a state hospital. As Cha had no health insurance, the doctor gave him two options: pay 1000 yuan (A$120) a day for treatment, the equivalent of his entire monthly income; or have his leg amputated. Fortunately, Cha received help from a Christian charity hospital, which was able to save his leg.

This case highlights the plight of an estimated 150 to 200 million rural–urban migrants who have moved to China's cities in search of work and better lives.

While their labour has fuelled China's economic growth, the majority of internal migrants never gain permanent residency in urban areas. For the most part, the lives of migrant workers are miserable. They have to live in makeshift shelters and eat the cheapest bean curd and cabbage. They have no insurance and their wages are often delayed. They are also discriminated against by urban people.

ACTIVITIES

Knowledge and understanding

1 Explain what the hukou system is. How did it restrict population movements?

2 Describe and account for change in China's population movement policy. What was the impact of the change?

3 Outline the urban population projections into the future.

4 Compare the opportunities for those who have an urban hukou and those new migrants who do not.

Geographical skills

5 Study Figure 15.5 and do the following tasks.

- a How many people will migration have added to cities by 2025?
- b Of the total urban population in 2025, what percentage will be migrants?

6 Study Figure 15.6 and do the following tasks.

- a In which year did China have the highest percentage of rural dwellers?
- b When did the rate of urbanisation start to accelerate?
- c What percentage of the population is expected to be urban dwellers by 2030?

15.3 Impact on Chinese cities

China is becoming an urban nation at a rate and scale never before seen. In just two decades, the lives of millions of Chinese people have been transformed. With increasing concentrations of people in industrial cities, the environment has become a casualty of the rural–urban migration. Huge areas of the countryside have become part of the rapidly advancing urban sprawl.

Rapid urban development

In China's eastern provinces are six of the world's top twelve economic 'hotspots': Shanghai, Beijing, Tianjin, Chongqing, Nanjing and Guangzhou. These cities are the favoured destinations of migrant workers. By 2015, China's cities are expected to make up 40 per cent of the global growth in the demand for residential and commercial floor space.

The government has built over a million new housing units in cities throughout China to deal with the rapid urbanisation. It has also provided assistance to workers to set up savings plans to purchase them. The skylines of cities and the countryside are filled with construction cranes, as high-rise apartments are built. **Urban sprawl** is consuming the east coast of China.

Impact on urban dwellers

Economic growth has delivered greater prosperity to urban dwellers. People living in Guangzhou in 2010 were earning four times the income that people earned in 1993. Urban residents also receive more benefits and social services from the government, such as health insurance, pensions and education. As city dwellers become more educated, they enter high-skilled jobs, where they earn even higher incomes. As a result, a **middle class** of wealthy consumers has emerged, and the numbers involved will lead to further economic growth. Figure 15.8 shows the rise in China's middle class.

Middle class

The increasingly wealthy middle class provides a huge market for consumer goods. Between 2000 and 2011, the number of automobiles per 100 urban households rose from less than one to more than 18; the number of computers from 8 to 80; the number of mobile phones from 16 to over 200; and the number of microwave ovens from 16 to 60.

15.8 The growth in China's middle class

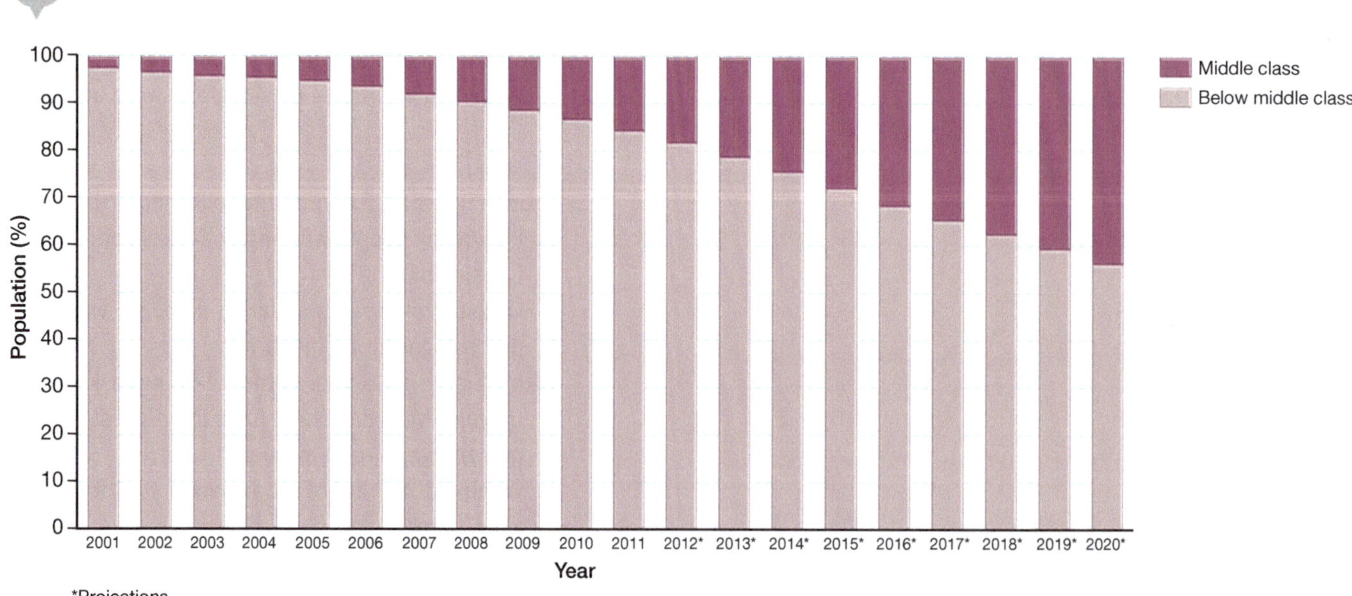

15.9 Pollution over Shanghai, 2013

Inequalities within cities

Migration has led to greater inequalities within cities. The incoming rural migrants are generally uneducated and low skilled, and find it difficult to get well-paid jobs. This has led to tension between the rich and the poor. Over one-third of urban crimes have been linked to young migrants. There has also been a disturbing number of suicides among young migrant workers unhappy with their wages and working conditions.

Environmental effects

As urban expansion continues to accelerate in China, the environmental consequences are alarming. A toxic, grey cloud often hangs over many of China's cities, as shown in Figure 15.9. Industrial wastewater and household sewage have contaminated rivers and coastlines. Water shortages and water pollution are becoming real threats. Air pollution has the potential to become the main cause of death in China. Heavy smog from factory emissions and car pollution can cause lung cancer and cardiovascular illnesses.

Some government initiatives to develop clean energy sources are underway. However, imposing controls on pollution would slow the economic growth to which the government is heavily committed.

DID YOU KNOW?

If current patterns continue, the number of middle-class Chinese will soar to 607 million by 2020, and spending by China's middle class will rival that of the United States of America.

ACTIVITIES

Knowledge and understanding

1. Describe where the most rapid urban development has occurred and will continue to occur.
2. Outline how the government has assisted with housing.

Applying and analysing

3. Create a PMI chart of the outcomes of China's urban expansion.
4. Write a short response to the following topic: *The greatest achievement of China has become its biggest burden.* In your response, consider the difficulty in balancing economic growth with environmental quality.

Investigating

5. Investigate the environmental issues of Chinese cities. Prepare an annotated visual display for a city of your choice entitled 'Choking in growth'. You should include:
   - a map of the city
   - an overview of the city—population size, demographics, main industries, growth rates
   - an outline of the main causes of the pollution
   - current solutions to the pollution problem. For example, have policies been introduced? Are they working? Why or why not? Provide alternative solutions.

15.4 China's rural towns and villages

Much of China's economic transformation has taken place in the special economic zones and big cities of eastern China. China's largely rural western provinces, in comparison, have not benefited from the economic transformation.

Promoting cities

For decades the Chinese government has focused on promoting industrial development in the urban and eastern parts of China. This is where the government has invested in transport and communications infrastructure. The poorest people in China are now concentrated in remote townships and villages in the western provinces. The educational, health and nutritional status of people in these provinces is well below that of people living elsewhere in China.

The Chinese government seeks to avoid social unrest, which could quickly take hold and threaten the stability of the vast country, by concentrating government spending in urban centres. In rural areas, discontent is less likely to escalate, as the rural settlements and their populations are quite scattered.

An exodus of migrants

The big cities offer a wider range of employment opportunities, which is an attractive prospect for young people living in rural areas. The potential for a more enjoyable, modern and adventurous lifestyle is also a major attraction.

City-based workers retain strong ties to their family. They travel home for important events, but return to their relatively well-paid jobs in the cities.

The flow of people is not always one way. During the 2008 global financial crisis, over 20 million workers returned to the countryside. Most returned to the cities as the world's economy recovered.

With so many young labourers leaving for the city, there are now many rural areas where it is mostly the elderly and women who are left to work on the farms. There are also many children left in the care of grandparents, while their parents work in the cities. An elderly man and his grandson are shown in Figure 15.10.

China's ageing population

In 1978, China introduced a new family planning policy, commonly known as the **One Child policy**. The policy permits families to have only one child. There have been changes to the policy since its introduction. The birth rate in China has fallen since the introduction of the policy.

Figure 15.11 illustrates China's falling birth rate in certain regions and an increasingly ageing population. Chinese officials are now discussing the emergence of

15.10 An old man and his grandson in Hebei Province, China

a '4–2–1' family structure developing throughout China as families have only one child. The '4' refers to two sets of grandparents, the '2' to the parents and the '1' to the only child. Traditionally, children look after their parents and grandparents in their old age. This will now be the responsibility of the one child—and people are living longer. China's ageing society and this new family dynamic provide China with many challenges.

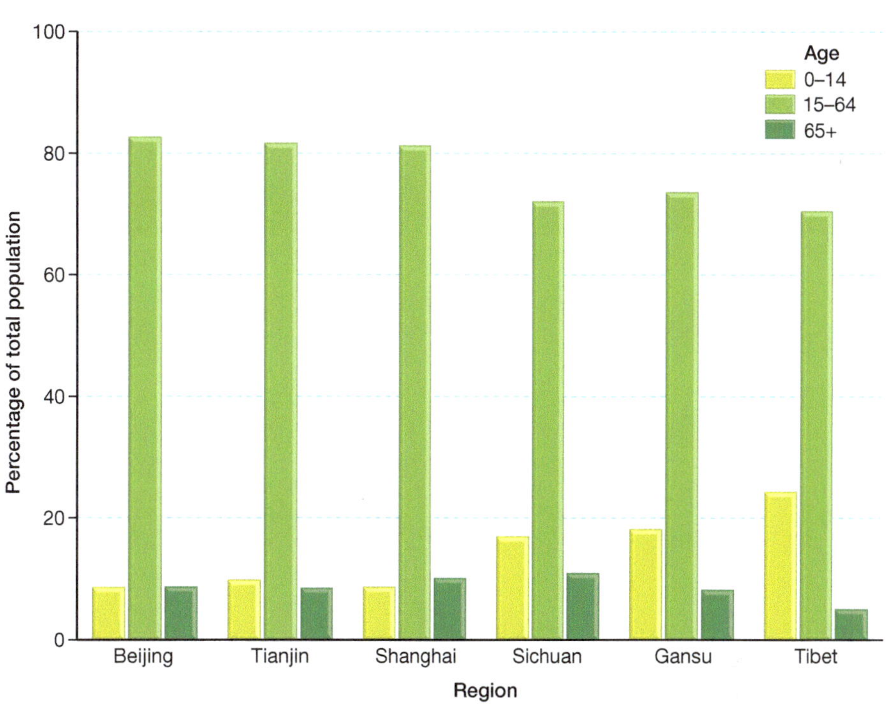

15.11 Population of selected Chinese regions, mid-2010

A widening gap between rural and urban

Half a billion people still live in rural areas, where almost 40 per cent of China's labour force work in agriculture. Some remote, inland regions remain much as they were 25 or even 100 years ago, with homes made of mud walls and earthen floors. Lack of adequate water supplies and sanitation services is common in many rural communities.

Despite China's economic transformation, the incomes of farmers are increasing more slowly than those of people working in the cities. The incomes of the rich have grown rapidly, which has widened the gap between rural and urban regions. The incomes of those living in the cities are two and a half times those of people living in rural areas.

Wide disparities are evident between the coastal provinces (where increasingly more people live in cities) and the inland provinces. This trend looks set to continue, as there are fewer opportunities in rural settlements. Rural schools have higher fees but fewer facilities than those in the city, so young people are disadvantaged. Many have to depend on farm labouring if they choose to stay in their village.

It is estimated that less than 30 per cent of the population in poor areas have safe drinking water and 22 per cent of rural residents cannot afford healthcare. The geographic distribution of health system coverage corresponds almost exactly with China's pattern of economic development, with the eastern provinces having the highest, followed by the central and then the western provinces.

ACTIVITIES

Knowledge and understanding

1 Explain why the Chinese government concentrates its spending in urban areas.
2 Analyse the movement of young people to and from the cities.
3 Describe and account for the widening gap between rural and urban areas.
4 Identify evidence for this widening gap.

Applying and analysing

5 Discuss the outcomes of the One Child policy and the creation of the '4–2–1' family structure.

Geographical skills

6 Study Figure 15.11 and answer the following questions.
   a List the region with the highest percentage of people aged 0–14 years.
   b List the regions with the lowest percentage of people aged 0–14 years.
   c Tibet, a remote rural area, has the lowest percentage of people over 55 years of age. Suggest reasons for this.

15.5 Case study: Shenzhen, China

Shenzhen, located in China's Guangdong Province, immediately to the north of Hong Kong, was China's first special economic zone (SEZs). It is also one of the world's fastest-growing cities and one of the greatest concentrations of economic activity on the planet.

15.12 Shenzhen today

15.13 Shenzhen's population, 1979–2010

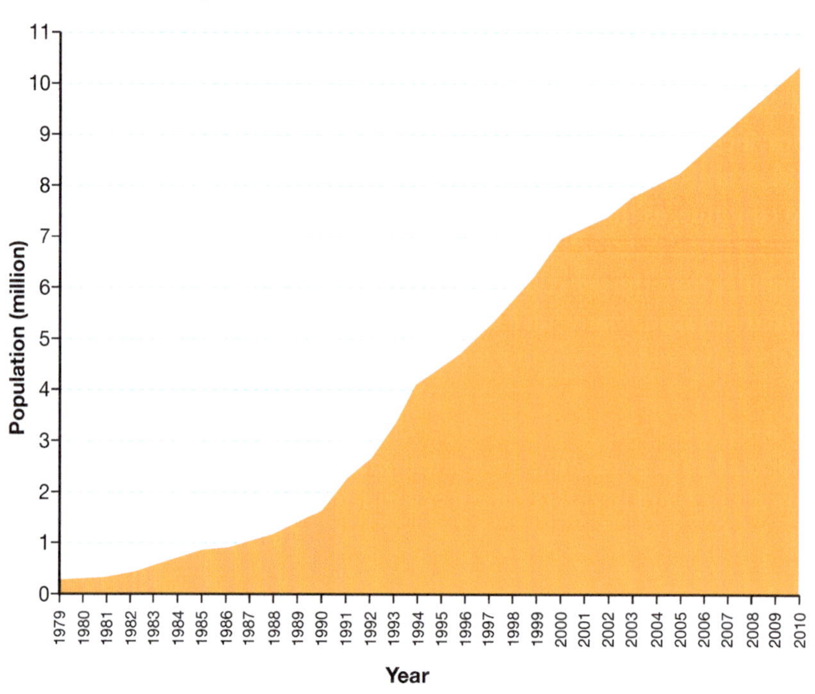

Population growth

In little more than three decades, Shenzhen, pictured in Figure 15.12, has emerged as an economic powerhouse and one of the largest cities in China. Shenzhen's population began to grow rapidly after the area was made one of China's five special economic zones (SEZs) in 1979, mainly due to its location close to Hong Kong. At the time, just 300 000 people lived in the area. By 1990, there were 1.2 million and by 2000, 7.1 million. In 2010, Shenzhen's population exceeded 10.3 million, as shown in Figure 15.13. More than 6 million of the city's residents are migrant workers who return home on the weekends and live in factory dormitories during the week.

Economic growth

China's SEZs were established as an experiment in capitalism. Chinese and foreign businesses—attracted by generous trade and tax benefits—have invested huge amounts of money in local industry. Foreign businesses have invested more than US$30 billion. While most investment has been in manufacturing, an increasing amount is being used to develop service-based industries. Industries include biotechnology/ pharmaceuticals, building/construction materials, chemicals production and processing, computer software, electronics assembly and manufacturing, instruments and industrial equipment production, medical equipment and supplies, research and development, and telecommunications equipment.

SPOTLIGHT

Foxconn's Shenzhen operations

Apple's iPod, iPhone and Mac Mini are all manufactured at Foxconn's Shenzhen plant. Foxconn (the Taiwan-based Hon Hai Precision Industry Co. Ltd) is the world's largest maker of electronic components (Figure 15.14). It is also China's biggest exporter and the largest private-sector employer. In addition to China, Foxconn has factories elsewhere in Asia, in Europe and in Latin America, which together assemble around 40 per cent of the world's electronics products.

Foxconn's largest factory is in the Longhua District of Shenzhen. Hundreds of thousands of workers (perhaps as many as 450 000) are employed at the Longhua Science and Technology Park, a walled facility sometimes referred to as 'Foxconn City'. Covering an area of 3 square kilometres, the vast complex includes fifteen factories and worker dormitories. It also has its own television station and fire brigade, and a downtown area complete with a grocery store, bank, restaurants, bookstore and a hospital. While some workers live in surrounding towns and villages, many live and work inside the complex. A quarter of the employees live in the dormitories, and it is not unusual for them work 12-hour shifts, six days a week.

15.14 Foxconn workers assemble about 40 per cent of the world's consumer electronics products.

Urban structure

The city is divided into ten districts—four city districts (Futian, Luohu, Nanshan and Yantian); two suburban districts (Bao'an and Longgang); and four 'new management' districts (Guangming, Pingshan, Longhua and Dapeng). Each has its own distinctive character. Futian is the city's major commercial district; Luohu is a popular shopping destination; Nanshan, Yantian, Bao'an, Pingshan and Longhua are dominated by industry, especially consumer electronics; and in Guangming are a number of ecological technology industries.

Economic specialisation

In Chinese cities there is a level of economic specialisation rarely found in the developed world. Datang, for example, makes 40 per cent of the world's socks; Songxia produces 350 million umbrellas each year; Wenzhou makes 70 per cent of the world's cigarette lighters; Qiaotou produces 70 per cent of the buttons used on clothing manufactured in China; and 40 per cent of the world's neckties are made in Shengzhou.

ACTIVITIES

Knowledge and understanding

1. Explain the role of government in the growth of Shenzhen.
2. Outline the economic importance of Shenzhen.
3. Describe the role of migrant workers in the city's economy.
4. Identify the city's main industries.

Applying and analysing

5. Discuss the ways in which the Foxconn example highlights the differences between Chinese cities and cities of the developed world.

Geographical skills

6. Study Figure 15.13. Using data from the graph, describe Shenzhen's population growth between 1979 and 2010.

CHAPTER 16

MANAGING AUSTRALIA'S CITIES

The growth of cities needs to be carefully managed if we are to minimise their environmental impact while enhancing their liveability. A more balanced distribution of Australia's population would be one approach, but this would require substantial expansion of existing regional centres or the building of new cities at great cost to accommodate increasing numbers of people. The latter has been done before—Canberra, Australia's capital, was a planned city and built from the ground up. Careful planning can make our cities more sustainable, on a city-wide scale or at the level of individual households.

In this chapter we examine the ways in which urban centres can be managed to achieve sustainability and enhance liveability.

KEY IDEAS

- To understand how urban centres can be better managed
- To investigate how planning can enhance the environmental sustainability and liveability of urban places

16.0 Perth, Australia

GLOSSARY

brownfield	a built-up area of land such as an old factory that is redeveloped
greenfield	an area of land that has not been developed
suburbanisation	the process in which people and businesses move out from the central areas of cities and into the suburbs
urban consolidation	increased population densities in existing urban areas
urban renewal	the process in which an unused area of urban land such as an industrial site is restored, redeveloped and improved

16.1 Managing growth of cities

In Australia, 75 per cent of the population live in just eighteen cities, each of which has a population of at least 100 000 people. Sydney and Melbourne now have populations of more than four million. Brisbane has more than two million and Perth has 1.7 million.

Australian cities

Australia's eighteen biggest cities are shown in Figure 16.1. Almost 75 per cent of all jobs in the country and about 80 per cent of all economic activity is located in these eighteen cities. With so many people, jobs and wealth in our cities, governments around the nation recognise the importance of carefully managing their growth.

In 2010, the federal government released an important report about Australian cities called *Our Cities—Building a Productive, Sustainable and Liveable Future*. Key issues included in the report were:

- providing employment opportunities
- providing important infrastructure, such as public transport, sewerage systems and roads
- providing enough affordable housing
- making our cities more sustainable by reducing consumption and dealing better with waste.

Planning for housing

As Australia's cities grow so too does the need for housing. The most common strategy used to increase the supply of new housing has been to develop housing estates on the edges of our cities. This is called **suburbanisation**. All Australia's cities and large towns now have complex suburban areas.

Economic costs

Suburbanisation continues to provide most new housing needed in Australia. Building new suburbs requires very detailed planning. Building the actual houses is the easy part. Complex infrastructure, such as sewerage, electricity, gas, telephone and internet lines, roads and public transport systems also need to be developed. There must also be plans for schools, shopping centres and other important community services such as parks, libraries and childcare centres.

16.1 Population of Australia's biggest cities, 2011

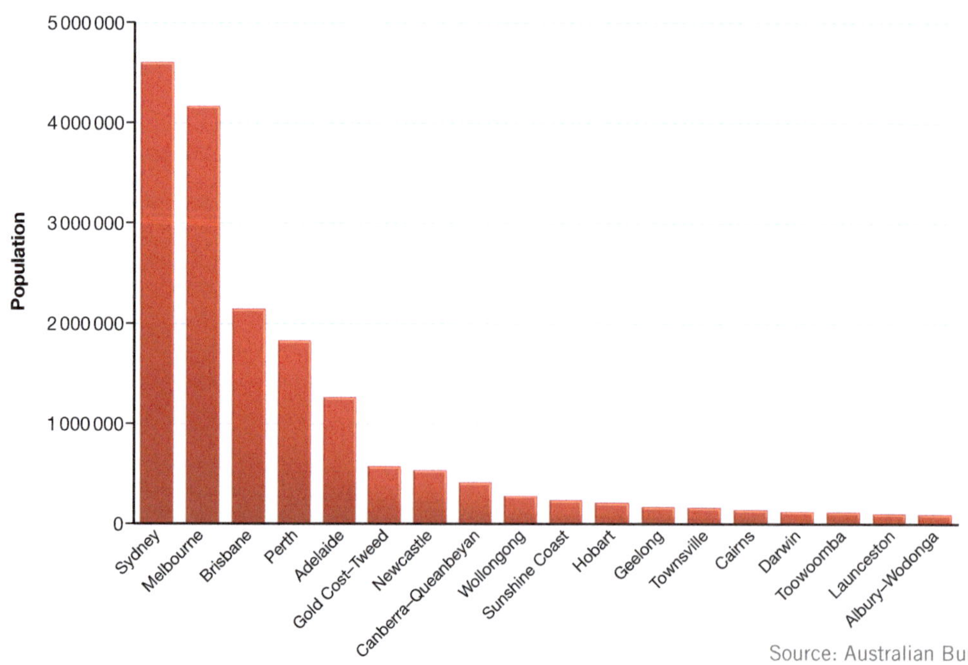

Source: Australian Bureau of Statistics, 2011 Census

16.2 The Palmview greenfield site, at Caloundra South, Queensland, will be developed for housing.

Environmental costs

The development of new suburbs typically involves clearing land that has been used for other purposes, such as agriculture or bushland. Geographers often call these areas 'greenfield sites', as they are undeveloped green zones (see Figure 16.2). In their natural, or near-natural state, these areas provide important habitats for native animals. These animals are displaced by suburban development. Suburban development also increases rates of water pollution; for example, oil and rubbish left on roads can be washed into stormwater drains. Introduced species of plants can result in weed infestations. Feral animals either kill native animals or compete with them for food.

SPOTLIGHT

Sydney's north–west rail link

To meet the needs of the rapidly growing population of Sydney's north-western suburbs, the New South Wales government is building an $8.5 billion, 23-kilometre rail link between Rouse Hill and Epping (Figure 16.3). Eight new railway stations will be built with parking for 4000 cars. The rail link includes 15 kilometres of tunnels between Bella Vista and Epping—the longest and deepest rail tunnels ever built in Australia.

16.3 Sydney's proposed $8.5 billion rail link is designed to reduce people's reliance on cars.

16.4 The Melbourne docks in the early 1900s (left) and today transformed into the Docklands redevelopment (right)

 SPOTLIGHT

Urban consolidation in Glenelg

Glenelg is a well-established beachside suburb in Adelaide. It is part of the Holdfast Bay City Council area and was first developed in 1839. This makes it one of the oldest urban areas in Australia. Like other major cities, Adelaide has had to diversify its housing choices to meet the changing needs of people and a growing population. As a result, Glenelg has become an area of urban consolidation.

A long-term development plan, released in April 2012, identified urban consolidation as a key feature of the area's revitalisation. Part of the plan is to further increase housing densities. This will be achieved through the development of medium-density housing, such as townhouses, as well as high-density housing (apartment towers). Some of these new apartments can be seen in Figure 16.5. An important part of the consolidation at Glenelg is to use existing sites, demolishing older buildings with no heritage value to make way for new buildings.

16.5 Urban consolidation is a feature of Glenelg in Adelaide.

Social costs

As our cities continue to expand, more and more people are forced to travel greater distances to work. In many of our cities, hundreds of thousands of people commute very long distances from outer suburbs every day. In many cases these suburbs are not well supplied by public transport, so vast motorway systems are needed and traffic congestion has become a major issue for Australia's large cities.

Commuting long distances has important social consequences: it affects people's health, as they spend so much time sitting in cars; it reduces the time they can spend with family and friends; and it costs families money in running a car. Traffic congestion also has environmental consequences. About 75 per cent of all Australians living in capital cities commute to work in cars, most commonly on their own. Cars are major contributors to air pollution in our cities. They also contribute significantly to greenhouse gas emissions, the cause of climate change.

Reducing urban sprawl

Urban consolidation involves the reuse of existing urban land for housing. Land that was once used for industrial purposes is redeveloped for housing. The Docklands redevelopment in Melbourne is a large residential and commercial building project built on waterside land that is no longer needed for its original purpose. The changes can be seen in Figure 16.4. This type of development is often called 'infilling' because it fills in gaps in the city that have been created by land no longer being used.

Another type of urban consolidation is to increase the density of housing in a residential area. Building density means the number of houses that occupy an area. Increasing building density involves building a few townhouses (medium-density housing) or even a multistorey apartment block with fifteen apartments (high-density housing) on a large block of land.

Urban consolidation means that more people can live in cities, reducing the need to build new suburbs further away from the city centre. Urban consolidation does mean, however, that some suburbs will change, and this can create tensions. For example, residents can become distressed when large apartment blocks are built where single-storey houses once stood.

Reducing traffic congestion

Building better infrastructure

Public transport infrastructure is extremely expensive to build but reduces traffic congestion more than building new roads or tunnels. A freeway can carry 2500 cars per hour, while a train can carry 50 000 people in the same time.

Motorways can cost billions of dollars and are always politically sensitive when their construction requires the demolition of homes or the loss of bushland. Tolls are now widely used to fund road infrastructure.

Railways are even more expensive to build and often require subsidisation by taxpayers—to charge the commuter the actual cost of providing the service would make it too expensive to use.

Using public transport

One of the big issues is encouraging people to use public transport rather than their cars. Public transport needs to be accessible, convenient, relatively cheap, safe and clean. Providing carparks at railway stations is one way to encourage the use of public transport. Increasing the cost of parking is a way of discouraging the use of cars.

The public is often divided on how best to meet transport needs of city dwellers, especially when funds are limited. Some argue that public transport should be the priority. Others argue that more motorways are the answer.

ACTIVITIES

Knowledge and understanding

1 Outline the key points to consider when managing cities.
2 Define the term 'suburbanisation'.
3 State what a 'greenfield' site is.
4 Explain the process of urban consolidation.

Applying and analysing

5 Create a PMI chart about suburbanisation.
6 Discuss the advantages and disadvantages of urban consolidation.
7 Produce a campaign plan to encourage people to use public transport rather than drive.

16.2 Impacts of large cities

As the population of a city grows, the city becomes like a giant magnet, drawing in more and more people and services from the places around it. Some cities grow so large that they dominate all other urban places in the country.

Australia

In Australia, the urban population is evenly spread, with five large cities—Sydney, Melbourne, Brisbane, Perth and Adelaide. However, the growth of these large cities is having a significant impact on regional areas. As each city's population grows, it draws in businesses attracted by the growing population. This then begins a cycle that leads to regional areas losing population, jobs and services to cities, as outlined in Figure 16.6.

16.6 Growing cities attract population and employment from regional areas.

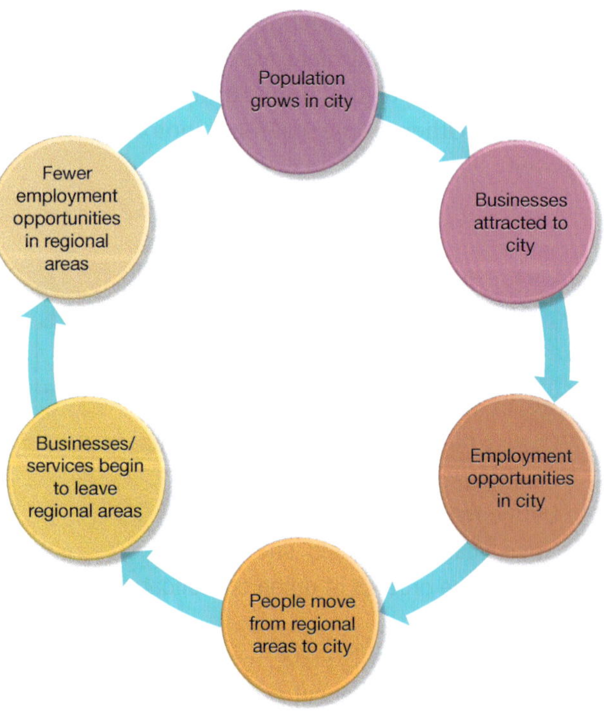

Rural Australia

The depopulation of regional areas is highlighted in Figure 16.7. The two population pyramids show the age distribution of people living in Melbourne and regional

16.7 Population pyramids for regional Victoria (top) and Melbourne (bottom)

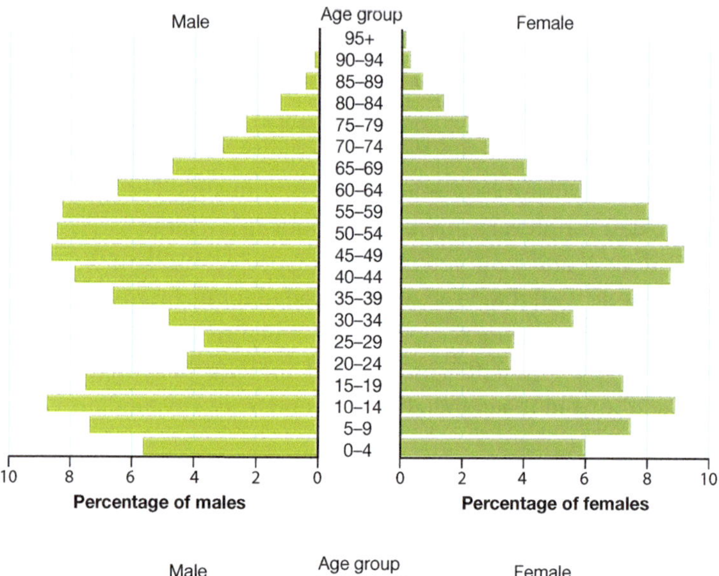

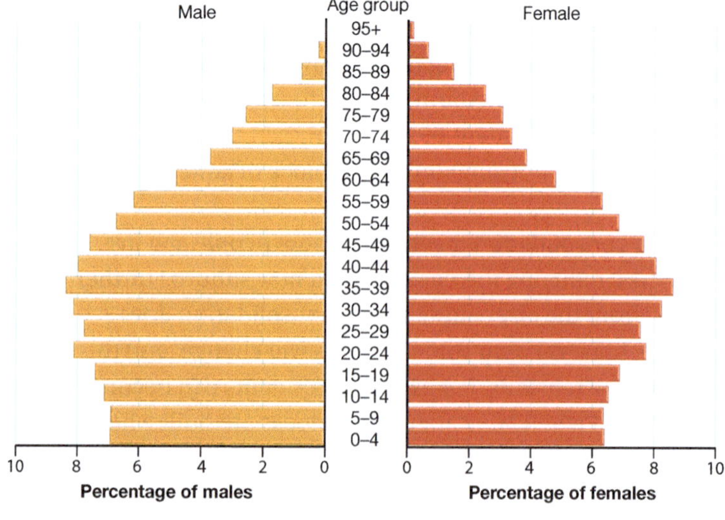

Source: Victorian Department of Planning and Community Development

Victoria in 2010. These pyramids show that while Melbourne has a fairly even distribution of population, in regional Victoria there is a dip in the percentages of people aged between 20 and 39 years. This is important because it shows that people of this age group have left regional Victoria. Most of those leaving do so to seek employment in Melbourne. A similar pattern can be found throughout Australia, the one exception being those regional areas that have a major mining industry, such as northern Western Australia and central Queensland.

SPOTLIGHT

Doctors in regional Australia

In many parts of regional Australia, there are severe shortages of doctors. This is having an impact on the health of those living there.

Figure 16.8 shows the number of doctors per 100 000 people in different parts of Australia. The table shows that there are many more doctors per 100 000 people in major cities than in regional areas. For example, in Sydney there are 343.8 doctors for every 100 000 people, compared to just 115.2 doctors in remote parts of the state. In other words, in Sydney, doctors make up about 0.34 per cent of the population but in remote areas they make up less than 0.01 per cent of the population.

16.8 The number of doctors per 100 000 people

Remoteness area	NSW	Vic.	Qld	SA	WA	Tas.	NT	ACT	Aust.
Major city	343.8	366.1	394.6	422.8	382.6			472.7	372.1
Inner regional	202.6	196.4	188.7	147.5	163.3	457.9			212.1
Outer regional	118.8	106.9	236.6	127.8	193.4	178.3	433.8		188.0
Remote/Very remote	115.2	265.3	161.8	126.1	221.1	101.6	385.8		216.4
Total	308.6	332.7	334.6	353.9	336.7	366.4	442.9	474.2	331.4

Source: Australian Medical Association, 2012

ACTIVITIES

16.2

Knowledge and understanding

1 Explain the difference in access to doctors for people living in cities compared with people living in other places in Australia.

2 List the reasons why people might be attracted to move to a city.

3 Explain what occurs when a region is depopulated.

Applying and analysing

4 Adopt the perspective of a young adult who lives in a regional part of Australia but is moving to a large city. Write a letter to a friend explaining your reasons for wanting to leave your town. Include in your response the economic and social reasons for your decision.

5 Study Figure 16.6 and write a short report explaining how cities can have an impact on the population of other places.

6 Study Figure 16.7 and do the following tasks.

  a What is the total percentage of males and females aged 20–24 living in:
  - regional Victoria?
  - Melbourne?

  b Using these graphs as stimulus, write a paragraph comparing the age structure of the populations of regional Victoria and Melbourne. Be sure to look for similarities and differences.

16.3 Case study: Canberra

As a capital city, Canberra is a very different type of city. In many countries the largest city is also the country's capital. In the case of cities such as London, Paris, Tokyo and Jakarta, the capital city is also the country's main commercial centre. However, Canberra is a city that was planned and purpose-built as a capital city, much like Washington, DC in the United States of America.

The city

Canberra is the most thoroughly planned Australian city. At Federation, in 1901, it was decided to build a capital city approximately midway between Sydney and Melbourne. Canberra is not a major commercial centre. It is dominated by government-related functions; its buildings include Parliament House and the headquarters of major government departments and agencies. Canberra is also home to the country's major national institutions, such as the Australian War Memorial, High Court, National Library, National Gallery and National Museum of Australia.

The current site was named the Australian Capital Territory (ACT) on 1 January 1911. An international competition was then held to find a design for the new capital. More than 130 architects from around the world submitted plans. American architect Walter Burley Griffin's design was ultimately successful, with his original plan for a city based on a series of 'zones' in a concentric (circular) design. The winning design is shown in Figure 16.9. It can still be seen today in Canberra, which is shown in Figure 16.10.

16.9 Walter Burley Griffin's original design drawings for Canberra. The concentric design of unique zones is clearly evident.

16.10 There are many important national institutions in Canberra, including Parliament House.

Functions

Parts of the city still have very clear functions. For example, the area around the Capitol in Griffin's original plan is where Parliament House is located, the Civic Centre contains the main shopping and commercial districts, and the area Griffin called the Government Group (now the Parliamentary Triangle) contains the headquarters of major government departments.

Bush capital

Canberra is sometimes known as the 'bush capital' because of its location and because it is surrounded by grazing properties and tree-covered mountains. Canberra takes up just 20 per cent of the ACT, and more than 50 per cent is designated as nature conservation areas. These areas are an important asset for the people of Canberra, providing them recreational areas very close to the city. In 2003, these areas were also a cause of enormous tragedy when a massive firestorm engulfed the city, killing four people and destroying more than 300 homes. A key part of the management of Canberra is now about planning for fires, as well as managing the growth of Canberra to minimise the impact of the growing city on the important nature conservation areas surrounding the city.

Population challenges

Dealing with a growing population is one of the key challenges facing Canberra. During the first decade of the twenty-first century, the population grew by about 50 000 to 367 800 people in 2011. Most of this growth has taken place in the northern part of the city, where new housing estates have been developed. For example, in the suburb of Gungahlin (Figure 16.11), about 10 kilometres north of the city centre, the population has more than doubled in the last 10 years. As the population of the region has soared there have been concerns that these new areas lack adequate infrastructure. The Gungahlin Community Council, for example, has called on the ACT government to consider building a new hospital in the area.

16.11 The population of Canberra suburb Gungahlin has more than doubled in the last 10 years.

Planning for the future

As the nation's capital, responsibility for urban planning and management is shared between the federal government and the ACT government. Over the years, there have been different plans developed for Canberra. The current plan is known as the National Capital Plan and sets out the future development of the city. Figure 16.12 shows how the land will be used in the future.

The National Capital Plan still includes the concentric design that Griffin developed. Another important feature is the natural bush spaces placed between the major urban areas. These areas, shown in pale green on the map, are areas that are not allowed to be developed for urban use. This land is sometimes set aside as nature conservation areas or may be used as parklands. These areas are an important way to manage the city's growth and to ensure that the character of Canberra as the 'bush capital' is retained.

16.12 Landuse for Canberra under the National Capital Plan

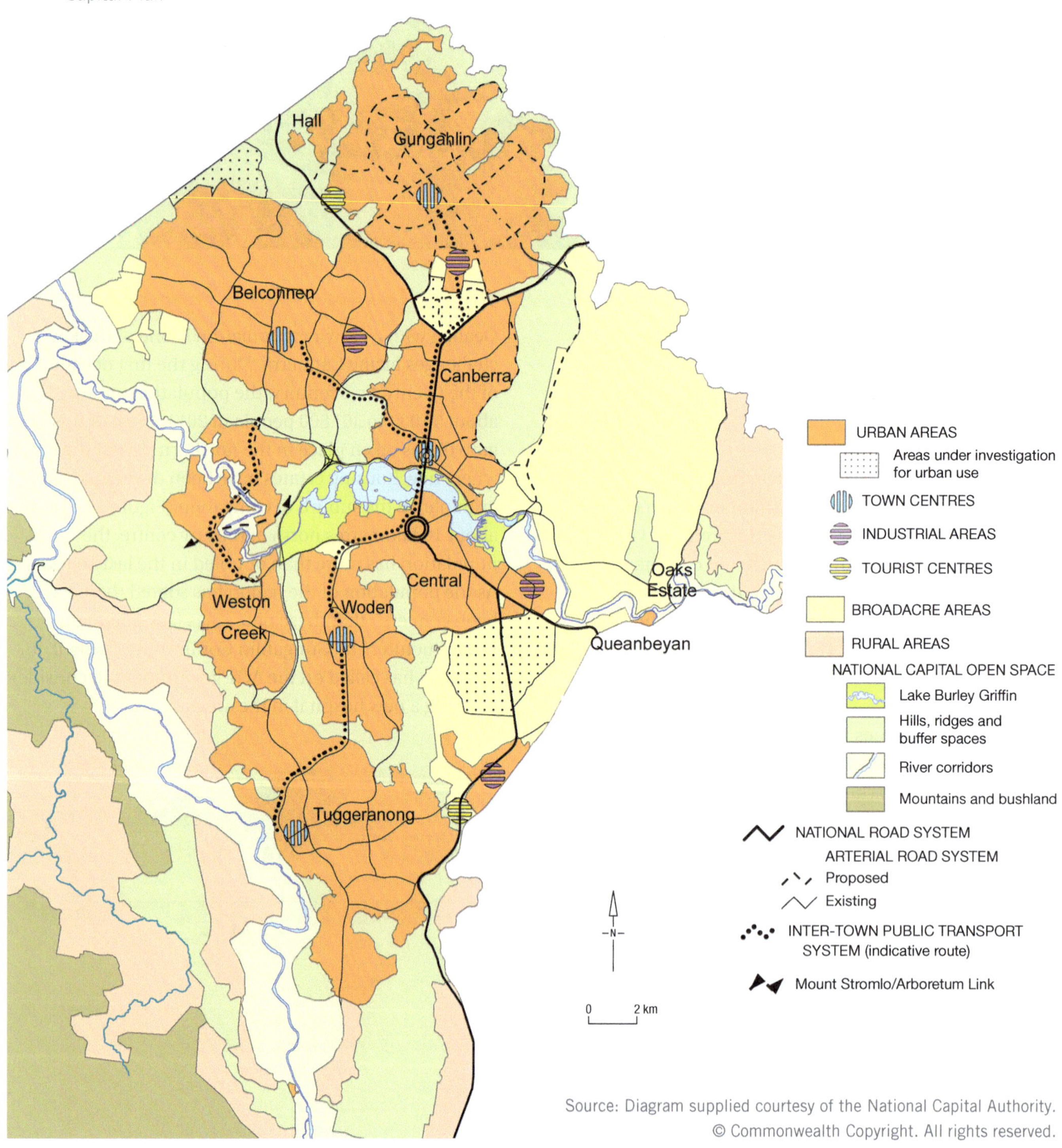

Source: Diagram supplied courtesy of the National Capital Authority.

SPOTLIGHT

The Molonglo Valley development

The Molonglo Valley is located about 10 kilometres to the south-west of the Canberra city centre. It was one of the areas most badly affected by the 2003 bushfires. A new housing estate being developed (see Figure 16.13) includes the area where many houses were destroyed. The development is being designed to incorporate sustainability. The new suburbs of Wright and Coombs have both received eco-design certification from the Urban Development Institute of Australia.

One of the key aspects of the design of Wright is the use of a grid system. By placing the housing in grids, the suburb becomes far easier to walk around, minimising the need for cars. Nearly every house will be no further than 800 metres from a reliable and regular public transport service. There are areas for high-density housing, such as apartments, medium-density housing, such as townhouses, and low-density housing—freestanding houses.

This mix of housing caters for various types of families and the size of the suburb can be reduced. This helps to make the suburb more sustainable by reducing the impact on surrounding bush and farmland, and making the suburb more bike- and pedestrian-friendly.

16.13 The Molonglo Valley development

ACTIVITIES

Knowledge and understanding

1 Explain why Canberra is sometimes called the 'bush capital'.
2 Describe the population growth of Canberra in recent years.
3 Outline the main features of the National Capital Plan.
4 Explain how the urban design of the suburb of Wright helps to make it more sustainable.

Applying and analysing

5 Study Figure 16.9—the original plan for Canberra developed by Walter Burley Griffin—and do the following tasks in groups.
   a Describe the main features of his design.
   b What features of his design do you think work best? Explain why.
   c What features would your group do differently? Explain why.
   d Examine the National Capital Plan in Figure 16.12. What similarities are there between this plan and that developed by Walter Burley Griffin?
6 Refer to Figures 16.10 and 16.12. Take on the role of an architect designing a capital city for a newly formed country in the Asia–Pacific region. Prepare a basic design for the main parts of the city. Along with your design, prepare a briefing sheet explaining your design decisions. You should describe the location of your city—near a river, on the coast or inland—and the shape of the land.

16.4 Case study: Urban renewal in Brisbane

Brisbane, the capital of Queensland, has a population of more than two million people, making it Australia's third largest city. It was first settled in 1824 and has developed into a significant business and tourism centre.

Urban renewal

Like all Australian large cities, Brisbane is experiencing a period of urban change as the city moves into the twenty-first century. A key aspect of this change is **urban renewal**; that is, existing parts of the city that have fallen into disuse, or at least underuse, are redeveloped and used for different purposes.

South Bank

The South Bank redevelopment, which lines the southern bank of the Brisbane River opposite the Central Business District, is an example of this renewal. For much of the twentieth century the South Bank district was used for industrial purposes, including shipbuilding, and contained a mixture of warehouses, poor-quality housing and small factories.

Urban renewal of the South Bank area began in the 1970s when it was selected as the location of a new cultural precinct for Brisbane. The construction of the Queensland Art Gallery, Museum and Cultural Centre in the area helped to kickstart the renewal. Most famously, South Bank was transformed to house Expo 88, the World Exposition, in Australia's bicentennial year (1988). After Expo ended, the temporary buildings were removed and the site was gradually transformed into the extensive parklands and popular cafe and restaurant precinct. These can be observed in Figure 16.14.

16.14 Urban renewal transformed an industrial area into the recreational parklands of South Bank.

The RNA Showgrounds redevelopment

Brisbane's newest urban renewal project is the redevelopment of the Royal National Agricultural and Industrial Association of Queensland Showgrounds in the suburb of Bowen Hills. Showgrounds are generally not used very often, often remaining empty for most of the year. This was the case with the RNA Showgrounds, which were only used for the EKKA Show (formerly called the Royal Queensland Show). EKKA runs for a two-week period each year—the rest of the time the showgrounds are empty.

This redevelopment, costing nearly $3 billion, is designed to transform the showgrounds into a major residential and cultural precinct. It is one of the most significant 'brownfield' developments ever undertaken in Australia. **Brownfield** developments are urban projects that take place on existing urban sites, whereas **greenfield** developments take place on areas that have not been previously developed, such as farmland.

The 22-hectare RNA site, when complete, will be home to about 3000 people, many of who will live in multistorey apartment buildings on the edge of the showgrounds. There will also be a major shopping centre, office towers and a range of other cultural facilities. It is expected that 15 000 people will be working in the area (see Figure 16.15).

The Royal International Convention Centre was one of the first sections of the redevelopment to be completed. It was opened in March 2013. The Convention Centre can hold up to 3000 people. The planners of the redevelopment hope the new centre will help Brisbane attract major conferences, which will bring in more tourists and tourist money.

16.15 (a) The RNA Showgrounds prior to the renewal of the site; (b) Artist's impression of the showgrounds once the renewal project is complete

a

b

ACTIVITIES

Knowledge and understanding

1 Define the term urban renewal.
2 Describe the transformation of the South Bank area of Brisbane.
3 Explain why the RNA Showgrounds were under-utilised.
4 Outline the redevelopment of the RNA Showgrounds.

Applying and analysing

5 Examine Figure 16.15. Describe the changes that are evident between the old site and the RNA redevelopment.
6 Take on the role of a real estate agent and prepare a promotional brochure for apartments in the RNA Showgrounds. In your brochure, consider the advantages of living in this environment. You may wish to access websites about the RNA redevelopment to assist you.

16.5 Case study: Upgrading Adelaide's CBD

Adelaide, the capital of South Australia, is one of the world's most liveable cities. There is, however, a need to enhance the city's liveability and sustainability. The redevelopment of the Rundle Mall, the city's premier retail strip, is part of the strategy to create of a more vibrant CBD.

The Rundle Mall

When closed to traffic in 1976, Adelaide's Rundle Street became Australia's first car-free pedestrian mall. In addition to its retailers, Rundle Street boasts some of the city's most famous architectural landmarks, such as the ornate Adelaide and Regent arcades. There are also several prominent sculptures in the mall. The best known is 'The Spheres', by artist Bert Flugelman—4 metres tall, it comprises two large stainless-steel spheres, each with a diameter of 2.15 metres, balanced one on top of the other. They are most commonly referred to as the 'Mall's Balls' or 'Rundle Mall balls'. The sculpture is a popular meeting place for visitors.

Today, the mall attracts more than 400 000 customers every week and 23 million visitors annually. It is, as a result, one of the most visited places in the state of South Australia. The mall offers the largest selection of shopping facilities in Adelaide, with three large Department stores, 15 arcades and centres, 700 retailers and more than 300 non-retail services and offices.

Revitalising the city's retailing heart

In 2011, Adelaide City Council decided to redevelop the Rundle Mall. The project was designed to reposition the mall as the primary retail precinct in Adelaide. The plan involved the transformation of the area into a unique, vibrant and attractive shopping experience in the heart of the city. The main retail competition comes from the large suburban shopping malls. The Adelaide shopping landscape is dominated by nation-wide retailers such as David Jones and Myer.

16.16 Adelaide's Rundle Street in 1938

The redevelopment of the mall involves the replacement of street paving, provision of new street furniture and lighting, removal of clutter and planting of mature trees. Public safety will be enhanced through the increased CCTV coverage throughout the mall. For those keen to go online, the mall will feature a number of wireless hotspots.

Figures 16.16 to 16.18 show how Rundle Mall has changed over time.

Rundle Street East

To the east, Rundle Street changes into one of Adelaide's most popular entertainment precincts, with its historic character and a contemporary life-style focus. The area boasts cutting-edge fashion stores and leading designer labels, funky gifts, homewares, jewellery and accessories. Also found there are some of Adelaide's best known cafes, restaurants and wine bars, most of which offer outdoor (alfresco) dining, making this strip one of Adelaide's favourite night spots for locals and tourists alike.

Rundle Street East is also the city's hub for the numerous annual festivals, events, street parades and parties, including the Adelaide Fringe, Adelaide Festival, Clipsal 500 V8 Car Race, Tour Down Under, East End Jazz Festival, several international film festivals and much more.

16.17 Rundle Mall, 2011

16.18 Artist's impression of the redeveloped Rundle Mall

Rundle Mall; HASSELL concept image

ACTIVITIES

Knowledge and understanding

1 State why Rundle Mall is important in terms of Australia's urban planning history.
2 Explain why the Rundle Mall is being redeveloped.
3 Outline the key features of the redevelopment.
4 Outline how Rundle Street East differs from the Rundle Street Mall.

Applying and analysing

5 As a class, discuss the advantages and disadvantages of traditional city centre-based retail strips compared with the large suburban shopping malls. What similarities are there between car-free pedestrian zones such as Rundle Street and the large suburban shopping malls?

Geographical skills

6 Study Figures 16.16 to 16.18.
  a Describe the changes you observe in the images.
  b In your opinion, have the changes been beneficial? Explain.
  c Would you change anything in the redevelopment plan?

Investigating

7 Identify and research an urban renewal project similar to that taking place in a place in a retail precinct near where you live. What is the nature and purpose of the development? What are its key features? Will it impact on other retail centres?

16.6 Sustainable cities

Our cities are important places. They provide most of the employment opportunities for Australians and they are where most of us live. Our cities also have an enormous impact on the environment. Reducing the impact that our cities have in the environment is a key part of the management of cities into the future.

Learning to live with nature

Bushland is a feature of all of Australia's large cities. Many suburbs are found close to the bush and have important bushland reserves, and even national parks within them. As our cities grow, the surrounding bushland is placed under increasing stress. Expanding suburbs lead to the clearing of bushland. Domesticated animals, especially cats, are introduced to the bush from surrounding suburbs and hunt native animals. Garden plants escape into the bush, taking over from native flora. Therefore, it is important to protect remaining bushland reserves and to carefully manage the impact of our cities on them, including the flora and fauna that live there.

Designing better houses

The way Australians live has a considerable impact on the environment. The average Australian household produces around 14 tonnes of greenhouse gas emissions each year. Australia's ecological footprint (a way of measuring the overall impact on the environment) is more than 2.8 times the world average.

The average size of new homes in Australia has increased by 40 per cent since the 1980s, making our homes now the largest in the world. During the same time Australian families have become smaller. The average size of a new home in Australia is about 240 square metres; in Britain an average new home is less than 80 square metres. Such big homes use huge amounts of energy to heat in winter and cool in summer.

As energy prices rise, greater emphasis is being placed on more sustainable housing designs. This, coupled with a gradual shift towards smaller homes, will, over time, help to reduce the environmental impact of our housing.

SPOTLIGHT

Protecting Brisbane's squirrel gliders

The squirrel glider (Figure 16.19) is a native species whose natural habitat extends along the east coast of Australia from the Victorian–NSW border to Far North Queensland. The glider is found in many locations throughout Brisbane, and the Brisbane City Council has developed an action plan to help protect it.

Loss of its natural habitat, as the suburbs of Brisbane continue to expand, is one of the main threats to the glider. There are also insufficient nesting sites. The glider nests in tree hollows but competition from introduced species, such as the common myna bird, is reducing the number of hollows available. Domestic cats are another problem for the glider. Hunting by cats is common and is a major cause of death for gliders all along the eastern coast.

The Brisbane City Council is taking a number of steps to help protect the glider, including purchasing bushland where the glider is common and protecting it from development. The council has formed partnerships with other organisations to purchase more bushland. The council has also supported research to discover more about the glider, its habitat and their needs. This has helped in the creation of education programs for local residents, as well as the identification of bushland areas for protection.

16.19 Brisbane City Council is actively managing populations of squirrel gliders around the city.

Passive housing design

Passive design uses the natural climate to heat and cool homes. The orientation of the building is a key part of passive design. For example, if you live on the east coast, having large windows facing westwards means that the hot afternoon sun will heat up the house. This is ideal if you live in a cool climate.

Shade is an important consideration. If you have large evergreen trees, such as gum trees, close to the house, you may need to use more energy to warm and light your home in winter. Deciduous (plants that lose their leaves in winter) vines provide shade in the hot summer months but allow warming sunshine in during winter.

Dense materials such as concrete, bricks and stone are good insulators, which means that they are good for maintaining more even temperatures year round. Roof and wall insulation are also very important for maintaining comfortable temperatures. Installing skylights is a very effective strategy for reducing the amount of energy needed to light a home. Figure 16.20 shows the features of an energy-efficient home.

In parts of Europe and North America, grass has been grown on rooftops for many years. The grass rooftops provide excellent insulation as well as habitat for insects. Some architects are beginning to use rooftop gardens in their designs in Australia (see Figure 16.21).

16.21 A rooftop garden on a home in Melbourne

16.20 An energy-efficient home

Well-designed shading can greatly reduce heat entering a home and reduce the running costs of cooling appliances.

External shading should be designed to let in the winter sun but totally shade the house from the summer sun. It can be overhanging eaves, horizontal shades or awnings and pergolas.

Solar panels capture the sun's energy and convert it into electricity that can be fed back into the grid.

Insulating homes is the best way to increase their energy efficiency. It prevents heat loss in winter and reduces radiant heat in summer.

Cross-ventilation occurs when two sides of the home are opened to allow air to flow from one end to the other. Encouraging airflow on summer evenings is an effective means of cooling the house and occupants at night.

In winter, draughts can account for 25% of heat loss. Reducing draughts is a cheap and cost-effective way of reducing heating and cooling costs. Heat is lost through cracks and gaps between walls, ceilings, floors, windows and doors, unnecessary vents, exhaust fans, unused fireplaces and skylights.

Landscaping can help to maintain comfortable temperatures. Breezes entering the home will be cooler if they have passed through gardens.

Daytime living areas should be located on the north side with large north-facing windows to capture the winter sun. Bedrooms and utility areas are best located on the south side.

Design open plan areas that can be closed off using doors. Zoning reduces the area you need to heat and cool.

Windows and other glazed surfaces can account for more heat gain and loss than any other surface. Choosing the right size windows and the right glazing material can significantly improve the efficiency of your home. Double-glazing reduces heat transfer.

North-facing windows should be full length to allow the heat from the winter sun in. East- and west-facing sides should have a minimum area of glass or none at all.

Managing our water

Approximately 21 per cent of all water consumed in Australia is in urban areas. The supply of water to our growing cities is one of the most significant urban management issues facing Australia. During the first decade of the twenty-first century Australia experienced one of the longest and most devastating droughts on record. Major cities were placed on severe water restrictions and, in some inland towns, water supplies went very close to running out. Scientists predict that climate change will mean that many parts of Australia will have even less water in the future.

Many cities have built or are planning to build desalinisation plants that turn salt water from the sea into fresh drinking water. However, these plants are expensive to build and run, and use huge quantities of energy to treat the water. Being more efficient in our water usage is a far more cost-effective way to manage our growing water needs.

Appliances such as dishwashers and washing machines now come with water rating labels to help consumer select products that use water in the most efficient way (see Figure 16.22). Many local councils and state governments now also require new homes to have water tanks installed to capture rainwater, which can be used to water gardens, wash cars and even flush toilets.

16.22 Water rating labels help consumers select water-efficient products such as washing machines.

Figure 16.23 shows the difference a water-efficient home can make. A conventional home typically uses about 330 kilolitres year (that's 330 000 litres!). This can be more than halved in a water-efficient home. Installing AAAA-rated appliances in the kitchen and laundry, and using water tanks, water-efficient showerheads and dual-flush toilets can make a huge difference to the amount of water used in our cities.

16.23 Comparison between a water-efficient and a conventional home

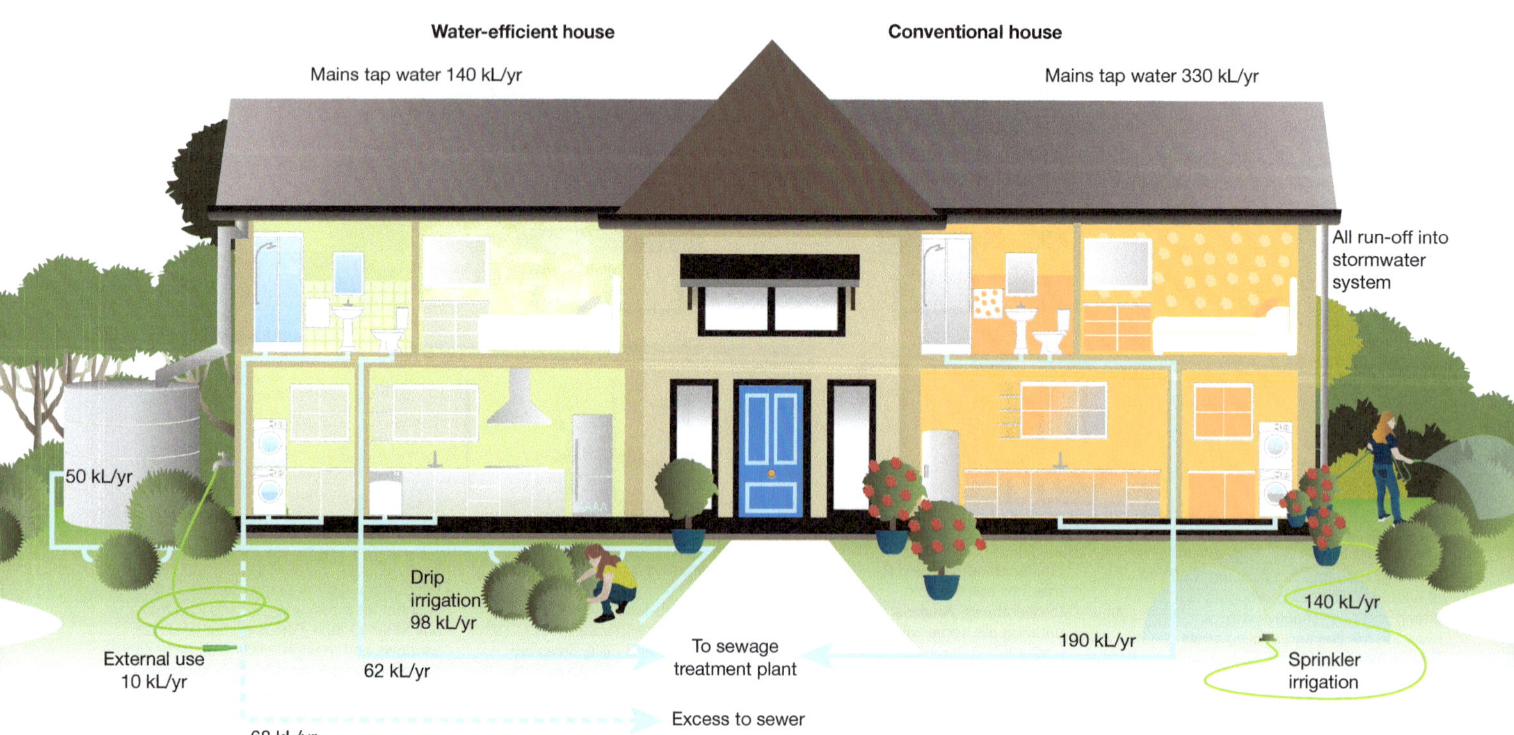

Source: Source: ACT Government www.thinkwater.act.gov.au/water_savingtips/visit.shtml

SPOTLIGHT

Cleaning up Hobart's stormwater

Every city has complex drainage systems to carry away stormwater after heavy rains. Even a small city such as Hobart has nearly 340 kilometres of major stormwater pipes, and many more smaller pipes. As stormwater flows through the city it picks up pollutants that then make their way into creeks, rivers and eventually the sea. In the past, after heavy rain, Hobart's Derwent River would be littered with the rubbish that had flowed from the stormwater system.

Hobart City Council has begun a program to install litter traps on major stormwater pipes (Figure 16.24). There are several different types of traps, from simple socks that go over the end of pipes as they reach creeks and trap the rubbish, to more complex floating traps that are used in major stormwater drains. Since these measures have been introduced, the level of litter reaching the river has been substantially reduced, improving the water quality and reducing the impact on the aquatic environment.

A floating litter trap used on the Hobart Rivulet to stop litter entering the Derwent River

ACTIVITIES

Knowledge and understanding

1 Outline the ways in which our cities affect the environment.
2 Explain how bushland is threatened by our cities.
3 Describe how Brisbane's squirrel gliders are being affected by the growth of the city.
4 Outline the strategies being used in Brisbane to reduce the impact on the gliders.
5 Explain the concept of passive house design.
6 State why is it so important that water is carefully managed in Australian cities.

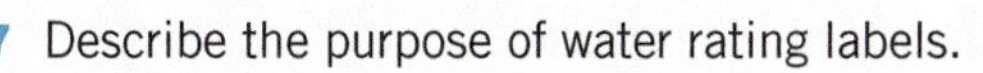

7 Describe the purpose of water rating labels.
8 Outline how Hobart City Council is improving water quality from stormwater drains.

Analysing and applying

9 Consider your house. Prepare a list of the features of your house that make it more sustainable and the things that make it less sustainable. What changes could you make to improve the sustainability of your house?

Review and reflect 3

Activity 1

Compare and contrast

Select one of Australia's capital cities and one capital city in the United States of America. Undertake research to find the following information.

a When was the city established and by whom?
b What is its population?
c What is the next largest urban centre in the same state or territory?
d Describe the site of the city.
e List its major industries, notable buildings and landmarks and any special events for which it is famous.
f Describe the environmental impacts of this city. Identify any progress that has been made in protecting environmental quality

Present an oral report to the class outlining the findings of your research. Include in your presentation a map of each city highlighting its major features.

Activity 2

Fast rail networks around the world

a Investigate a fast rail network from two of the following countries. Select one country from Asia and one from Europe.
    - *Asia:* Japan, China
    - *Europe:* Spain, Germany, France and Italy

b Prepare a report recommending the building of a fast rail network in Australia. Your report must include:
    i a map of the proposed route
    ii an indication of how fast the train will travel
    iii the economic, social and environmental benefits and costs of the project.

Activity 3

A sustainable house

Working with a partner, prepare a design for a sustainable house for your suburb. Remember to consider passive design features to reduce energy use as well as water-saving strategies. Present your design as an annotated visual display. Highlight each sustainable feature and explain why it is sustainable.

Activity 4

Design a new mining town

Imagine you are a young architect submitting a plan to a major mining company for the design of a new mining town in regional Western Australia.

a Create your plan (map) of a new mining town for 1500 people. Design a symbol for each feature in your plan. Draw a map outline on an A3 sheet of paper. Include the following features on your map:
    - housing estate (calculate the total area of the estate required based on population)
    - houses
    - townhouses
    - ten shops
    - F–12 school
    - multifaith worship centre
    - community centre
    - sports facilities
    - hospital
    - power station
    - water supply dam
    - conservation area
    - hotel
    - town hall/offices
    - minor roads that join areas of the town
    - major roads that exit/enter town.

b Create a key for your symbols and give your town a name.
c Think of two other features you would add to your town. Create symbols for them and add them to your map. Do not forget BOLTSS.

Activity 5

Impact on Chinese cities

a Read the article about housing inequality in China, and then answer the questions that follow.

> In recent decades, Chinese cities have experienced profound social, economic and spatial transformations. In particular, Chinese cities have witnessed the largest housing boom in history and unprecedented housing privatisation. As a result, urban residents have enjoyed spectacular housing improvement with better housing quality, higher residential mobility and higher rate of home ownership than before. Dominated by public rental housing only three decades ago, China now is a country of home owners, with more than 70 per cent of urban residents owning homes—higher than in many developed countries. Moreover, more than 15 per cent of urban households own multiple homes. Yet, this spectacular housing success is not shared by all social groups. Housing inequality is rising rapidly, and residential segregation is increasingly prevalent in previously homogeneous Chinese cities. Low-income groups such as migrants have to live in crumbling shacks in so-called 'urban villages', while the new *nouveaux riches* live in exclusive gated villa communities that are on a par with upmarket gated communities in the West. The residential landscape in Chinese cities is becoming increasingly polarised, which is reshaping the social, economic and political landscape in Chinese cities, and is challenging our perception of Chinese cities.
>
> Source: *Housing inequality in China*, edited by Si-ming Li and Youqin Huang, Routledge, 2013

i What do urban residents enjoy now as a result of the social, economic and spatial changes in Chinese cities?

ii Are good quality homes available to all people in Chinese cities? Explain your answer.

iii What similarities do Chinese cities share with cities in developed countries?

b In parts of China there are whole cities that have been built but are not occupied. Investigate why these cities are not occupied when there is a housing crisis.

Activity 6

Natural gas in Australia

a Use the internet to investigate how natural gas is obtained from the earth.

b Investigate how one country other than Australia produces and uses natural gas. You could include the following headings in your slideshow, report or poster: the level of natural gas reserves; the level of production; how much is exported and how much is consumed domestically; how the country addresses the sustainability of natural gas resources.

c Write a blog on one of the following topics. Research and include appropriate illustrations and links.

- Natural gas and climate change
- Environmental issues of natural gas
- Leaking natural gas
- Natural gas vehicles
- Safety concerns with natural gas

or another topic approved by your teacher.

Index

A

D

E

F

G

H

N

O

P

Q

R

S

T

U

V

Acknowledgements

We thank the following for permission to reproduce copyright material. The following abbreviations are used in this list: t = top, b = bottom, l = left, r = right, c = centre.

AAP: AAP Image/Lyn Durham, p. 309b; AP/Kyodo News, p. 79t; Bill Bachman, p. 29; Andrew Brownbill, p. 118t; David Hancock, p. 268; Emilio Morenatti, p. 255; David Wethey, p. 77.

Adelaide City Council: With thanks to Adelaide City Council and Hassell, p. 315b.

Airview Aerial Photography: p. 275.

Alamy Ltd: AA World Travel Library, p. 233; Aerial Archives, p. 36l; AfriPics.com, p. 68l; Arcaid Images, p. 309t; Bill Bachman, p. 168; Bon Appetit, p. 41; CTR Photos, p. 315t; CulturalEyes-AusGS, p. 261tr; Danita Delimont, p. 62; Nikki Edmunds, p. 111; Rob Francis, p. 23b; Global Warming Images, p. 108; Glyn Genin, p. 236; Richard Green, p. 228; Blaine Harrington III, p. 72br; Martin Harvey, p. 193r; Paul Kingsley, p. 261br; Frans Lanting Studio, p. 27b; Lebrecht Music and Arts Photo Library, p. 27t; Keith Levit/Design Pics Inc., p. 130b; Lou Linwei, p. 196; Ern Mainka, p. 25; Bruce Miller, p. 147; Tom Mueller/imagebroker, p. 37t; NASA Archive, p. 2; Rolf Nussbaumer Photography, p. 130t; Michael Oakes, p. 156t; Ingo Oeland, p. 199b; Andrea Pistolesi/Tips Images, p. 194; Radius Images, p. 197b; redbrickstock.com, p. 313t; Ingo Schulz/imagebroker, p. 188r; Bryan Simpson, p. 304tr; David South, p. 159c; StockShot, p. 100; Keith Taylor, p. 59; Universal Images Group/DeAgostini, p. 28; David Wall, pp. 44bl, 95b, 101b, 135, 145b, 198bl, 266, 304b; Rob Walls, p. 258; John Warburton-Lee Photography, p. 44br; Darwin Wiggett/All Canada Photos, p. 52tl; Michael Willis, p. 300.

Auscape International Photo Library: Jean-Marc La Roque, p. 89; Wayne Lawler, p. 68r.

Australian Geographic: David Dare Parker, p. 281.

Beebe, Sam: Ecotrust, CC by 3.0, p. 7.

Bureau of Meteorology (BOM): p. 113 (all); map courtesy of the National Climate Centre, Australian Bureau of Meteorology, Melbourne, Australia, pp. 117t, 117b.

Chapman, Chantelle: www.chantellechapman.daportfolio.com, p. 257.

Comfort, Geoff: p. 311.

Corbis Australia Pty Ltd: Noah Addis, p. 221; Alaska Stock, p. 73b; Barrett & MacKay/All Canada Photos, p. 143;Bettmann, p. 82; Corbis, p. 191b; Corbis RF pp. 151tl, 151tr, 151br; Edward S. Curtis/Christie's Images, p. 33l; Cameron Davidson, p. 248l; Tim Fitzharris/Minden Pictures, p. 31t; Werner Forman, p. 33r; Walter Geiersperger, p. 16; Robert Glusic, p. 250; Jon Hicks, p. 263; Fritz Hoffmann/In Pictures, p. 218; Imaginechina, p. 186; Colin Monteath/Hedgehog House/Minden Pictures, p. 19tl; Steve Parish, p. 87b; Radius Images, pp. 20, 44tr; Jose Fuste Raga p. 72bl; Nick Rains, 87t; Michael Reynolds/epa, p. 201l; Claude Robidoux/All Canada Photos, p. 31b; Qilai Shen/In Pictures, pp. 293, 299; George Steinmetz, p. 201r; Torleif Svensson, p. 19tr; Darwin Wiggett/All Canada Photos, p. 122; Gerhard Zwerger-Schoner/imagebroker, p. 72tl.

Department of the Army, U.S. Army Corps of Engineers, Omaha District, p. 175b.

Department of Sustainability, Environment, Water, Population and Communities: © Commonwealth of Australia 2012, pp. 47, 157.

Department of Primary Industries: This map is © State of Victoria, Department of Primary Industries. Reproduced with permission, p. 88t.

Dreamstime: pp. 18tr, 19bc, 19cl, 26, 36r, 37b, 39, 53tc, 53b, 54, 63 (all), 69, 70t, 70b, 80l, 98t, 125, 127, 129, 138, 155, 159t, 182, 199t, 213l, 248t, 248b, 248bc, 248tc, 249br, 249bcr, 249tcr, 249tr, 249bl, 252, 261bl, 298, 312.

Design Architect's Adrian Smith and Gordon Gill: Rendering © Adrian Smith + Gordon Gill Architecture, p. 227b.

Fairfax Photo Sales: Craig Abraham, p. 119t; *The Age* News, p. 114; Joe Armao, p. 227t; Matt Davidson/*The Age*, p. 118b; Pat Scala, pp. 115, 270.

Fisher and Fisher: p. 177r.

Fotolia: fanfan, p. 52br.

George Schlegel lithographers: p. 247t.

Geoscience Australia: Commonwealth of Australia (Geoscience Australia) 2013, p. 237l; © OzCoasts (Geoscience Australia) 2012. The Commonwealth has provided this image for general educational purposes only and not for commercial use. The Commonwealth does not guarantee its accuracy and disclaims liability for any loss whatsoever arising from its use for any purpose, p. 95tr.

Getty Images: AFP, cover, p. 74, p175t; Franz Aberham/StockImage, p. 192; Ted Aljibe, p. 256; Alexandro Auler/LatinContent, p. 99t; Auscape/UIG, p. 46; Auscape/UIG,

Torsten Blackwood, p. 84b; Anders Blomqvist, p. 213r; Adam Burton, p. 136; Peter Cade, p. 34; ChinaFotoPress, pp. 99b, 104; Paul Cowell Photography, p. 56; Danita Delimont, pp. 72tr, 139; Grant Dixon/Lonely Planet Images, p. 145t; Ruth Eastham & Max Paoli, p. 188bl; Gamma-Rapho via Getty Images, p. 38t; Guenter Fischer, p. 71; Christopher Groenhout/Lonely Planet Images, p. 198br; Peter Harrison, p. 222; Peter Harrison/Photolibrary, Photography by Claire Chao, p. 189; Hoberman Collection/Universal Images Group via Getty Images, p. 188tl; Bay Ismoyo/AFP, p. 214; Paul Kane/Getty Images News, p. 235; Frans Lemmens/Lonely Planet Images, p. 191t; Frans Lemmens/Lonely Planet Images, p. 193l; Hiroyuki Matsumoto, p. 247b; Ted Mead, p. 128; Morales/age fotostock, p. 316; Michel Porro/AFP, p. 291; Aizar Raldes/AFP, p. 98; Mark Ralston/AFP, p. 215; 2011 Jorge Silva, p. 97; Penny Tweedie, p. 276; UIG via Getty Images, p. 230; Guy Vanderelst, p. 295; Frank J Wicker, p. 178; Milton Wordley, p. 197t; Ariadne Van Zandbergen/Lonely Planet Images, p. 51.

Glenvale School: pp. 116, 117c.

Hamilton, A.: A reflection on Aboriginal People in Central Australia, from A. Hamilton, 'Coming and Going: Aboriginal Mobility in North-west South Australia, 1970–71' in *Records of the South Australia Museum*, Vol 20, pp. 47–57, 277.

Hammer, Michael: p. 156b.

Hobart City Council: p. 319.

Hoeper, Frank: Bremen, Germany, p. 81l.

Kleeman, Grant: pp. 48, 149.

Lend Lease Corporation: pp. 264, 313b.

Lock the Gate Alliance: p. 283.

Map Illustrations: p. 92.

McDougall, Garry: p. 40.

NASA Images: NASA Earth Observatory/Jesse Allen, p. 84tr; Courtesy of the NOAA Coastal Services Center, p. 85b; Robert Simmon, p. 91l; National Imagery and Mapping Agency (NIMA) of the U.S. Department of Defense (DoD), the German and Italian space agencies, and the Jet Propulsion Laboratory, p. 93b; courtesy of nasaimages.org, pp. 91r, 246, 251.

National Archives of Australia: (35772), p. 308.

National Capital Authority: Diagram supplied courtesy of the National Capital Authority. © Commonwealth Copyright. All rights reserved, p. 310.

National Library of Australia (NLA): Joseph Lycett. Working the Land. nla.pic-an2962715-s20. National Library of Australia, p. 43.

National Oceanic & Atmospheric Administration (NOAA): NOAA Center for Tsunami Research, p. 79b.

Nest Architects: Photographer: Nic Granleese, p. 317t;

News Limited Images (Newspix): Chris Scott, p. 133; Jason Edwards, p. 225.

Nicholson Cartoons: Cartoon by Nicholson from *The Australian*: www.nicholsoncartoons.com.au, p. 21.

NSW Department of Lands: p. 203.

NSW Government Land & Property Information: © LPI, Department of Finance and Services 2013, www.lpi.nsw.gov.au, p. 169 (all).

NSW Rural Fire Service: p. 119b.

Pearson Australia: Alice McBroom, 265, 318t.

Photodisc: p. 151bl.

Picture Desk, The: 20th Century Fox/Bazmark Films/ The Kobal Collection, p. 23t.

Routledge: Housing Inequality in China, Youqin Huang, Si-ming Li, Routledge, p. 321.

Science Photo Library SPL: David Parker, p. 61; Martin Bond/Science Photo Library, p. 65;

Shutterstock: pp. 18tl, 18bl, 18br, 19bl, 19cr, 19br, 24, 30, 38b; 44tl, 52tcl, 52tcr, 52tr, 53bc, 53t, 53c, 73tl, 73tr, 80r, 101t, 107, 158, 159b, 160, 166, 206; Joseph Kim, 261tl.

Strudwick, Shane: Discover Murray River, p. 154.

Skyworks: Ken Rae, p. 88b.

Smithsonian Institution, The, National Portrait Gallery: Smithsonian American Art Museum, Washington, DC/Art Resource, NY, p. 32.

SNR Uvac: Mr Vojislav Jovanovic, p. 162.

Spiire: p. 177l.

Stanic, Mick: www.flickr.com/photos/splatt, p. 132.

State Library of South Australia: B 9358, p. 314.

State Library of Victoria: Pictures Collection, State Library of Victoria, p. 304tl.

Sunshine Coast Regional Council: p. 303t.

Swisstopo: Reproduced by permission of swisstopo (BA13007), p. 93t.

Thredbo Resort: p. 134.

Transport for NSW: p. 303b.

Troppo Architects: p. 267.

Tweed River Entrance Sand Bypassing Project: p. 95tl.

UNESCO Publishing: p. 50.

Wilder, Altus: p. 103.

Wildlight: Jaime Plaza, p. 45.